KU-030-070

The body at work

The body at work

BIOLOGICAL ERGONOMICS

Edited by
W.T. SINGLETON
Professor of Applied Psychology, University of Aston, Birmingham

CAMBRIDGE UNIVERSITY PRESS
Cambridge
London New York New Rochelle
Melbourne Sydney

Published by the Press Syndicate of the University of Cambridge
The Pitt Building, Trumpington Street, Cambridge CB2 1RP
32 East 57th Street, New York, NY 10022, USA
296 Beaconsfield Parade, Middle Park, Melbourne 3206, Australia.

First published 1982

Printed in Great Britain at the University Press, Cambridge

Library of Congress catalogue card number: 81–18096

British Library Cataloguing in Publication Data

The body at work.
1. Human mechanics
I. Singleton, W.T.
618'.7 QP303

ISBN 0 521 24087 5

CONTENTS

CONTRIBUTORS

Dr P.R. Boyce
The Electricity Council Research Centre
Capenhurst
Chester CH1 6ES

Dr. S. Cole
Ministry of Defence
Army Personnel Research Establishment
c/o Royal Aircraft Establishment
Farnborough
Hants.

Dr D.R. Davies
Applied Psychology Dept
University of Aston
Birmingham B4 7ET

Dr D. Grieve
Royal Free Hospital School of Medicine
Dept of Anatomy
8 Hunter Street
London WC1N 1BP

Dr D. Jones
Dept of Applied Psychology
UWIST
Penylan
Cardiff

Dr D.McK. Kerslake
Lime Tree Cottage
Chandlers Lane
Yateley
Camberley
Surrey GU17 7SP

Dr S. Pheasant
Royal Free Hospital School of Medicine
Dept of Anatomy
8 Hunter Street
London WC1N 1BP

Professor W.T. Singleton
Applied Psychology Dept
University of Aston
Birmingham B4 7ET

Dr W.R. Withey
Admiralty Marine Technology Establishment (P.L.)
Fort Road
Alverstoke
Gosport
Hants. PO12 2DU

PREFACE

The term 'work' is used here as a convenient shorthand for purposive human activity. It includes not only labouring for monetary gain but all similarly demanding activities such as war, sport, games, hobbies and housework.

A person in action is subject to various external and internal constraints. The external constraints arise from the nature of the specific task; the internal constraints are more general, they can be studied systematically and the results should be applicable to a wide range of people and situations. These studies of general constraints on human activity are collectively called ergonomics. There are two separable aspects: constraints arising from bodily functions and constraints arising from mental functions. This book is restricted to the former, hence the title – *The Body at Work: Biological Ergonomics.* To encompass ergonomics a complementary book would be required which would deal with 'the mind at work: psychological ergonomics'.

Ergonomics is a difficult subject to comprehend because it is multidisciplinary, because the boundaries are not agreed, because it is part science and part technology and, not least, because it is still immature. As explained in Chapter 1 it has a history which goes back less than forty years.

Nevertheless, in this period and even before it in a more haphazard fashion, there have been a large number of respectable research workers and practitioners who have made a variety of significant contributions. Many of these contributions will no doubt be developed further in the future but they are unlikely to have been totally misconceived. This book is intended to be a record of these contributions; it is a description of the finally established knowledge in ergonomics. Some of this knowledge is based on the physical sciences (physics and chemistry), some on the biological sciences (mainly anatomy and physiology) and some on psychology (mainly human performance studies). They have in common that the concepts are secure and the evidence is reliable – in short they qualify as science.

One reason for grouping these topics together is that, because they are to do with man at work, they can make a contribution to the more effective design of work and the work situation. This is the technological aspect of ergonomics and this book is also intended to enable the student to apply this knowledge to further studies of man at work. The purpose of such application is to make the work and the worker more effective, which might emerge as increased safety or productivity or in the longer term as improvements in health and well-being.

These are self-evidently valid objectives but they are not the fundamental reason why ergonomics exists as a separate discipline. The underlying entity of ergonomics is the style of thinking of ergonomists. They do not think like psychologists or physiologists or engineers or doctors. Most ergonomists at the present time come from one or other of these older disciplines but when they practise as ergonomists they have this unique approach which is scientifically interdisciplinary, oriented towards practice and yet focussed on the person working. I hope that this book also communicates this new philosophy.

University of Aston in Birmingham W. T. Singleton
November 1980

1 INTRODUCTION

W. T. Singleton

1.1 Origin of ergonomics

UK and USA

There have been activities in many countries which might be mentioned as the disparate origins of ergonomics. This book inevitably concentrates mainly on the work which has been recorded in books and journals written in English. The origin of this work is in the wartime and immediate post-war period in the UK and the USA; the term ergonomics was coined by K.F.H. Murrell in 1949. The Ergonomics Research Society emerged in 1949–1950 from various meetings held in London, Oxford and Cambridge. The intention was to facilitate the exchange of ideas and expertise between the many disciplines which had made a contribution to the increased effectiveness of human performance during the Second World War and to extend the application of these principles beyond the military sphere. The first important symposium was held in Birmingham in 1951 'Human Factors in Equipment Design' and the second at Cranfield in 1952 on 'Fatigue'. These proceedings emerged in print in reverse order as Floyd and Welford (1954) and (1953) respectively.

The disciplines involved included anatomy, physiology, psychology, industrial medicine, industrial hygiene, design engineering, architecture and illumination engineering (Murrell, 1965), but the Society settled down for the first ten years as a forum for two academic groups: the anatomists/physiologists and the experimental psychologists. This was because the active members and authors of papers were working mainly from universities. The emphasis was on the development of theory and methodology which in principle was relevant to industrial problems although the experience and existing application was restricted largely to military operations.

The senior academics who supported the Society from its inception and became its first two honorary members were Sir Wilfred Le Gros Clark, Lees Professor of Anatomy at Oxford, and Sir Frederic Bartlett, Professor of Experimental Psychology at Cambridge. Members of their departments and the associated research units are prominently represented in the early proceedings. The Medical Research Council sponsored the Climate and Working Efficiency Research

Unit in Oxford and the Applied Psychology Research Unit in Cambridge. The MRC was responsible during the war (and remains so up to the present time) for the Personnel Research Committees of the Royal Navy, the Royal Air Force and the Army. It was activities under the aegis of these committees which could be regarded as the origin of ergonomics. Another Unit at Cambridge in the immediate post-war period was that for research into problems of ageing sponsored by the Nuffield Foundation. The head of this unit, A.T. Welford, was the first chairman of the Society appointed in 1954 and the founding editor of the journal *Ergonomics* which first appeared in 1957. Two other academics worthy of mention who encouraged the Society in its earliest days were G. Crowden, Professor of Applied Physiology at the London School of Hygiene and Tropical Medicine, and T.U. Mathew, Lucas Professor in the Principles of Engineering Production at Birmingham. The early history of the Society is given in detail in Edholm and Murrell (1973).

The 1951 membership list also includes Paul Fitts, Leonard Meade and Clifford Morgan from the USA. The corresponding activity in America first emerged in print as 'Lectures on Men and Machines' sponsored in 1947 by the Special Devices Centre, Office of Naval Research. In the foreword, Commander Wylie USN pays tribute to the 'remarkable group of four young men' who wrote the lectures: C.T. Morgan, A. Chapanis, W.R. Garner and F.H. Sanford. The subtitle was 'An Introduction to Human Engineering', but these titles were changed to *'Applied Experimental Psychology: Human Factors in Engineering Design'* when the lectures were converted into a book (Chapanis, Garner and Morgan, 1949). Another important book of the same period was *Human Factors in Undersea Warfare* (Lindsley, 1949) produced by the Panel on Psychology and Physiology, Committee on Undersea Warfare, National Research Council; the chairman of the panel was D.B. Lindsley. Although this latter has a balanced physiological and psychological content the former is predominantly psychological with some engineering.

American learned society meetings in this field originated in the annual Human Engineering Conferences held by the Office of Naval Research beginning in 1953 and the annual meetings of the Flight Safety Foundation which began in 1948. Division 21 (Engineering Psychology) of the American Psychological Association was organised in 1957 and so also was the Human Factors Society of America (McFarland, 1971). The *Human Factors Journal* started publication in 1958.

Ergonomics was built on a balanced foundation of anatomy, physiology and psychology but the American counterpart – human engineering or human factors – was more of an alliance between psychology and engineering. In both cases the psychology was essentially human experimental psychology (Grether,

1968). Although the pre-war industrial psychology in the USA (Viteles, 1933; Maier, 1946) and the occupational psychology in the UK (Chambers, 1951; Blain, 1970) had acknowledged the importance of the experimental method, in fact the knowledge and expertise available at the time was mainly from tests and surveys. From this basis there was considerable contribution to the war effort in selection and job allocation (Vernon and Parry, 1949), but this was separate from the contribution of the experimental psychologists to equipment design, no doubt because these activities fit so differently within military organisations. The potential link between the two – training – was an area where psychologists made little contribution until after 1945. There were extensive innovations in training during the war but they were in industry rather than the military services because the services had adequate supplies of fit able young people and it was in industry where there were labour deficiencies in numbers and skills. Thus, training at this time emerged from the methods aspects of time and motion study rather than from psychology (Shaw, 1952; Seymour, 1968). Again there was corresponding activity in America, particularly expressed through the Training Within Industry (TWI) movement, which began in 1940.

The Anglo-Saxon countries had emerged from the war relatively unscathed but almost immediately got involved in the so-called cold war. This resulted in continued military expenditure including that on associated research activities such as human factors. The generous expenditure on research led to the development of new thinking in military units (Taylor, 1957), in university departments (Fitts, 1951) and in that unique American institution the consulting research organisation, such as the RAND Corporation, Dunlap and Associates, Bolt, Beranek and Newman, the American Institute for Research, and the Systems Development Corporation. The proceedings of the Fourth Annual Meeting of the Human Factors Society in 1960 appeared as a book (Bennett, Degan and Spiegel, 1963). There are more than fifty authors and by origin they divide almost equally into three sections: the universities, industrial and military organisations, and consulting organisations, Correspondingly, more than half the twenty or so authors in the seminal book (Gagné, 1962) are from consulting organisations. Training was now emerging as a key problem because the high-calibre war-time personnel had now returned to civil life. Conscripted personnel were only available for limited periods yet the complexity of equipment continued to increase. This was also the period of rapid development of civil aviation and the associated aviation medicine (McFarland, 1953) which overlapped with ergonomics but was never quite regarded as part of it. By 1960 there was considered momentum in ergonomics and human engineering. There were innovative research units in the universities, the military establishments and in industry.

In the UK there was sympathetic interest from those concerned with policymaking at government level; it was not difficult to get Ministers of the Crown to open and address ergonomics meetings. National funds were channelled into the subject through the Department of Scientific and Industrial Research and the Medical Research Council. Industrial representatives were prepared to attend meetings and to discuss methods of implementation.

The picture was even more encouraging in the USA. Kraft (1958) had carried out a survey which demonstrated that there had been rapid growth in the subject from 1954 and there were forty-three companies with human factors research programmes – although these were concentrated in the aircraft and supporting industries (Wood, 1958). Kraft (1962) carried out another survey on a wider industrial basis which indicated that there were 137 companies with such programmes with a total staff of at least 1300. This growth had been stimulated by the Department of Defense edicts of 1954/7 (e.g. Chaillet, 1967) that all military equipment design should have a human factors input, and by the expansion of missile and man-in-space programmes during 1957/61.

Europe

The Ergonomics Research Society quickly attracted strong participation from Continental Europe including E.H. Christensen from Sweden, G.C.E. Burger from the Netherlands and G. Lehmann from West Germany. Their contribution was essentially medical, stimulated by the need to develop work physiology in Europe at a time when there was much hard physical work and limited food and fuel supplies. This was a useful counterbalance to the American and British innovative activity from experimental psychology. Although there was extensive British research in climate physiology and in biomechanics (Edholm, 1967), the Scandinavians developed functional anatomy and applied it to postural problems of chair and workspace design as well as to diagnosis and therapy. The centres of expertise in work physiology were in Germany and Holland, respectively the Max Planck Institute at Dortmund and the Institute of Preventive Medicine at Leiden.

The re-emergence of European industry proceeded with more co-operation and intercommunication than in pre-war days; productivity comparisons were made between European countries and between Europe and the USA. The USA was regarded as the model and there were productivity missions across the Atlantic from a wide range of industries. This activity was organised and encouraged by the European Productivity Agency (EPA) set up in 1953 as a subsidiary of the post-war Organisation for European Economic Co-operation (OEEC). EPA Project 335 was concerned with 'Fitting the Job to the Worker'. Characteristically this started by sending a mission to America in 1956; the American project advisor was H.S. Belding of the University of Pittsburgh, and the secretary was K.F.H. Murrell, who had recently joined the University of Bristol after an innovative period as Head of Ergonomics in Tube Investments, a British Steel Company. The team contained representatives from the main European countries including F.H. Bonjer from the Netherlands Institute of Preventive Medicine at Leiden, B.G. Metz, Director of the Centre for the Study of Applied Physiology of Work in Strasbourg, B. Schulte, a method engineer from Siemens Schuckertwerke A.G., and W.T. Singleton, an applied psychologist from the British shoe industry, all of whom have remained active in the ergonomics field.

The team defined 'fitting the job to the worker' as 'applying knowledge of human capacities to work design' and said that this 'branch of learning' was known as human engineering in the USA and ergonomics in Europe (Murrell, 1958). They found that most of this kind of work depended on military or aviation support with very little in American industry outside the traditional industrial medicine and hygiene. There was considerable discussion as to the relationship between industrial hygiene and ergonomics, the consensus being that environmental hazards such as noise and heat are part of ergonomics but toxic hazards are not. There were also discussions as to how far problems of diet needed to be considered, but given the rapidly improving European situation and the virtual absence of this problem in America it did not seem to be very important; this aspect emerged again later as ergonomics spread to developing countries (Christensen, 1964).

The American tour was followed by a technical seminar for European experts in Leiden in 1957 (Murrell, 1958) and a tripartite international conference for employers, workers and experts in Zurich in 1959 (Metz, 1960). These meetings were well attended and resulted in further meetings in the separate countries. A conference was organised by the Department of Scientific and Industrial Research in the UK in 1960 (DSIR, 1961). Again there was encouraging attendance by industrial representatives and papers from research workers which demonstrated that ergonomics work was being developed in the shoe industry (Singleton, 1958), the Post Office (Conrad, 1960), the steel industry (Sell, 1971) and in road transport (Grime, 1958). There was one paper from abroad in which A.Y. Wisner talked about the physiological work in Renault which had begun in 1954, with particular emphasis on seat design. There was also ergonomics work going on in the electrical industry by this time, notably by J.C. Jones in Amalgamated Electrical Industries (AEI) and B. Shackel in Electrical and Musical Industries (EMI). The DSIR followed up this meeting by producing in 1962 a series of twelve booklets by specialist authors under the general title *Ergonomics for Industry;* these were later reproduced in the journal *Applied Ergonomics* and eventually as a separate handbook (Shackel, 1974).

A further consequence of the Leiden meeting was the setting up of the

International Ergonomics Association which held its first meeting in Stockholm in 1961 (Welford, 1962). The President was S. Forssman of Stockholm and the Secretary was E. Grandjean of Zurich.

By the 1960s the world was recovering rapidly from war. Books on ergonomics began to appear in other languages (e.g. Grandjean, 1963; Lomov, 1964; Luthman, Aberg and Lundgren, 1966; Faverge, Leplat and Guiguet, 1966; Montmollin, 1967). The development of ergonomics with particular reference to the formation of societies is reviewed by Sell (1969). He considers that Europe is relatively strong in physiological studies while in the USA the emphasis has been on psychological problems associated with aerospace and defence. The appearance of Japan on the ergonomics scene is too recent to evaluate but the orientation seems to be towards applications. He notes that there are still very few papers reporting studies which use techniques from more than one basic discipline. Fowler (1969) takes an American look at Europe mainly from the point of view of research interests; he considers that there are more physiologists involved compared with the USA and that Europe also has more emphasis on the application of human factors to industrial problems.

Klinkhamer (1970) reviewed the progress of ergonomics in the Netherlands. He mentioned four centres of research: the Netherlands Institute of Preventive Medicine which worked extensively on dynamic muscular load, the Institute of Perception at Soesterberg, the Institute for Perception Research at Eindhoven and the Laboratory for Ergonomic Psychology in Amsterdam. There were also ergonomics groups working in industry at Philips in Eindhoven, Netherlands Steel in Ijmuiden, Netherlands Railways and the Dutch State mines. The industrial groups were all interdisciplinary, involving technologists, physicians, psychologists and work-study engineers (Dirken and Klinkhamer, 1974). There was also a group at Wageningen, concerned with the ergonomics of agriculture. Thus the Dutch are relatively strong in production rather than product ergonomics and compared with other European countries they make more extensive use of experimental psychologists. In Belgium and Luxembourg the situation at this time was similar to that in Holland although on a smaller scale, with some ergonomics work in industry through the industrial physicians and some also through psychologists working from the universities of Leiden, Liége and Brussels.

In France also it was the occupational physicians who first took up ergonomics and applied it particularly to transport systems and to the steel industry. The physiologists working from the universities – notably B.G. Metz in Strasbourg and A.Y. Wisner in Paris – took up the problems of ergonomics of transport systems and the steel industry and developed interdisciplinary teams of research workers to deal with problems such as heat and shift-work which focus on the measurement of stress (Parrot, 1971). The more informational aspects of stress were studied by a group involved in air traffic control systems (Leplat and Bisseret, 1966).

In Germany there was a corresponding development of ergonomics from the strong foundations of work physiology but with a greater involvement of work study/industrial engineering from the REFA organisation. Here also air traffic control problems stimulated the change of orientation from physiologically based stress research to a more psychologically based approach (Rohmert, 1973). Germany was unique in Europe in the 1960s in following the American tradition of control-engineering type modelling of the human operator based on the behaviour of pilots (Bernotat and Gartner, 1972). In East Germany the industrial engineering approach was gaining strength at this time (Hacker, 1973) and there was some orientation towards a systems approach (Timpe, 1969). In Scandinavia in the 1960s the strong anatomy/physiology approach continued to develop but with broadening to the interdisciplinary aspects of real problems and the introduction of the systems approach (Ivergard, 1969; Saric, 1970).

In the USSR the context for the development of engineering psychology was the same as that in the USA, namely early work on aviation and rapid post-war development stimulated by the defence, atomic energy and space programmes. The universities involved include Leningrad, Moscow, Kharkov and Kiev (Lomov and Berrane, 1969). There was some emphasis on a cybernetics approach and there seems to have been a considerable involvement by mathematicians as well as the more orthodox team approach of physiologists, psychologists and physicians. There were several Soviet books published around 1970 which were collections of translations of typical papers from Europe and the USA.

Krivohlavy (1969) mentions departments of engineering psychology in three universities in Czechoslovakia: Prague, Brno and Bratislava. The emphasis here seems to have been more on industrial problems, following the work study tradition with safety as the dominant criterion and with an unusual incorporation of aesthetics as part of ergonomics. In Rumania also, work in the field seems to be essentially production ergonomics (Seminara, 1975), with supportive research in a few university centres of engineering psychology. There is close integration within work psychology generally between ergonomics and personnel psychology. In Bulgaria there is a considerable tradition of physiological research and this was used as the foundation of ergonomics work with characteristic emphasis on physiological indicators of stress and fatigue (Tsaneva, 1972). There was considerable official interest in ergonomics expressed through its incorporation in five-year plans and organisations founded in the 1960s such as the Centre of Industrial Design which has a department of ergonomics (Seminara, 1976). Poland is another centre of ergonomics in Eastern Europe with long-standing interests in production ergonomics, e.g. the design of machine tools, and in ergonomics as an aspect of occupational medicine with concern for health and safety. In Yugoslavia, again ergonomics developed in close association with safety, and the journal *Ergonomija* is published by the Safety Institute in Belgrade.

In the northern Mediterranean countries from Portugal across to Turkey ergonomics began to emerge in the 1960s, particularly in Italy from the usual twin foundations of industrial engineering/work study and occupational health. In Japan (Oshima, 1969, 1970) ergonomics work in the 1960s appeared to centre on transport, particularly railways, but also with some unusually early development of consumer ergonomics for kitchens, bicycles and clothing.

In general most non-English-speaking countries are much more knowledgeable about ergonomics in the English-speaking countries than the other way around. This might have been justifiable in the immediate post-war period when most of the research and application was in the USA and Western Europe, but it was getting more difficult to defend even in the 1960s and today it is most unfortunate that language, cultural and ideological barriers reduce the flow of information to the West. The two United Nations bodies with an interest in ergonomics: the World Health Organisation (WHO) and the International Labour Office (ILO) have invested considerable effort in establishing communication between the communist and the capitalist countries and between the developed and the developing countries. The communist countries have one considerable advantage over capitalist countries in that there is much less emphasis on the need to justify ergonomics in economic terms. If the authorities can be convinced that ergonomics is necessary for health and/or safety purposes then it can be introduced through administrative machinery without the need for elaborate validation by criteria such as productivity.

The developing countries

In the 1960s the United Nations organisations began to devote more and more of their efforts to the developing countries; their mission in the health field (WHO) and in the labour field (ILO) became one of absorbing technical knowledge from the developed world and transferring it to the developing world. Ergonomics came within this context and in 1967 the WHO organised an inter-regional course in ergonomics for developing countries in Bombay (Singleton and Whitfield, 1968). This course was later published in book form (Singleton, 1972). At first sight it might seem that these countries have many more urgent problems than the introduction of ergonomics, but it can be argued that their main under-used resource immediately available is manpower, and ergonomics is concerned with the proper utilisation of manpower. They acquire machinery and systems designed in the developed countries and they need to be able to identify the ones most appropriate to their needs and to decide on any changes required as a result of their different situation (e.g. different-sized working populations and different climates). In some countries excellent occupational health services already exist and these can be used as the foundation for ergonomics development. There are some different problems from those studied by ergonomists in the developed world, the most obvious being the plethora of hard physical work in harsh environments on restricted diets. This leads of course to the importance of considering working periods and shifts appropriate to the situation (Maule and Weiner, 1975). Lippert (1967) deplores the shortage of Western literature on ergonomics for developing countries but on the other hand Daftuar (1971) is able to quote about forty references to work done in India. Thompson (1972) and Singleton (1973) outline some of the problems of applying ergonomics to these countries.

In Chapanis (1975) a number of different issues in relation to developing countries are discussed in detail, in particular anthropometric differences between populations, problems of technical communication between countries and the organisation of multinational experiments as a basis for the standardisation of such different design issues as keyboards and the content of car driving tests. Shackel and Van Nes (1975) mention the ergonomics work now getting under way in Brazil; not surprisingly this is in the transport and steel industries. There are serious accident and safety problems and the use of hand tools is also being studied.

1.2 Structure and content of ergonomics

Introductory discussion

Ergonomics has two parts:

1. an interdisciplinary research activity based on the human sciences of anatomy, physiology and psychology;
2. an operational activity which sometimes functions independently but more usually finds expression through one of the two established technologies of medicine and engineering.

The purposes of particular projects vary from an academic need to know, to direct support for an urgent practical decision. The criteria vary from health through safety to operational effectiveness.

At first sight the above would seem to be all that needs to be said about what ergonomics is, but, for a variety of reasons, there has been a continuous and sometimes bitter debate concerning ends, means, responsibilities and boundaries. The controversy contains a mixture of the following issues:

1. Should the anatomical/physiological approach dominate the psychological/engineering approach or vice versa?

2. Should the application be in association with medicine or with engineering?
3. Should there be a profession of ergonomists or should this title merely indicate a specialism by an individual whose first loyalty is to one of the older disciplines?
4. Should the emphasis be on research or on application?
5. Should the criterion of excellence be academic impeccability or operational success?
6. Should the fundamental method be the laboratory-based experiment or the field-based survey and analysis?
7. Should applied ergonomics concentrate on problems of production or problems of products?

Put in this bald fashion it would be agreed immediately by most people that there cannot be specific answers to such general questions: it always depends on contextual factors such as the objectives of the particular study, the relevance of particular kinds of evidence, the expertise of the particular investigators and so on. Nevertheless, it is a matter of charming naivety, refreshing iconoclasm, boring self-examination, irritating pugnacity or professional insecurity, depending on one's mood, that the unfailing characteristic of any long meeting about ergonomics is at least one session about 'why and whither ergonomics?'. This tendency towards obsessive introspection is a standard characteristic of all new and interdisciplinary topics, motivated partly by interprofessional rivalry and partly by the genuine need to define the new field.

'When the Ergonomics Research Society was first formed, two secretaries were appointed, one psychologist and one physiologist, because neither side would trust the other; there was no president for the same reason' (Murrell, 1967). The proceedings of the first industrial conference (DSIR, 1961) has an introduction by Edholm about the physiological contribution and another by Welford about the psychological contribution. There are separate sessions about 'The Place of Ergonomics in Industry', 'Ergonomics and Production' and 'Ergonomics and Products'. The final session, inevitably about 'The Future of Ergonomics', contains discussions of the scope and application of Ergonomics, and comments by perceptive industrial people on the apparent divergence of approach between physiologists and psychologists, on the relationship between work study and ergonomics and on the need for ergonomics to justify its existence in economic terms.

Anatomy, physiology and psychology

The distinctions between anatomy, physiology and psychology cannot be made with any exactness. Broadly, anatomy is concerned with the structure of the organism, physiology with the way it functions and psychology with the way it behaves, that is adapts, survives and pursues aims in relation to the external world. These distinctions are hazy, however, in that one kind of anatomy highly relevant to ergonomics is functional anatomy, which is very close to physiology. The older morphological anatomy is the origin of anthropometry – the measurement of man which is the foundation of workspace and living space design. Work physiology is a specific topic based on man as a source of energy, but environmental physiology is essentially about adaptation, which overlaps with psychology. Psychology and anatomy share many design problems: for example machine controls involve anatomical issues of physical dimensions and forces and psychological issues of identification and timing characteristics. The extent of the overlaps and the complementarity of the different points of view are the real justification for a science of ergonomics as distinct from a technology.

Differences between physiological and psychological approaches are essentially differences in criteria. These are discussed in considerable detail in Singleton, Fox and Whitfield (1971). Before the conference on which this book is based, the organisers raised three questions:

1. For what kind of ergonomics problems is it more appropriate to use physiological criteria, psychological criteria, or a combination of the two?
2. When both physiological and psychological criteria are used in an experiment, how well do they agree?
3. When principles of equipment design are based on findings achieved in experiments using physiological criteria, do they agree with principles arrived at using psychological criteria?

In answer to the first question it was concluded that the measures taken in either science are highly dependent on specialist interpretation and thus that the skills of the investigators determine the utility of the technique. In relation to the second question it has to be accepted that sometimes there is agreement and sometimes disagreement, depending on the context, because the physiological measures depend on the state of the man and the psychological measures often reflect the total situation. The remedy is to use both. The third question conceals a more basic question as to whether individual well-being or system efficiency is the objective.

Although it is a negation of the total concept of ergonomics to divide it into parts depending on different human sciences, one cannot ignore the fact that many ergonomists have specialised in one of these disciplines both in terms of their original training and in terms of their interests. However, it is increasingly difficult to do research on an ergonomics topic such as stress without using concepts and techniques from more than one discipline, and it is impossible to

Table 1.1. *Contributions of the human sciences to ergonomics*

	Anatomical ergonomics	Physiological ergonomics	Psychological ergonomics
Origin	Morphological and functional anatomy	Whole body physiology	Experimental psychology
Content	Anthropometry; biomechanics	Work physiology; environmental physiology	Sensory, perceptual and cognitive psychology
Models of the human operator	Structure of linked masses Specialised muscle groups about rotating joints	Heat engine using food and oxygen as fuel; sensitive, adaptive but fragile mechanisms	Communication device; skilled practitioners
Practical expertise	Workspace dimensions; seating; lifting; application of forces	Measurement of energy expenditure; measurement of severe environments	Presentation of information; assessment of performance
Current research	Functioning of the low back; optimal postures	Static work; combinations of stressors	Information coding; total information load; human error
Current measurement problems	Dynamic anthropometry; gait measurement	Strain as a consequence of stress	Human skill; human reliability
Common problems	Evaluative criteria for stress, fatigue, effort, comfort and satisfaction		

solve a practical problem such as a workspace design with experts from only one discipline. The present situation is summarised in Table 1.1.

Ergonomics and work study

In a symposium on engineering psychology (Geldard, 1957) an industrial consultant, Ungerson, commented severely on papers by Singleton and Simister and by Seymour with the implication that the same or better results could have been obtained by any competent methods study engineer.

The relationship between work study and ergonomics can only be understood by looking at the history and practice. Work study has been applied on a very wide scale in post-war industry, mainly using time study for the measurement of work necessary for the application of piece-work methods of payment. The complementary study of working methods by motion study was regarded in principle as a necessary preliminary to time study but in practice relatively little effort was devoted to it. All the techniques were developed in the early years of this century and have changed little, partly because they were so effective and partly because the theoretical background was weak (Singleton, 1979). The concepts of effort and fatigue as used by the time study engineer have not proved susceptible to development. The concepts of motion economy have been restructured and extended by the application of principles of anthropometry and biomechanics. In relation to heavy physical work and environmental stress, physiological methods are much more exact and more theoretically defensible. The contribution of the experimental psychologist became important as technology changed the emphasis in so many industrial tasks from energy and manipulation demands to the sensory and perceptual demands of display scanning and inspection tasks (Table 1.2).

However, these differences should not have been sufficient to result in the manifest lack of rapport between practitioners in the two subjects as a result of which both have been the losers. Work study has not yet taken up the opportunity to establish itself on a more secure conceptual foundation and ergonomics has not been able to extend its application widely to production problems. The work study man has learned the difficult art of selling himself and his recommendations to a sceptical management, but to achieve this aim he has concentrated on the criteria of cost reduction and productivity improvement. The ergonomist has maintained strong research interests at the expense of his reputation as an efficient solver of real problems.

Production and product ergonomics

For the production ergonomist the key operator is the production worker; for the product ergonomist the key operator is the product user. There

Table 1.2. *Comparison of work study and ergonomics*

	Work study	Ergonomics
Origin	Shop-floor production problems	Military equipment design
Favoured procedure	Generalisation from observation and experience	Inferences from theory and experimental evidence
Validation	Before/after comparisons	Experimental methods
Special techniques	Estimation by human observers	Acquisition of reliable evidence about behaviour
Advantages and limitations	Required training relatively short (typically a few months)	Required training extensive (typically several years)
	Scientific basis relatively weak but strong on validation by success in practice. Cheap and rapid in execution	Scientific basis relatively strong but weak on short term cost-effectiveness. Slow and expensive in execution
	Sensitive to industrial criteria, skilled in marketing the profession	Relatively insensitive to industrial criteria, hitherto ineffectual in marketing the profession
	Can have long-term disadvantages in excessive emphasis on economic criteria	Has long-term importance for health and safety

Modified from Singleton (1972).

is inevitably overlap if the product, e.g. a machine-tool or a fork-lift truck, is used in the production of other products, but the two are sufficiently different to warrant some separate discussion (Table 1.3).

The production ergonomist is usually faced with problems in the form of on-going processes within which he must make some modifications. His colleagues are work study engineers, production engineers, industrial engineers and occupational health personnel. His primary objectives are likely to be productivity and safety although he will also consider long-term health and satisfaction. The product ergonomist is, or should be, involved throughout the creative process of designing and making something new. His colleagues are industrial designers, design engineers and user-preference specialists. His primary objectives are likely to be safety and efficiency, although he also will consider long-term health and satisfaction. The kinds of operators and their expected level of training vary with the particular system. The production ergonomist is normally dealing with trained and experienced personnel but there may be limited potentiality for further training so that, on grounds of acceptability, he may have to consider operatives with fixed skills. The product ergonomist may have to work within the capacities and abilities of a very wide user population.

Product ergonomists are to be found in industries where design is critical either because the product is so widely used, e.g. chairs and clocks, or because it is complex and misuse might be disastrous, e.g. aircraft and power generation. Production ergonomists are found in industries where efficiency depends heavily on individual worker-skills, e.g. shoe-making and metal working, or where there are unusual environmental hazards, e.g. mining, steel and glass-making. In general ergonomists are not found in small firms; they are in nationalised industries and co-operative research and development institutes where they can serve very large operator populations.

In principle one would expect the production ergonomist to be a close colleague of the industrial engineer and the production engineer. In practice the situation is highly variable because the functions of these engineers vary so widely. In the UK the production engineer may be a specialist in production technology only, e.g. metal working or moulding; insofar as he has a management function he often turns to economics rather than ergonomics as a broadening discipline. Some of the university departments of production engineering have extensive interests in ergonomics but others do not.

Similarly the product engineer or designer is still at the early stage of incorporating ergonomics within his expertise. In the UK there is another related specialist called an industrial designer who has a *beaux arts* rather than an engineering background; he specialises in styling the product although he may also consider function and to this extent he would regard ergonomics as part of

Table 1.3. *Comparison of production and product orientations of ergonomics*

	Production ergonomics	Product ergonomics
Related disciplines	Work study, safety engineering, industrial engineering, occupational health	Design engineering, industrial design, user-preference study
Population considered	Workers	Users
Procedure	Assessment and modification of production machines and processes	Participation in design decisions about new systems
Criteria	Productivity and safety	Safety and efficiency

Table 1.4. *Attributes of ergonomics as a science and as a technology*

	Science	Technology
Origin	Combined approach of anatomy, physiology and psychology to man at work	Man as an integral component of a working system
Methodology	Provision of data and principles on human limitations	Design of human tasks as system functions
Favoured techniques	The laboratory experiment	Task analysis; simulation
Special problems	Assessment of work-load and stress; expression of variability and contextual dependence of human functions	Allocation of function between men, machines and procedures; training versus work design

his responsibility. For many products the ideal design team contains engineers specialising in mechanisms, industrial designers specialising in appearance and ergonomists specialising in relationship to the user, with all of them using a common functional language.

Intermediate discussion

These brief accounts should suffice to clarify why the questions posed in the introductory discussion are too general and too academic. Ergonomics did develop from military and academic roots which are fundamentally different from industry in concern for costs, in time scale and in the characteristics of operator populations. Military personnel are fit young men and most academic experiments have relied on this population, but the industrial population covers a much wider range of age and fitness, as well as differences in sex, culture, attitudes and motivation. Most industrial personnel think in terms of careers, trades and professions. There are limits to what is acceptable or tolerable which have no origin in biology but which are none the less potent because they depend on traditions and demarcations. They change between industries and geographical areas, which makes the concept of the universal human operator even more abstract and meaningless.

Nonetheless these more elusive social limitations only obscure biological limitations, they do not eliminate them. Susceptibility to damaging the back may change with age, sex and previous experience but the principles of avoiding damage to the back by considerations of posture and lifting techniques can be stated independently of these variables. Similarly, although there are differences in sizes and proportions between populations these differences are small compared with the mistakes about dimensions which can be made by the insensitive designer, and in any case such differences are measurable. Thus, even if we assume for the moment that ergonomics is a discipline based in academic research, there is still a responsibility to communicate what is known for the ultimate benefit of the human operator in the widest sense. What to communicate, how to do it and to whom are serious questions which are quite central to ergonomics. Broadly one can try to communicate either data and principles or sensitivity and methods of finding out. The application can be done through specialists or as a subdiscipline within other technologies such as medicine, engineering and industrial design (Table 1.4).

Data and principles

The accumulation, collation and storage of data relevant to people at work was one of the earliest achievements of ergonomics. The first large-scale attempt was the Tufts bibliography (1960). A similar project was started in the

UK by the Department of Scientific and Industrial Research at their Warren Springs Laboratory, and later transferred to the Department of Engineering Production in Birmingham. It appears in journal format as *Ergonomic Abstracts* and provides a search service in relation to specific enquiries. Such comprehensive data banks require a general structure and listings of headings for human factors data. There have been many attempts to summarise and present this kind of information in the form of manuals. Two of the earliest were Woodson (1954) and Murrell (1957). The military services organised considerable efforts in this direction. Morgan *et al.* (1963) was finally published after a preparation phase of more than ten years, in the UK the Medical Research Council first produced a manual in 1960, and this eventually became a joint services manual (MRC, 1970). There have been specialist manuals designed for particular industries, e.g McFarland *et al.* (1954) and Kellerman, van Wely and Williams (1963). More detailed data sets became available from the space programme (Webb, 1964).

It is very difficult to estimate the value of these publications. They are tedious and expensive to prepare, partly because the available data is so widely scattered in the literature but mainly because most of the data are not in the form in which they can be used directly by designers. A considerable process of translation and recoding is required. This varies from relatively simple issues such as changing the medical terminology favoured by anthropometrists into nouns, and more usually into schematic diagrams, which an engineer can be expected to understand, to much more difficult conceptual issues originating in the different objectives and requirements of generators and users. For example, there is a very extensive literature on reaction-times but this is based on the theoretical exploration of man as an abstract communication device and decision maker; it is difficult, if not impossible, to extrapolate from these data to standard ergonomics problems such as time required to scan a battery of dials or time required for a driver to brake in response to the behaviour of the car in front of him. Data in research literature are very rarely expressed in the form of probability functions, so that the designer has no possibility of combining human factors data with other systems data so as to arrive at an optimal cost/value solution to a given problem.

Most human characteristics have very complex contextual dependencies which are not readily expressible in tabulations of numbers or even in multivariate equations. The result is that about the only established numerical knowledge about the human operator is his physical dimensions and exertable forces (McFarland *et al.*, 1954; Dreyfus, 1960), and even for these there are complications due to differences not only between populations of different sex, age, geographical location, economic status, nationality and so on but also because dimensions change in time. In particular they increase as standards of living rise.

Most of the contents of these manuals are general recommendations and principles stated in verbal terms and it is difficult to answer critical design questions such as 'how much does it matter if this recommendation is not followed?'. Thus, the order of difficulty of producing a human factors manual is quite different from that of producing an engineering standards manual. However, there is obviously considerable value in the publication of data in the form of standards on such varied problems as the size of children's desks and the human tolerance of vibration (Whitfield, 1971).

Sensitivity and methods

An alternative approach is to attempt to inculcate in designers an attitude appropriate to the full consideration of human factors. This can be done in several ways including the presentation of short courses. Such courses can be extremely valuable if they concentrate either on a particular issue common to many industries, such as dial design, or on a particular topic, such as the ergonomics of machine tools.

A more document-based way of achieving comprehensive sensitivity is to provide the designer with checklists which he can run through in relation to any problem (Burger and de Jong, 1962). A general checklist has been designed for the International Ergonomics Association (IEA), and is reproduced in Edholm (1967). Again it is difficult to assess how effective this approach has been in practice. How far such lists are used and whether they achieve the desired result has never been evaluated. They are useful as the basis of exercises on ergonomics courses and eventually they might be the origin of regulations about ergonomics if a given list can be accepted for a given purpose as a compulsory requirement in a design appraisal. Ergonomics standards are another way of ensuring that checks are made on designs during the design process. The IEA has taken the initiative in this field and there is extensive work in progress on the formulation of international as well as national standards (Metz, 1976).

The concept of conveying methodology, that is telling the customer not what the answer is but rather how to find out, has interesting potentialities which have not been explored. The difficulty is that no one has yet developed a comprehensive set of crude and approximate but simple and cheap techniques for finding solutions to ergonomics problems. Most of the available techniques are intended for use in the acquisition of evidence which can be reported in scientific journals with appropriate confidence in reliability. If a designer had a way of finding out which would only take a few hours he would be much more likely to use this than to go and search for partially relevant data in a library. On the other hand the prospect of setting up an experiment or simulation which will take weeks is enough to persuade most designers to make a guess at the answer, and improvements on uneducated guesses should not be difficult. For example, techniques of using hardboard mockups and fitting trials with say three subjects across the expected range have much to commend them.

Methods of evaluating machines or systems are also important. They are, of course, essential in the military situation where there is rarely any opportunity to operate the system completely in the situation it has been designed to meet, but they are also useful and necessary when, for example, a large company is considering investing in a new machine on a wide scale, or when an established machine has a poor safety record. Such procedures are rather like the use of checklists in that the evaluation must be systematic and exhaustive. Such practice is bound to increase as ergonomics-type evidence becomes accepted and even required in courts of law when questions of allocating responsibility between say the designer, the manager/owner and the operator arise.

Ergonomics, medicine and engineering

The only feasible vehicle for the widespread application of ergonomics is through other disciplines. This is not a question of principle or theory; it is simply a matter of the numbers of professionals available. Given the length of training for a specialist ergonomist and the number of training places available, the only possibility for the widespread utilisation of ergonomics during this century is through the established technologies. The relationship of ergonomics and work study has already been discussed. There are other possibilities, such as the teaching of ergonomics in schools and technical colleges, but this is likely to be restricted to reasonably well understood and circumscribed topics such as the ergonomics of lifting.

Medicine would seem to have considerable potential because the medical practitioner has some basic knowledge of biological sciences and he is experienced in the art of combining personal observation with fragmented factual evidence. However, there are difficulties: medical biology is different in orientation from ergonomics biology. The medical practitioner knows about structural rather than functional anatomy, his physiology is rather too cellular in emphasis and his psychology is restricted to psychopathology. Perhaps most important of all, his emphasis is entirely towards the person as an isolated entity or at best towards the person in a domestic context. He has little acquaintance with or even interest in the work context. Clearly these comments do not apply to occupational health personnel but these make up only a very small proportion of the total medical profession. Nevertheless, although small, this medical specialism is increasing and its members are closest to the ergonomist in objectives and experience. If there is an established occupational health department in an institution it can provide a highly appropriate environment for the growth of ergonomics in that institution.

Engineering has its own difficulties as a partner. The lack of biology is perhaps not as important as it might appear at first sight: providing there is enough interest it can be acquired readily enough. Ergonomics as a whole is not foreign to the thinking of aeronautical, electrical and control engineers because they are accustomed to dealing in dynamic functional concepts. Unfortunately this is less true of mechanical and civil engineers, although newer graduates in these disciplines are much improved on the more traditional exponents. Even so the engineer's natural reaction to something as complex and untidy as the human operator is either to regard him as infinitely capable or to design him out of systems as something which adds unquantifiable erratic behaviour. The concept of the man as an essential partner to the machine with his own attributes and limitations is not readily acceptable to someone whose training has concentrated entirely on hardware. Admittedly the good engineer involved in the performance of real systems does realise that he cannot avoid human problems and he may go to an ergonomist for help. There are then further difficulties because he would like to be given the unambiguous numerical answers to which he is so accustomed on matters such as strengths of materials; the probabilistic context-dependent statements which he does get do not endear him to ergonomics as a supporting discipline. Again there are exceptions, and individuals from all branches of engineering have made their contribution to ergonomics as a science as well as a technology. A design department will eventually accept ergonomics, and even a specialist ergonomist, if sufficient effort is made to provide the right material and to present it in a manner which is as close as possible to the way the engineer thinks.

These comments, summarised in Table 1.5, are not intended to be patronising or disparaging. It would be quixotic for someone from a new discipline to comment adversely on the procedures of technologies with centuries of achievement as evidence of their success. Rather the intention is to indicate why co-operation, although possible, is never straightforward and why progress is bound to be slow.

Final discussion

There have been other typologies, such as the separation of military, industrial, domestic and sport ergonomics, but there is no difference in theory or methodology across this particular spectrum. Although the typical operator changes and there are differences in emphasis of objectives – e.g. in the military situation the effects on long-term health are given minimal consideration, in the domestic situation there is a dominant emphasis on safety – there are no differences which justify treating the topics as separate disciplines or even approaches.

Table 1.5. *The place of ergonomics between medicine and engineering*

	Medicine	Ergonomics	Engineering
Objective	Prevention and amelioration of disease	Preservation of health and improvement of working efficiency	Making physical devices which further human objectives and support human functions
Supporting sciences	Anatomy, physiology, biochemistry	Anatomy, physiology and psychology	Physics, chemistry and materials science
Specialised knowledge	The human body and its malfunctions	Human advantages and limitations *vis à vis* machines	Characteristics of physical materials and making of engines
Specialised techniques	Clinical diagnosis	Task analysis	Structural analysis, energy analysis
Specialised applications	Occupational health	Work design	Machine design
	Minimisation of accident and long-term health risks	Allocation of function and interface design	
	Reduction of environmental hazards for heat, cold, light, noise and vibration		

Another typology is to separate the man/environment, man/machine and man-in-system relationships. These are different in supporting disciplines and appropriate methodology but usually a specific real problem involves all three relationships – which, of course, is why ergonomics exists as an integral discipline.

The way forward for ergonomics would seem to be to seek friends and converts in other more-established fields and to establish respectability by achievements in practice (ergonomics as a technology) and by generating and structuring reliable knowledge about man at work (ergonomics as a science). This book attempts to support both these aims by presenting what is known in an orderly manner and by relating this to how to apply knowledge in helping to deal with real problems.

There is one final question to be mentioned: the key problems of criteria and evaluation. What exactly is the ergonomist trying to do and how can he find out how successful he has been? These are so closely related as to be inseparable: each determines the other and the fundamental one is the criterion or the criteria. It was stated at the beginning of this section that criteria vary from health through safety to operational effectiveness. Unfortunately all of these terms are relational rather than absolute and ergonomics does not pretend to deal entirely with them, only to contribute towards them. However, even if we cannot define perfect health, safety and effectiveness the other contributors to these ends – medicine and engineering – cannot either, and yet they have achieved acceptability, their successes are acclaimed and their failures are dismissed as merely unfortunate. It seems that criteria for success of the profession need to be discussed separately from criteria for success of particular projects.

Criteria for the success of the profession have already been mentioned: an established range of knowledge and an exclusive methodology. Given sufficient time, society as a whole will eventually accept that for this sort of problem one goes to a specialist with this kind of expertise and most of the time one gets this kind of success. For ergonomics do we know the kind of expertise and the kind of success that is expected? This obviously depends to some extent on the problem. For military systems the criterion is operational effectiveness, for civil aviation systems it is safety, for industrial systems it has tended to be productivity. This last is the one which could do with further examination, particularly as societal attitudes to work and workers are currently changing rapidly. Perhaps the ergonomist should approach a potential employer not with any claim to improve productivity but rather to improve safety and reliability and to maintain the long-term health of the employees. The employer should invest in ergonomics not to increase his profits but because it is his inescapable responsibility to his employees.

This approach, even if it is no more than a change of emphasis, has a number of fundamental implications. It means that ergonomics should depend more on the law to bring the minority of recalcitrant employers into line; it means that ergonomics standards are more important; it means that, other things being equal, ergonomics is closer to medicine than to engineering. In terms of evaluation it means that credibility depends more on a sound theoretical background than on immediate evidence of success for particular projects. Paradoxically, although the *raison d'être* of ergonomics is its practice as a technology, its evaluation is more dependent on its progress as a science. This is not to suggest that in any way the academics are more fundamental than the practitioners: both are needed to develop the subject as a science and as a technology.

1.3 Biological and psychological ergonomics

The distinction between biology and psychology is difficult to sustain with precision; indeed psychology is sometimes defined as a branch of biological science and both are concerned with behaviour. Nevertheless, it is useful to distinguish between studies where description and prediction are based on an understanding of bodily mechanisms, and those based on functional concepts to do with the mind – 'the organised totality of psychical structures and processes' as James Drever put it. In studying behaviour in relation to work we obviously need both, sometimes separately and sometimes in combination. Of the total set of issues to do with man at work, some are best approached through biology and some through psychology, although there will usually be overtones of or insights from the complementary view. The need for eclecticism can complicate understanding because the same words are used with different connotations in different disciplines. Consider for example the title of the last chapter in this volume, 'Hearing and Noise'. 'Hearing' implies essentially a biological mechanism, but 'Noise' for the physicist means a particular form of energy disturbing the air while for the psychologist it means a sensation and an attitude (not wanted) to that sensation. These ambiguities and obscurities can only be overcome by careful definitions and awareness of particular contexts.

If we consider the broad issue of a man working, clearly his primary requirement is energy. The following chapter deals with the source of energy needed for human activity and the properties and limitations of the mechanisms which acquire, process, store and distribute this energy. The effective application of energy can only be through the controlled activities of the musculo-skeletal system. These mechanisms also have their properties and limitations which are relevant to man and work; their study comes under the heading of biomechanics which is the title of the third chapter.

These and all other bodily mechanisms only function optimally within narrow ranges of environmental parameters. Performance can be disturbed and ultimately the body can be damaged in various ways by vibration and acceleration; these matters are dealt with in the fourth chapter. As a warm-blooded creature man is highly susceptible to variation in temperature, and his reaction to heat and cold and his adaptation to changes in this climate variable are described in the fifth chapter.

Man receives much of the information he needs to direct and control his behaviour through the two highly specialised distance receptor systems: vision and hearing. The last two chapters deal respectively with these two systems, exploring not only the biological mechanism but also the associated physical phenomena of light and sound. The end results in discriminations and conscious awareness are described partly in these terms but mainly through the direct measurement of environmental effects on performance.

This is not the whole of ergonomics. There are other important variables to do with human performance generally and variables such as learning, fatigue, individual differences, aspirations and so on which are best approached through psychology rather than biology. However, the material based on biology is perhaps more definite and secure although no more important than the rest, and it is presented here as an integrated subset of ergonomics knowledge.

References

Bennett, E., Degan, J. and Spiegel, J. (1963) *Human Factors in Technology.* New York: McGraw-Hill.

Bernotat, R.K. and Gartner, K.P. (1972) *Displays and Controls.* Amsterdam: Swets and Zeitlinger.

Blain, I. (ed.) (1970) *Occupational Psychology: Jubilee Volume, 44.* London: NIIP.

Burger, G.C.E. and de Jong, J.R. (1962) Aspects of ergonomics job analysis. *Ergonomics,* 5, 185–201.

Chaillet, R.F. (1967) Human Factors requirements for the design of US Army material. In *The Human Operator in Complex Systems,* ed. W.T. Singleton, R.S. Easterby and D. Whitfield, pp. 178–86. London: Taylor and Francis.

Chambers, E.G. (1951) *Psychology and the Industrial Worker.* Cambridge: Cambridge University Press.

Chapanis, A. (1975) (ed.) *Ethnic Variables in Human Factors Engineering.* Baltimore: Johns Hopkins University Press.

Chapanis, A., Garner, W.R. and Morgan, C.T. (1949) *Applied Experimental Psychology.* New York: Wiley.

Chapanis, A., Garner, W.R., Morgan, C.T. and Sanford, F.H. (1947) *Lectures on Men and Machines.* Baltimore: Johns Hopkins University Press.

Christensen, E.H. (1964) *Manual Work: Studies on the Application of Physiology to Working Conditions in a Sub-tropical Country.* Geneva: ILO.

Conrad, R. (1960) Ergonomics in the Post Office. In DSIR: Ergonomics in Industry. London: HMSO.

Daftuar, C.H. (1971) Human factors research in India. *Human Factors,* **13,** 345–53.

Dirken, J.M. and Klinkhamer, H.A.W. (1974) Education and training in ergonomics in the Netherlands. *Ergonomics,* **17,** 709–15.

Dreyfus, H. (1960) *The Measure of Man.* New York: Whitney Library.
DSIR (1961) *Proceedings of Conference on Ergonomics in Industry.* London: HMSO.
Edholm, O.G. (1967) *The Biology of Work.* London: Weidenfeld and Nicolson.
Edholm, O.G. and Murrell, K.F.H. (1973) *The Ergonomics Research Society: A History 1949–1970.* Monograph available from the Secretary, Ergonomics Society.
Faverge, J.M. (1966) *L'Ergonomie de processes industriels.* Brussels: Brussels University Press.
Faverge, J.M., Leplat, J. and Guiguet, B. (1966) *L'Adaptation de la machine a l'homme.* Paris: University Press of France.
Fitts, P.M. (1951) Engineering psychology and equipment design. In *Handbook of Experimental Psychology,* ed. S.S. Stevens, pp.1287–1340. New York: Wiley.
Floyd, W.F. and Welford, A.T. (1953) *Symposium on Fatigue.* London: H.K. Lewis.
Floyd, W.F. and Welford, A.T. (1954) *Human Factors in Equipment Design.* London: H.K. Lewis.
Fowler, R.D. (1969) An overview of human factors in Europe. *Human Factors,* **11,** 91–4.
Gagné, R.M. (ed.) (1962) *Psychological Principles in System Development.* New York: Holt, Rinehart and Winston.
Geldard, F. (convenor) (1957) Engineering psychology: a BPS symposium. *Occupational Psychology,* **31,** 209–56.
Grandjean, E. (1963) *Physiologische Arbeitsgestaltung.* Munich: Otto Verlag. (Published in English in 1969 as *Fitting the Task to the Man: An Ergonomic Approach.* London: Taylor & Francis.)
Grether, W.F. (1968) Engineering psychology. *American Psychologist.*
Grime, G. (1958) Research on human factors on road transport. *Ergonomics,* **1,** 151–62.
Hacker, W. (1973) *Allgemeine Arbeits - und Ingeniur-psychologie* Berlin: UEB.
Holmes, K.S., Floyd, W.F. and Conrad, R. (1960) Ergonomics in the Post Office. In DSIR: *Ergonomics in Industry.* London: HMSO.
Ivergard, T. (1969) *Information ergonomi.* Stockholm: Raben and Sjogren.
Kellerman, F.Th., van Wely, P.A. and Williams, P.J. (1963) *Vademecum: Ergonomics in Industry.* Eindhoven: Philips.
Klinkhamer, H.A.W. (1970) Ergonomics around the world: the Netherlands, *Applied Ergonomics,* **1,** 166–7.
Kraft, J.A. (1958) A follow-up survey of human factors research in aircraft, missiles and supporting industries. *Human Factors,* **1,** 23–5.
Kraft, J.A. (1962) The 1961 picture of human factors research in business and industry in the USA. *Ergonomics,* **5,** 293–9.
Krivohlavy, J. (1969) Engineering psychology: Czechoslovakia. *Human Factors,* **11,** 87–90.
Laitakari, L.K. von (1962) Biotechnologie in Finnland. (In German.) *Ergonomics,* **5,** 301–10.
Laner, S., Crossman, E.R.F.W. and Hellon, R.F. (1960) Ergonomics in the steel industry. In DSIR: *Ergonomics in Industry.* London: HMSO.
Lehmann, G. (1953) *Praktische Arbeits Physiologie.* Stuttgart: George Thieme.
Leplat, J. and Bisseret, A. (1966) Analysis of the processes involved in the treatment of information by the air traffic controller. *The Controller,* **5,** 13.
Lindsley, D.B. (ed.) (1949) *Human Factors in Undersea Warfare.* Washington: National Research Council.
Lippert, S. (1967) Ergonomics needs in developing countries. *Ergonomics,* **10,** 617–26.
Lomov, B.F. (1964) *Ingenieur Psychologie.* Berlin: UEB Verlag. (Originally published in Russian in 1963).
Lomov, B.F. and Berrane, C.M. (1969) The Soviet view of engineering psychology. *Human Factors,* **11,** 69–74.
Luthman, G., Aberg, N. and Lundgren, N. (eds.) (1966) *Handbok i ergonomi.* Stockholm: Almquist and Wiksell.

McFarland, R.A. (1953) *Human Factors in Air Transportation.* New York: McGraw-Hill.
McFarland, R.A. (1954) *Human Factors in Highway Transport Safety.* Boston, Mass: Harvard School of Public Health.
McFarland, R.A. (1971) Ergonomics around the world: the United States of America. *Applied Ergonomics,* **2,** 19–25.
McFarland, R.A., Damon, A., Standt, H.W., Moseley, A.C., Dunlap, J.W. and Hall, W.A. (1954) *Human Body Size and Capabilities in the Design and Operation of Vehicular Equipment.* Boston, Mass: Harvard School of Public Health.
Maier, N.R.F. (1946) *Psychology in Industry.* Boston: Houghton Mifflin.
Maule, H.G. and Weiner, J.S. (1975) Some observations on the teaching of ergonomics. In *Ethnic Variables in Human Factors Engineering,* ed. A. Chapanis. Baltimore: Johns Hopkins University Press.
Metz, B. (ed.) (1960) *Fitting the Job to the Worker: Zurich.* Paris: OEEC.
Metz, B. (1976) Ergonomics and standards. *Ergonomics,* **19,** 271–4.
Montmollin, M. de (1967) *Les systèmes hommes-machines.* Paris: University Press of France.
Morgan, C.T., Cook, J.S., Chapanis, A. and Lund, M.W. (1963) *Human Engineering Guide to Equipment Design.* New York: McGraw-Hill.
MRC (1960) *Human Factors in the Design and Use of Naval Equipment.* MRC-RNPRG 60/962. London: Medical Research Council.
MRC (1970) *Human Factors for the Design of Naval Equipment.* London: Medical Research Council.
Murrell, K.F.H. (1957) *Data on Human Performance for Engineering Designers.* London: Engineering.
Murrell, K.F.H. (1958) *Fitting the Job to the Worker: USA and Leiden.* Paris: EPA/OEEC.
Murrell, K.F.H. (1965) *Ergonomics.* London: Chapman and Hall.
Murrell, K.F.H. (Convenor) (1967) Why ergonomics? *Occupational Psychology* **41,** 17–24.
Oshima, M. (1969) Ergonomics in the design of consumer products in Japan. *Ergonomics,* **12,** 701–12.
Oshima, M. (1970) Ergonomics around the world: Japan. *Applied Ergonomics,* **1,** 70–2.
Parrot, J. (1971) The measurement of stress and strain. In *Measurement of Man at Work,* ed. W.T. Singleton, J.G. Fox and D. Whitfield, pp. 27–34. London: Taylor and Francis.
Rohmert, W. (1962) *Untersuchungen über Umselermüdung und Arbeitsgestaltung.* Berlin: Beuth Vertrieb.
Rohmert, W. (1973) *Psycho-physische Belastung und Beanspruchung von Fluglotsen.* Berlin: Beuth Vertrieb.
Saric, I. (1970) *Manniska-Maskinsystem.* Stockholm: P.A. Radet.
Schmidtke, H. (ed.) (1971) *Ergonomie.* Munich: Hanser.
Sell, R.G. (1969) Ergonomics around the world. *Applied Ergonomics,* **1,** 62–4.
Sell, R.G. (1971) The Human Factors Section of BISRA 1950–1964. *Applied Ergonomics,* **2,** 226–9.
Seminara, J.L. (1975) Human factors in Romania. *Human Factors,* **17,** 477–87.
Seminara, J.L. (1976) Human factors in Bulgaria. *Human Factors,* **18,** 33–44.
Seymour, W.D. (1968) *Skills Analysis Training.* London: Pitman.
Shackel, B. (ed.) (1974) *Applied Ergonomics Handbook.* Guildford, Surrey: IPC.
Shackel, B. and Van Nes, F.L. (1975) Ergonomics in Brazil. *Applied Ergonomics,* **6,** 43–4.
Shaw, A.G. (1952) *The Purpose and Practice of Motion Study.* London: Harlequin Press.
Singleton, W.T. (1958) Production problems in the shoe industry. *Ergonomics,* **1,** 307–13.
Singleton, W.T. (1960) Application in machine design. In DSIR: *Ergonomics in Industry.* London: HMSO.
Singleton, W.T. (1972) *Introduction to Ergonomics.* Geneva: World Health Organisation.
Singleton, W.T. (1973) Ergonomics: comparative problems in India and Europe. *Behaviormetrics,* **3,** 57–66.

Singleton, W.T. (1979) *The Study of Real Skills*, vol. 2, *Compliance and Excellence*. Lancaster: MTP Press.

Singleton, W.T. and Whitfield, D.C. (1968) The organisation and conduct of a WHO inter-regional course on ergonomics for developing countries. *Human Factors*, **10**, 633–40.

Singleton, W.T., Fox, J.G. and Whitfield, D.C. (1971) *Measurement of Man at Work*. London: Taylor and Francis.

Taylor, F.V. (1957) Psychology and the design of machines. *American Psychologist*, **12**, 249–58.

Thompson, D. (1972) The application of ergonomics to developing countries. *Applied Ergonomics*, **3**, 92–6.

Timpe, K.P. (1969) Engineering psychology in the German Democratic Republic. *Human Factors*, **11**, 81–6.

Tsaneva, N. (1972) Fatigue at work. Paper given at the Fourth Scandinavian Conference on Ergonomics, Helsinki.

Tufts (1960) *Handbook of Engineering Data*, 2nd edn. Medford, Mass,: Tufts College.

Vernon, P.E. and Parry, J.B. (1949) *Personnel Selection in the British Armed Forces*. London: London University Press.

Viteles, M.S. (1933) *Industrial Psychology*. London: Cape.

Webb, P. (ed.) (1964) *Bioastronautics Data Handbook*. Washington, DC: NASA.

Welford, A.T. (ed.) (1962) Proceedings of the First International Congress on Ergonomics. *Ergonomics*, 5.

Whitfield, D. (1971) British Standards and ergonomics. *Applied Ergonomics*, **2**, 236–42.

Wood, C.C. (1958) Human factors engineering: an aircraft company Chief Engineer's viewpoint. *Ergonomics*, **1**, 294–300.

Woodson, W.E. (1954) *Human Engineering Guide for Equipment Designers*. Berkeley: University of California Press.

2 THE PROVISION OF ENERGY

W.R. Withey

2.1 Existing knowledge

Every living cell in the body requires energy to function (to do work): if it is a nerve cell it requires energy to generate nervous impulses; if it is a muscle cell it requires energy to contract. The energy for all these processes is provided by the internal chemical processing (metabolism) of the food we eat. During metabolism the food is broken down (catabolised) by combination with oxygen from the inspired air (oxidation) in a series of closely regulated steps. Energy is liberated and water and carbon dioxide produced:

$$\text{food} + \text{oxygen} \longrightarrow \text{energy} + \text{carbon dioxide} + \text{water}$$

Obviously this simple outline neglects the detailed aspects of the system, but it will serve as a guide to the factors which are important in the processes which ultimately enable the body to function.

Composition of the diet

Since food is converted to energy, and since energy cannot be created or destroyed (First Law of Thermodynamics), it follows that in order to obtain sufficient energy to maintain an active life the body must be supplied with sufficient food. Too small an intake of food yields insufficient energy and the deficit must be made up from body stores of carbohydrates and fats. If these are depleted too severely (as in chronic starvation) then the vital organs will cease to function effectively and death will ensue. If too much food is eaten, then that food which is not needed to provide energy for body functions will be stored as fat, and the person who overeats will become obese. Maintaining a correct energy balance is therefore an essential part of providing the energy for life. Before discussing energy balance, though, it is necessary to consider how the *type* of food we eat is important.

All food has three main components: carbohydrates, fats and proteins. In any one food the amount of each of these is fixed, but foods vary enormously in their composition. For instance, sugar is 100% carbohydrate, butter is mostly (80%) fat, and most meats contain a large percentage of protein. For a detailed discussion of the composition of individual foods see Paul and Southgate (1978). It is essential that the diet contains adequate amounts of each of these main components, otherwise the building processes (anabolism) necessary in the body for growth, reproduction, repair of tissues, etc., cannot take place. However, a satisfactory diet not only contains carbohydrates, fats and proteins in adequate amounts, but also contains them in an optimum ratio of about 60% carbohydrate, 20% fat and 20% protein. Any major deviation from this composition can lead to metabolic disorders. For instance, a diet rich in carbohydrate may contain insufficient protein to ensure a supply of components for anabolism, while a fat-rich diet can lead to poor absorption of the dietary components from the intestine.

It is essential that the diet contains, in addition to carbohydrates, fats and proteins, minute amounts of other chemicals: minerals, vitamins and essential amino acids. However, their role in supplying energy for body functions is indirect so only brief mention of them will be made.

Minerals. Some elements such as calcium, phosphorus, iodine, iron, magnesium and zinc are essential to the body. The daily requirements vary with age, body weight and gender but typically range from about 100 μg of iodine to about 1 g of calcium. The role of each of the minerals required by the body is highly specific and their absence leads to serious ill-effects. For instance, lack of calcium leads to bone abnormalities whilst iodine deficiency causes goitre.

Vitamins. Vitamins are small molecules important in the intermediary metabolism (see below) of carbohydrates, fats and proteins. About a dozen vitamins (or groups of vitamins) are known and they fall into two different groups: water-soluble vitamins (vitamins C, B_2 (riboflavin), B_1 (thiamine), etc.) and fat-soluble vitamins (vitamins A, D, E and K). Symptoms of vitamin deficiency can range from the relatively minor (e.g. night blindness caused by vitamin A deficiency) to the fatal (e.g. scurvy caused by vitamin C deficiency).

Essential amino acids. Amino acids are the building-blocks of protein. Of the 21 types commonly found in animal protein 10 cannot be synthesised by man and must therefore be taken in with the diet. Those 10 are thus known as the essential amino acids. The role of each acid is specific, but none is directly related to energy production in an otherwise well-fed person.

For a further discussion of these dietary components see Harper, Rodwell and Mayes (1977). Their concentration in different foods is given in detail by Paul and Southgate (1978).

Energy balance

Each main component of the diet yields a constant amount of energy: 1 g of carbohydrate or protein produces about 4 kilocalories (kcal) of energy,

1 g of fat produces about 9 kcal. (One calorie is the quantity of heat required to raise the temperature of water from 15 to 16°C, and is equivalent to 4.18 joules (J): see Table 2.5.) Thus it is possible by ascertaining the composition of a diet to estimate its energy value. For example, a typical daily menu for a fairly active male might be 350 g carbohydrate, 100 g fat, 100 g protein. This would yield approximately:

$$(350 \times 4) + (100 \times 9) + (100 \times 4) \text{ kcal} = 2700 \text{ kcal } (11\,300 \text{ kJ})$$

This daily intake is adequate to sustain life indefinitely. However, if the daily intake of carbohydrate were reduced to 200 g without a corresponding increase in the other components, the daily energy intake would fall to 2100 kcal (8800 kJ), and this would be too small a supply of energy to keep the person active. The deficit of 600 kcal would be obtained by the catabolism of body fat and the person would lose weight. Such a weight loss would ultimately be catastrophic.

Similarly if the intake of carbohydrate were to be doubled to 700 g the diet would then provide 4100 kcal (17 100 kJ) and the person would convert the daily surplus of 1400 kcal to fat which would be stored. It is essential, therefore, to maintain an energy balance. A surplus of only 100 kcal (418 kJ) each day could lead to a gain of more than 6 kg of fat in a year. What is surprising is that in most people the mechanism which balances energy intake and output is so finely tuned that body weight remains essentially constant for very long periods. This indicates that the appetite is under very close control. For a discussion of factors influencing energy balance and obesity in man see Garrow (1978).

Biochemistry of dietary components

Carbohydrates

All carbohydrates are composed of carbon, hydrogen and oxygen. There are four major classes:

Monosaccharides, with the general formula $C_nH_{2n}O_n$. There may be three, four, five, six or seven carbon atoms in the monosaccharide unit. The most important monosaccharides are the hexoses (six carbon units) glucose, fructose and galactose:

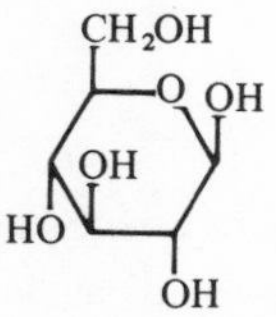

Glucose

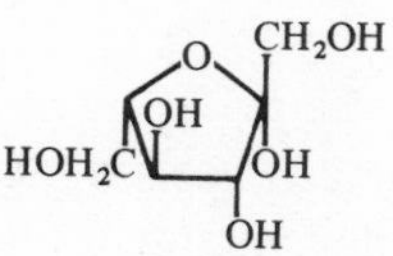

Fructose

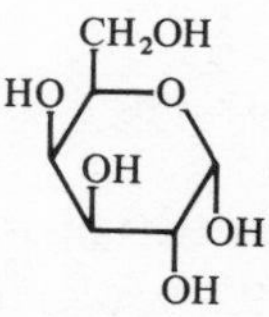

Galactose

It is not appropriate to discuss here the complex chemistry of monosaccharides. For a fuller discussion see Harper *et al.* (1977).

Disaccharides, which are composed of two monosaccharides. For example, sucrose is composed of one unit of glucose and one of fructose:

CH_2OH O CH_2OH O OH HO O HO CH_2OH OH OH

This can be broken down by enzymes in the body to its constituent monosaccharides.

Oligosaccharides, which are composed of between three and six monosaccharides.

Polysaccharides, which are composed of more than six monosaccharides. Typical polysaccharides are starch and inulin, which are found in vegetable matter; and glycogen, which is found in muscle and liver.

Fats (lipids)

The exact nomenclature of fats is still under review, but for general purposes dietary fats can be considered as being either fatty acids or their derivatives (triglycerides, phospholipids, etc.) or steroids such as cholesterol.

Fatty acids, with the general formula $C_nH_{2n+1}COOH$ (where n can be up to 23) are termed 'saturated' fatty acids, i.e. there are no double bonds between carbon atoms. Another group of fatty acids is termed 'unsaturated' and these have fewer hydrogen atoms, giving general formulae of the type $C_nH_{2n-x}COOH$, where x is 1, 3, 5 or 7 etc., depending upon the degree of unsaturation. Fatty acids combine with glycerol to form an important group of compounds called triglycerides. (R represents a fatty acid):

$CHOC(=O)-R$
$CHOC(=O)-R$
$CHOC(=O)-R$

Triglycerides are important in the diet because they yield more energy upon oxidation (9.3 kcal/g) than do carbohydrates or proteins. Fatty acids also combine with a phosphoric acid residue

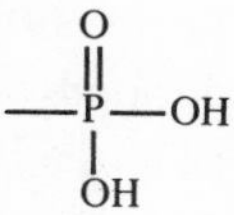

$$-\overset{\overset{\displaystyle O}{\|}}{\underset{\underset{\displaystyle OH}{|}}{P}}-OH$$

to form phospholipids, which are essential components of all cell membranes.

Steriods are a group of molecules which contains 19 carbon atoms. A typical member of the steroids is cholesterol, which is found in the diet, in many animal

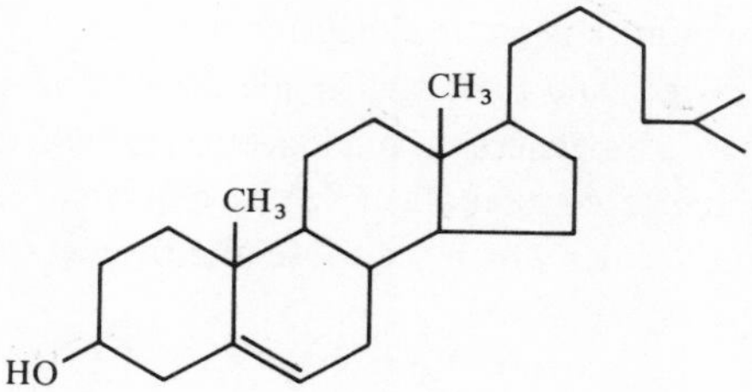

CH_3
CH_3
HO

tissues and is important in a wide range of cellular reactions. Its role in the provision of energy is small and the steroids will not be discussed further.

Proteins

Proteins are a group of molecules (some are the largest found in the body) which are composed of chains of sub-units called amino acids. The architecture of proteins is complex. Their primary structure is formed by them linking together in a chain:

– amino acid – amino acid – amino acid – amino acid –

The chain may be tens of thousands of amino acids in length. About 20 different amino acids are commonly found in proteins, so the possible number of different proteins is astronomical. The general formula of an amino acid is

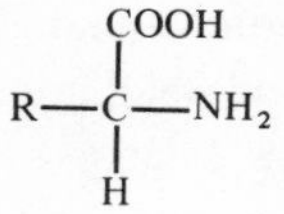

$$R-\overset{\overset{\displaystyle COOH}{|}}{\underset{\underset{\displaystyle H}{|}}{C}}-NH_2$$

where R may be a simple moiety such as $-CH_3$ or a complex one such as

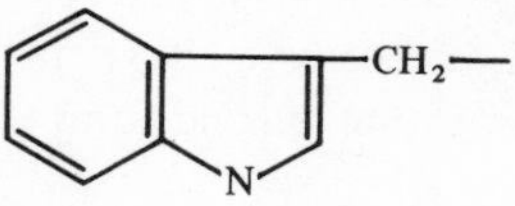

CH_2-
N

Proteins do not remain in simple chains as shown above, but coil into a helical structure as the result of the formation of hydrogen or disulphide bonds between amino acids. This coiling is termed their secondary structure. The tertiary structure is formed by the further folding and coiling of the chain, and some proteins have a quaternary structure formed when two or more molecules link together.

Proteins are essential for growth and repair processes in the body, but from an energy-producing viewpoint are the least important group of dietary components and so will be discussed only in the context of their direct role in the intermediary metabolism of carbohydrates and fats. Any amino acids which are surplus to the body's immediate requirements for anabolic processes are deaminated (have the $-NH_2$ group removed) and then further metabolised to produce energy by two main routes: (1) by ketogenesis to oxaloacetic acid which is an important intermediate in carbohydrate metabolism, (2) by gluconeogenesis, i.e. building up of new glucose molecules which are subsequently metabolised to yield energy.

Production of energy from dietary components

Digestion of food

The purpose of the digestion of food, which takes place in the gastro-intestinal tract, is to produce from the complex molecules ingested, simple, small molecules which can be absorbed from the large intestine or small intestine into the blood. These small molecules (intermediary metabolites) are then transported to the sites of their final degradation to yield energy.

Food digestion begins in the mouth where the food is masticated. Enzymes contained in the saliva begin to digest starches to smaller chains of sugar molecules. In the stomach (the contents of which are strongly acid) the proteins are attacked by pepsin enzymes to yield some amino acids and small chains of amino acids (polypeptides). The semi-solid food mass (chyme) then passes into the duodenum where it is neutralised by secretions from the pancreas to provide a suitable environment for the pancreatic enzymes to continue the catabolism of proteins. Lipase enzymes also degrade triglycerides to mono- and diglycerides and fatty acids, and other enzymes attack other lipids. In the small intestine the process of metabolism of carbohydrates to glucose, fructose, galactose and other monosaccharides is completed; fats are broken down to fatty acids and glycerol; and polypeptides are completely digested to amino acids. The formation of intermediary metabolites is thus complete (Fig. 2.1).

The intermediary metabolites are absorbed through the walls of the intestines, enter the bloodstream and are transported to cells where they are used. Any unabsorbed material leaves the large intestine in solid form as faeces.

Intermediary metabolism of carbohydrates

Blood glucose (a six-carbon molecule) has several possible fates, depending upon body needs:

1. Conversion to glycogen (glycogenesis) in the liver. This is a readily reversible process which allows the storage of sugar molecules in a form available to yield energy at times of demand. The breakdown of glycogen to glucose is called glycogenolysis.
2. Breakdown to three-carbon molecules by a route called the Embden–Meyerhof pathway.
3. Direct breakdown via the hexose monophosphate shunt to three-carbon molecules.

The three-carbon (triose) molecules formed by these processes are further metabolised to pyruvic acid (pyruvate) which then undergoes one of two types of

Fig. 2.1. Pathways of intermediary metabolism of carbohydrates.

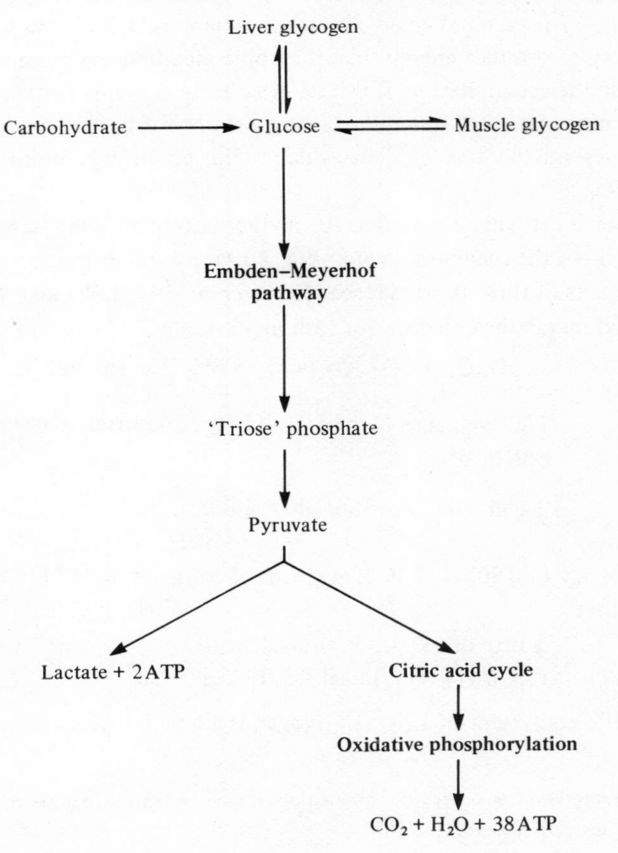

degradation depending upon the amount of oxygen available to the cells. If adequate oxygen is present, the three-carbon pyruvate molecule is oxidised to carbon dioxide and water in a complex series of events. First, it is converted to acetyl-coenzyme A which is metabolised via the citric acid (tricarboxylic acid or Krebs) cycle, in which one pyruvate molecule yields three carbon dioxide (CO_2) molecules and five pairs of hydrogen atoms. In this process one molecule of the high-energy compound adenosine triphosphate (ATP) is formed. Next, in a process called oxidative phosphorylation, the hydrogen atoms are oxidised by the components of the cytochrome chain to form water molecules, which releases energy for the generation of more ATP. This whole process of glucose (or glycogen) metabolism in the presence of oxygen is called aerobic metabolism. It produces overall 38 molecules of ATP for each molecule of glucose oxidised.

If there is inadequate oxygen to allow oxidative phosphorylation, the glucose molecules are broken down by an anaerobic process which forms lactic acid (lactate). In this process there is a net gain of only two molecules of ATP (or about 15 kcal of energy) for each molecule of glucose metabolised, and is thus a much less efficient way to produce energy than is aerobic metabolism. It is also self-limiting because the accumulation of lactate effectively prevents further metabolism by that route. To break down the lactate a supply of oxygen is needed and this supply represents the 'oxygen debt' which builds up during continuous exercise.

Thus in summary it will be seen that the metabolism of carbohydrates to produce energy follows the scheme shown in Fig. 2.1 (for a full discussion of the biochemical details of these processes see Harper *et al.*, 1977). We may therefore write the overall metabolic equation for carbohydrates as:

$$C_6H_{12}O_6 + 6O_2 \longrightarrow 6CO_2 + 6H_2O + \text{'energy'}$$

or

$$180 \text{ g glucose (1 mole)} + (6 \times 22.4) \text{ litres oxygen} \longrightarrow 686 \text{ kcal}^*$$

thus

$$1 \text{ g glucose} \longrightarrow 686/180 = 3.8 \text{ kcal}$$

this uses

$$(1/180) \times (6 \times 22.4) \text{ litres of oxygen} = 0.747 \text{ litres}$$

therefore

$$1 \text{ litre of oxygen is equivalent to } (1/0.747) \times 3.8 \text{ kcal} = 5.09 \text{ kcal}$$

Thus the calorific equivalent of 1 litre of oxygen is about 5.1 kcal.

* The standard free-energy of combustion of glucose (various figures may be found in standard texts).

Since 38 molecules of ATP are formed for each molecule of glucose oxidised, the amount of energy available as ATP is $38 \times 7.3^* = 277$ kcal, giving an overall efficiency for glycolysis of $277/686 \times 100 = 40\%$. This figure represents the minimum likely value, but some authors quote an efficiency of 60% or more. Lehninger (1971) and Becker (1977) discuss this in greater detail. The energy which is not trapped as high-energy chemical bonds in ATP is released as heat. It is this heat which maintains body temperature.

Intermediary metabolism of fats

Fat absorbed from the intestine is usually deposited in the body as a triglyceride unless it is required immediately. When it is needed to supply energy, the triglyceride is first hydrolysed to glycerol and free fatty acids (FFA or non-esterified fatty acids (NEFA)). The glycerol becomes phosphorylated to triose phosphate and is metabolised to carbon dioxide and water, with production of ATP by the carbohydrate pathways previously described.

The FFA are metabolised by the process of β-oxidation to give acetyl-coenzyme A which then enters the citric acid cycle. Thus the central role of the citric acid cycle in producing energy in the form of ATP is readily apparent. The energetics of FFA metabolism are such that one six-carbon molecule of FFA yields 44 molecules of ATP when it is completely oxidised (compared with 38 molecules of ATP for a six-carbon sugar). The process of intermediary metabolism of fats can be summarised as shown in Fig. 2.2.

Fig. 2.2. Pathways of intermediary metabolism of fats.

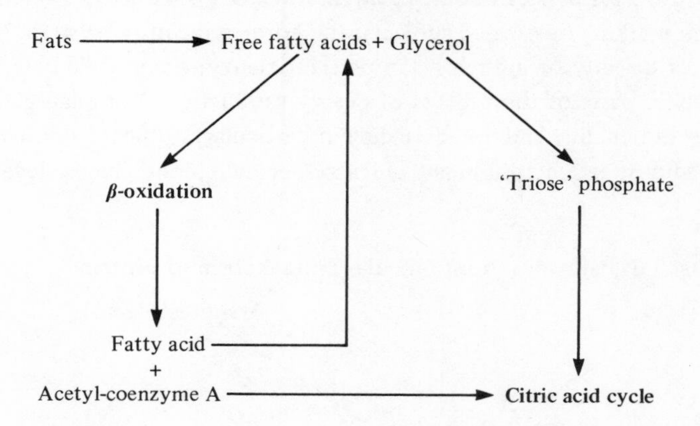

* The standard free-energy of hydrolysis of ATP to ADP (adenosine diphosphate).

Intermediary metabolism of protein

The amino acids absorbed from the gastro-intestinal tract form a pool upon which the body can draw for synthesis of new proteins. Any amino acids surplus to this requirement can be used to supply energy. The exact pathway of intermediary metabolism varies between the amino acids. Usually the first step is removal of the amino ($-NH_2$) group by a deamination process. In the case of alanine this deamination leads to the formation of pyruvate which is metabolised as previously described. Similarly, aspartic and glutamic acids are deaminated to give compounds which themselves form a part of the citric acid cycle. All the amino acids have comparable pathways of catabolism (see Fig. 2.3) and thus play an important role in the production of energy in the form of ATP synthesis.

Control of intermediary metabolism

At a simple level, the control of ATP synthesis can be regarded as being related to need. If the substrate is available to fuel the various pathways of intermediary metabolism then ATP synthesis continues. If, however, the products of the various reactions accumulate because they are not being used in energy-demanding processes, then the sequence of metabolism will slow and may even reverse. Control is also exerted at other levels. For instance, each of the very many steps in each metabolic pathway is enzyme-controlled. It is, therefore, essential that these enzymes are available. In some cases coenzymes or metal ions are required too. The rate of many of the enzyme reactions is variable according to complex interactions occurring within the cell.

There is also a major set of controls on intermediary metabolism which is important in working or stressed subjects: the hormonal control system. Many hormones act directly or indirectly on particular enzyme-controlled steps to control specific parts of the process of energy production. For example, catecholamines (adrenaline and noradrenaline in particular) influence the release of FFA from adipose tissue; insulin and glucagon regulate blood glucose level by

Fig. 2.3. Pathways of intermediary metabolism of protein.

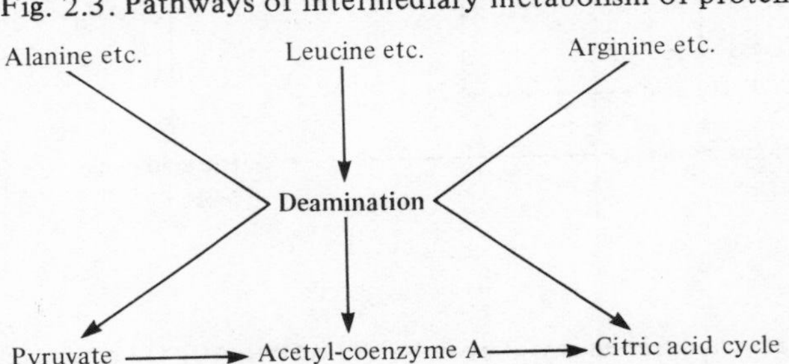

controlling glycogenesis and glycogenolysis; the glucocorticoids from the adrenal cortex increase the release of FFA from the tissues and raise blood glucose levels.

It is apparent, then, that the production of energy from the components of the diet is under a complex series of controls. It is not possible to consider any one metabolic pathway, or one substrate, in isolation. It will be readily appreciated, however, that what has been described above could not occur in an efficient manner in the absence of oxygen. Therefore we must now examine the way in which oxygen reaches the tissues and the importance this has in allowing us to assess the energy requirements of various activities.

Oxygen uptake

The production of energy-rich compounds such as ATP requires oxygen. This comes from the air which is breathed in (inspired). The process of oxygen transfer from the air to the cells is called respiration. Two distinct phases of respiration are recognised: first, external respiration whereby oxygen is taken up and carbon dioxide is lost from the body; and secondly internal respiration, the process of the carriage of gas in the blood and the exchange of oxygen and carbon dioxide between the blood and the cells.

External respiration

The lungs are compliant organs which lie in the closed thoracic (chest) cavity. The outer surface of the lung is 'stuck' to the inner surface of the thoracic wall by the forces of surface tension, and thus when the thoracic wall

Fig. 2.4. Processes involved in external respiration.

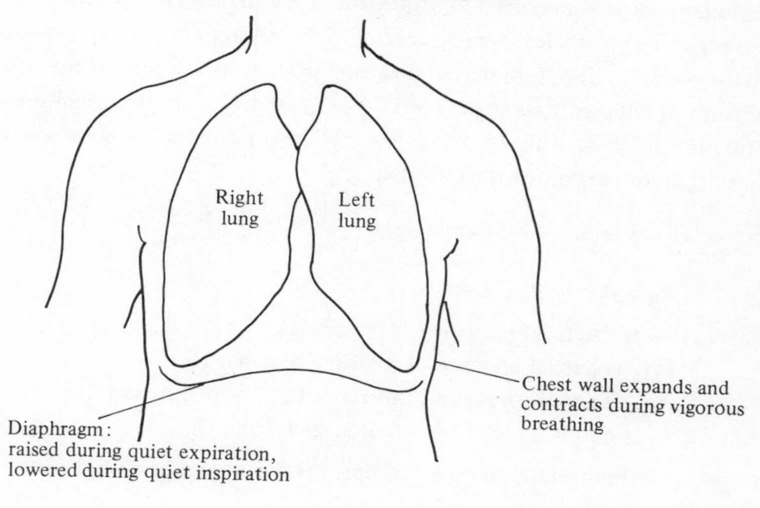

expands so do the lungs. The size of the lungs is also controlled by the position of the diaphragm, a thin muscular sheet which forms the floor of the thorax. In normal, quiet breathing the diaphragm is raised and lowered under control from the central nervous system. This cyclic change in the position of the diaphragm is responsible for the movement of air in and out of the lungs (Fig. 2.4).

So we can now describe the process of external respiration in greater detail. Under control of the respiratory centre in the brain, the diaphragm is lowered. This increases the volume of the thorax and thus of the lungs and therefore lowers the pressure inside the lungs (according to Boyle's law pressure of a gas is proportional to its volume), causing air to enter them through the mouth. The volume of air entering the lungs is called the tidal volume, and at rest is about 0.5 litres. During expiration the diaphragm is raised, the thoracic volume is decreased, the thoracic (hence lung) pressure rises and air flows out of the lungs. This process is repeated about 15 times each minute at rest, giving a total volume breathed each minute (minute volume) of about 7.5 litres.

During exercise, minute volume is very much greater – up to 75 litres or more. This is brought about by an increase in both respiratory frequency (30 breaths or more per minute) and tidal volume (which can rise to as much as 3 litres). Whereas at rest tidal volume is controlled almost entirely by movement of the diaphragm, in exercise the chest wall expands and contracts under the control of the muscles between the ribs (intercostal muscles). The muscles themselves are controlled by the respiratory centre in the brain.

Movement of oxygen into the blood

It is the function of external respiration processes to get oxygen into the lungs so that it may move into the blood by the process of diffusion: a movement of molecules from an area of high partial pressure to an area of low partial pressure. The term partial pressure means that fraction of the total gas pressure which a particular gas exerts. For example, in air at normal pressure (760 mm Hg) 21% of the air is oxygen, 78% nitrogen and 1% other gases (carbon dioxide, argon, krypton, etc.), so that:

$$\text{total pressure (760 mm Hg)} = \left(\frac{21}{100} \times 760\right) + \left(\frac{78}{100} \times 760\right) + \left(\frac{1}{100} \times 760\right)$$

or $$P_B = P_{O_2} + P_{N_2} + P_x$$

where P_B is the total pressure (760 mm Hg);
P_{O_2} is partial pressure of oxygen (160 mm Hg);
P_{N_2} is partial pressure of nitrogen (593 mm Hg); and
P_x is partial pressure of other gases (8 mm Hg).

In the lung, the situation is more complex. First, the air is always saturated with water vapour. At normal body temperature (37 °C) this gives a partial pressure of water vapour about 47 mm Hg. Secondly, the lung always contains 40 mm Hg partial pressure of carbon dioxide. Thus in the lung

$$P_B = P_{O_2} + P_{N_2} + P_{H_2O} + P_{CO_2}$$

(P_x has been omitted because of its very small value)

where, normally, P_{O_2} is about 100 mm Hg, P_{N_2} about 570 mm Hg and P_{H_2O} and P_{CO_2} as given above.

In the blood entering the alveolar capillaries (the capillaries surrounding the small air sacs in the lungs) the partial pressures are different again. This is because the capillaries contain mixed venous blood, i.e. 'used' blood returning from the organs that is replete in carbon dioxide and deplete in oxygen. For mixed venous blood P_{O_2} is about 40 mm Hg, P_{CO_2} about 46 mm Hg and P_{N_2} and P_{H_2O} the same as the lung values. Table 2.1 summarises the position so far.

We are now in a position to understand why oxygen enters the blood of the alveolar capillaries: it does so because the partial pressure of oxygen in the mixed venous blood is very much lower than that in the lung (alveolar) air. It therefore diffuses from the lung and enters the blood. Simultaneously, the carbon dioxide in the mixed venous blood diffuses out into the alveolar air along its partial pressure gradient. Thus the overall process occurring in the alveolar capillaries is for the blood to be enriched with oxygen (the P_{O_2} rises from 40 to 95 mm Hg) and depleted of carbon dioxide (the P_{CO_2} falls from 46 to 40 mm Hg).

The carriage of oxygen to the tissues is the process of internal respiration.

Internal respiration

Once oxygen has diffused into the blood from the alveolar air it dissolves in the plasma. However, since 100 ml of blood can dissolve only about 0.3 ml of oxygen, and this is clearly an inadequate amount to keep the body supplied with oxygen without an enormous throughput of blood by the heart, a further series of reactions occurs. The oxygen enters the red blood cells and

Table 2.1. *Air and blood gas partial pressures*

	Partial pressure (mm Hg) of gases in			
	Inspired air	Lungs	Oxgenated blood	Mixed venous blood
Oxygen	160	100	95	40
Carbon dioxide	0.2	40	40	46
Nitrogen	593	570	570	570
Water vapour	0	47	47	47

here combines with the haemoglobin they contain to form oxyhaemoglobin. Normal blood contains about 15 g haemoglobin/100 ml and this can combine with about 20 ml of oxygen: thus haemoglobin increases the oxygen-carrying capacity of the blood by nearly 70-fold. The diffusion and combination reactions are much more complex than this simple description supposes, but for the purposes of understanding the role of oxygen in the supply of energy such a description is adequate.

In the tissues the reverse process occurs: oxyhaemoglobin dissociates to release oxygen. The amount of oxygen released depends upon a number of factors such as the body temperature, the pH of the blood and the partial pressure of carbon dioxide in the tissues. The best way to visualise these effects is diagrammatically by means of the oxyhaemoglobin dissociation curve (Fig. 2.5).

Oxyhaemoglobin dissociation curve

The amount of oxygen with which haemoglobin combines is dependent upon the factors illustrated in Fig. 2.5. The S (sigmoid) shape of the curve arises because of the shape and mode of action of the protein chains which form the haemoglobin molecule, and it ensures certain significant points. First, the flat part of the curve at high (about 100 mm Hg) P_{O_2} means that even if the P_{O_2} of alveolar air (hence alveolar capillary blood) is somewhat lower than 100 mm Hg, the haemoglobin is almost fully saturated with oxygen. Secondly, the steep part of the curve below about 80 mm Hg means that even small

Fig. 2.5. Oxyhaemoglobin dissociation curve.

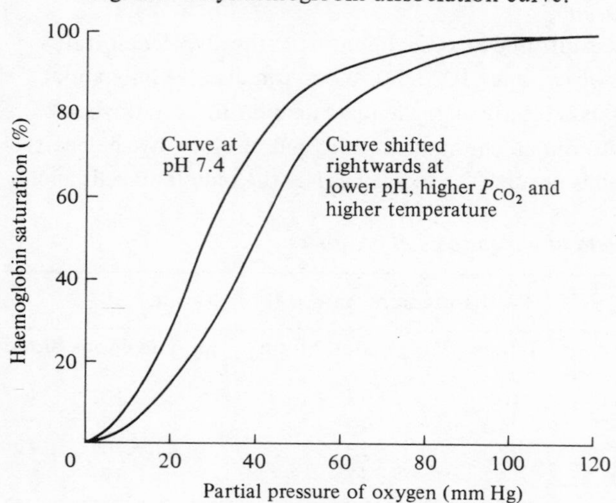

changes in P_{O_2} in the tissues lead to large changes in saturation, i.e. large amounts of oxygen can be released from oxyhaemoglobin.

The rightward shift of the curve at higher values of P_{CO_2} and lower (more acidic) values of pH and higher temperatures (the Bohr effect) facilitates the release of oxygen from oxyhaemoglobin. In the working muscles P_{CO_2} is greater than in the lungs, pH is more acidic because of the formation of acidic metabolites such as lactate and pyruvate, and tissue temperature is higher. In these conditions more oxygen is released from the oxyhaemoglobin than would be the case if the dissociation curve remained static.

So the carriage of oxygen in the blood is achieved by a mechanism which firstly ensures that the maximum possible amount of oxygen is collected at the lungs; and secondly that the maximum amount of oxygen is lost from the blood in the tissues. This release of oxygen occurs in an area of low partial pressure of oxygen (because the cytochrome system is using up oxygen). Movement of oxygen into the cells to the site of its utilisation is, therefore, once again along a partial-pressure gradient.

Carriage of carbon dioxide

We saw previously how the metabolism of the intermediate products of carbohydrate, fat and protein metabolism led to the formation of carbon dioxide. Since this is occurring in the cells, it follows that there will be a high partial pressure of carbon dioxide in the cells (it has been estimated at about 50 mm Hg). And since oxygenated blood has a P_{CO_2} of about 40 mm Hg there is a partial pressure gradient for carbon dioxide from the tissues to the oxygen-rich blood which reaches them from the lungs. Thus whilst oxyhaemoglobin is dissociating to release oxygen to the tissues, the tissues are giving up carbon dioxide to the blood. (Note how the high P_{CO_2} will also assist the release of oxygen by the Bohr effect.)

The carbon dioxide enters the plasma where some of it remains as 'dissolved' carbon dioxide. But most of it enters the red blood cells where the reaction

$$H_2O + CO_2 \rightleftharpoons H_2CO_3 \rightleftharpoons H^+ + HCO_3^-$$

is catalysed by an enzyme called carbonic anhydrase. The HCO_3^- (bicarbonate) ion thus formed diffuses out of the red blood cell and is carried in the plasma. Other ionic shifts occur to maintain the electrical neutrality of the red blood cell. A third way in which carbon dioxide is transported is in combination with haemoglobin as carbamino compounds, but this mechanism accounts for only a small fraction of overall carbon dioxide carriage.

In the same way that the presence of carbon dioxide influences the oxygen-carrying capacity of the blood, the presence of oxygen influences the blood's capacity to carry carbon dioxide. This is best illustrated by reference to Fig. 2.6. This shows that in the presence of high P_{O_2} (such as in the alveolar capillaries) the curve shifts to the right, i.e. the oxygenated blood can contain less carbon dioxide, so facilitating its release. Conversely, in the tissues where P_{O_2} is low, the amount of carbon dioxide which the blood can carry is increased.

Quantitative aspects of metabolism

It is apparent, then, that the metabolism of food to produce energy-rich molecules such as ATP uses up oxygen and releases carbon dioxide. It is possible to write exact equations to represent these overall reactions, for example:

$$C_6H_{12}O_6 + 6O_2 \rightarrow 6CO_2 + 6H_2O \text{ (for carbohydrate)}$$
$$2C_{57}H_{110}O_6 + 163O_2 \rightarrow 114CO_2 + 110H_2O \text{ (for fat)}$$

To metabolise one molecule of this six-carbon sugar, six molecules of oxygen have been taken up. Similarly 163 molecules of oxygen have been used in catabolising two fat molecules. Thus to release energy from these compounds definite volumes of oxygen are taken up and definite volumes of carbon dioxide liberated. The ratio of carbon dioxide output to oxygen uptake is termed the respiratory quotient (RQ). If carbohydrate is being metabolised the RQ is 6/6 = 1; if fat is being used the RQ is (in the above case) 114/163 = 0.7.

Fig. 2.6. Carbon dioxide dissociation curve.

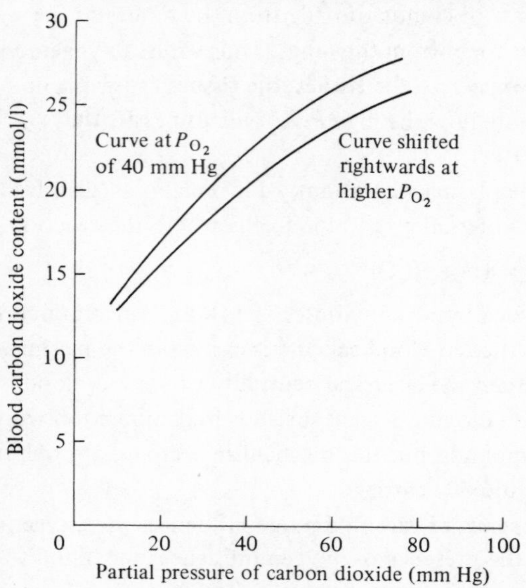

As we have seen, the calorific equivalent of 1 litre of oxygen is 5.1 kcal if glucose, a carbohydrate, is metabolised. Similar calculations for fats give values of 4.7 kcal. Thus it is possible to draw up a relationship between RQ and energy yield per litre of oxygen uptake. Some typical values are given in Table 2.2. It follows, therefore, that if we wish to measure the amount of energy being used in any activity we can measure the amount of oxygen being taken up. This figure can then be converted to an approximate energy equivalent. The value so obtained will be approximate because the basis of the calculation is the value for the free energy of combustion of the metabolite. This varies firstly with substrate: e.g. glucose gives 3.69 kcal/g, sucrose 3.96 kcal / g, starch 4.1 kcal/g, protein 5.65 kcal/g and fat 9.4 kcal/g. Secondly, the amount of 'available' energy (the amount of energy available from the substrate before metabolism) is lower than this free energy. The widely used values are 4 kcal/g for carbohydrate, 9 kcal/g for fat and 4 kcal/g for protein. The value for protein is so much lower than expected because the nitrogenous parts of the molecule play no part in energy production. Overall, however, the use of these values does allow an estimate of energy output from measured values of oxygen uptake.

Basal metabolic rate

If oxygen uptake is measured in fasting subjects at complete rest, it is found to be about 0.25 1/min. The actual value depends upon age, height and gender. However, taken as an average value, this amount of oxygen uptake represents about 1.2 kcal/min of energy required to maintain this basal metabolism. Over 24 hours this amounts to $1.2 \times 60 \times 24 = 1328$ kcal. That is, if a man lay at rest all day, his body would still need about 1300 kcal of energy. This is the energy required by the body to carry out all resting functions: for the heart to pump blood, for the brain to function, for the respiratory muscles to perform work, etc. For much of the time, however, the body is not at rest but is moving. Hence the energy requirement is much greater – approximately 2500 kcal/day (depending on size, gender and activity).

Table 2.2. *Calorific equivalent of oxygen*

Respiratory Quotient	Energy yield (kcal) per litre of oxygen uptake
0.7	4.69
0.8	4.80
0.9	4.92
1.0	5.09

Oxygen cost of various activities

It is possible to measure the energy requirement (oxygen cost) for any activity and so build up a picture of the daily calorific requirements of various life-styles. Table 2.3 shows the oxygen cost of some typical daily activities. Exceptionally some activities (lumberjacking, cross-country skiing) may demand oxygen uptake in excess of 3 litres/min.

It is possible to use such figures to estimate the calorific requirement of almost any pattern of rest and activity. Typical results of such calculations show values of about 2000 kcal for a fairly sedentary life, 3500 kcal for a moderate industrial job and 5000 kcal for a heavy industrial job. The energy used in excess of basal metabolic rate is mainly used in muscular exercise, and it will be instructive to look at how the muscles use up energy.

The use of energy by muscles

Most of the muscles in the body are striated (skeletal) muscle, so called because of their striped appearance under a microscope. They are composed of many long, thin cells called muscle fibres. Each fibre is composed of a number of fibrils and each fibril is, in turn, composed of a series of protein filaments arranged geometrically (see Fig. 2.7). The large filaments are a protein called myosin, a part of which is an enzyme, which splits ATP to provide energy for the contraction process. It is the formation and breakage of the bonds which uses energy, and which results in a shortening (contraction) of the muscle fibres and hence of the whole muscle.

All striated muscles are controlled by nerves. When a nervous stimulus arrives at the muscle, calcium ions are released from the cytoplasm of the muscle cells. These ions remove an inhibition on the actin molecules and allow them to form chemical bonds with myosin, using ATP to supply the energy. The actin–myosin interaction results in a relative movement of the protein filaments, i.e. a short-

Table 2.3. *Oxygen cost of various activities*

Activity	Oxygen cost (l/min)
Sleeping	0.25
Sitting at rest	0.34
Typing	0.46
Ironing	0.88
Gardening	1.16
Digging	1.78
Cycling (about 13 m.p.h.)	2.00
Stair climbing	2.40

From Webb (1973).

ening of the muscle. The release of energy from ATP involves the reaction

$$ATP \rightarrow ADP + \text{phosphate} + \text{'energy'}$$

This conversion results in a depletion of the muscle reserves of ATP, which are very small. The ADP must, therefore, be re-formed and this is achieved by the transfer of a phosphate group from another 'high-energy' molecule, creatine phosphate, to the ADP. Creatine phosphate is re-formed from ATP when muscular activity ceases.

If the muscles continue to contract, as in prolonged physical exercise, the creatine phosphate and ATP stores are replenished by the catabolism of blood glucose or muscle glycogen by aerobic glycolysis. Free fatty acids (FFA) are used to supply energy 'at rest', i.e. when there is little muscular activity, or during recovery from exercise. The way FFA are used to provide energy has been described above (see p.37).

Oxygen debt

In what has just been described it has been assumed that the supply of oxygen to the tissues is adequate. At times, however, this is not the case. In these circumstances aerobic glycolysis and FFA metabolism via the citric acid cycle cannot continue and the pyruvate formed as an intermediate metabolite is converted anaerobically to lactate:

$$CH_3COCOOH \rightleftharpoons CH_3CHOHCOOH$$

This lactate accumulates first in the cells and then in the blood (lactate levels may rise from about 10 mg/100 ml blood at rest to 200 mg/100 ml during prolonged, severe exercise). The accumulation of lactate in the muscles, or the absence of sufficient ATP, causes muscular fatigue in which the muscles will fail to contract or will contract only weakly even though they are being stimu-

Fig. 2.7. Schematic representation of the arrangement of actin and myosin filaments in striated muscle.

lated. When exercise ceases oxygen uptake remains elevated so that the lactate can be oxidised to pyruvate by the removal of hydrogen ions which are then transferred to oxygen atoms by the components of the cytochrome chain.

Thus if one sets out to measure the oxygen cost of a particular activity it is essential to consider the possible presence of an oxygen debt. In these circumstances oxygen uptake rises at the onset of the activity, remains elevated and only falls some time after the activity has ceased. In order for the total oxygen cost of the activity to be measured oxygen uptake should therefore be followed until it has reached pre-exercise levels.

Limitations of energy supply

It will be apparent that in order to maintain basal metabolic rate and to supply the energy needs of muscular activity, an adequate supply of ATP is demanded. This, in turn, points to several possible points at which the supply of energy might be disrupted.

Inadequate diet

The exact influence of inadequate diet depends upon the magnitude and duration of the deprivation. Daws *et al.* (1972) have shown that calorie restriction for 10 days led to no change in the oxygen uptake required for certain standardised activities. However, Keys *et al.* (1950) reported that after 24 weeks of calorie restriction the maximum work output of the subjects (maximum aerobic capacity) was reduced by about 30%. This made any activity more demanding during starvation than when the subjects were eating a diet that supplied their calorie requirements.

These results were found in otherwise healthy subjects. In subjects exposed to prolonged starvation or other dietary deficiencies there is a massive and progressive decrease in maximum aerobic capacity caused by skeletal muscular atrophy and myocardial (heart muscle) dysfunction. No such changes are found in acute calorie restriction and it seems as if in this case the mechanism limiting energy expenditure is dehydration and general metabolic disorder, including acidosis and low haemoglobin levels in the blood.

Deficient diet

In some cases a diet may have sufficient calories but be deficient in other ways. For example, if essential amino acids, vitamins or minerals are absent, signs of these deficiencies will become apparent. The time-course of the onset of symptoms depends upon the particular chemical which is missing. For example, inadequate vitamin C will lead to reduced aerobic capacity after about 90 days. The mechanism of such deficiency is not known but it may be related to a possible role of vitamin C in the oxidation–reduction reactions in the cytochrome chain of intermediary metabolism.

Iron deficiency may also exert limitations on the supply of energy because of its intimate role in the carriage of oxygen by haemoglobin. Each molecule of haemoglobin contains four ferrous ions which are active in coupling with molecular oxygen. Because of the constant turnover of body proteins it is essential to have an adequate supply of iron to maintain haemoglobin levels in the red blood cells. Loss of iron from the diet reduces aerobic work capacity because the delivery of oxygen to the tissues is reduced.

Metabolic defects

One example will serve to show how a metabolic disorder can influence work capacity. A common characteristic of a group of diseases called lipidoses is a defect in one or more of the enzymes responsible for the metabolism of fats. If these cannot be adequately catabolised the source of free fatty acids, so essential in prolonged severe exercise, is also inadequate. This leads to a reduced capacity to do work because the substrate of the energy-producing reactions is limited.

Limitations of cardiac output

If subjects are exposed to progressively more severe exercise it is found that heart rate eventually reaches a maximum value (Fig. 2.8), which represents the approximate point at which cardiac output is maximal. There are very wide inter-individual differences in maximum heart rate, and further large changes

Fig. 2.8. Relationship between heart rate and oxygen uptake. Note how the linear relationship is lost at high values of oxygen uptake as heart rate reaches a plateau.

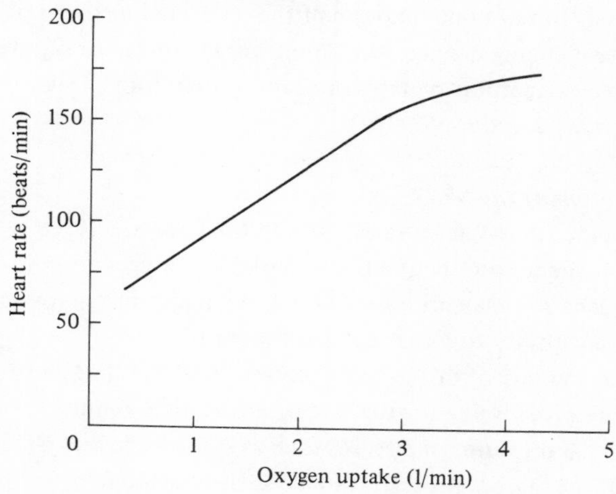

with age and with gender. A full discussion of the characteristics of these differences is given by Astrand and Rodahl (1977). If cardiac output is maximal and a greater demand for work is placed upon the subject, energy production cannot increase to cope with the demand. In such circumstances anaerobic metabolism will increase until it too is limited and the available energy will also reach maximum values.

Blood flow disorders

As we have seen, the efficient metabolism of dietary components to produce energy relies upon an adequate oxygen supply to the tissues. In turn, this depends upon an adequate blood flow to the tissues. In diseases such as atheroma, or cardiac insufficiency, blood flow through skeletal muscle may be low. When these muscles are called upon to do work the tissue oxygen concentration therefore rapidly falls below that at which aerobic metabolism can continue. Since anaerobic metabolism can only continue for a very short time, exercise is soon limited by the absence of adequate ATP and creatine phosphate. Such limitations on work capacity are often accompanied by pain, as in the vascular disease intermittent claudication.

Reduced oxygen-carrying capacity of the blood

Even if blood flow through a tissue is adequate, the capacity of that tissue to produce energy may be limited by a low level of oxygen in the blood. Strictly, it is not the partial pressure of the oxygen which is important, but its oxygen content. Fig. 2.9 shows the oxyhaemoglobin dissociation curve of blood with reduced oxygen-carrying capacity. It can be seen that even at high partial pressures of oxygen (such as in the alveolar capillaries) the haemoglobin is only half saturated, i.e. the blood contains only about half the 'expected' amount of oxygen. In times of increased energy demand the blood will be unable to supply the requisite amount of oxygen; aerobic metabolism cannot therefore be sustained and the aerobic capacity is reduced.

Reduced oxygenation of the blood

Even if 'normal' blood flows at 'normal' rates in the tissues, oxygen may be a limiting factor on energy production because of poor oxygenation of the blood. Such poor oxygenation may arise for at least two major physiological reasons: impaired diffusion capacity or impaired lung function.

In the former condition the rate of diffusion of oxygen from the lungs into the blood is impaired. The normal value for the rate of diffusion is about 20 ml/min per mm Hg partial pressure. In diseases such as sarcoidosis this rate may be diminished by 90% or more. This results in a less-than-adequate amount of oxygen entering the blood with a consequent diminished aerobic capacity.

In impaired lung function insufficient air may be moved into the lungs to carry out normal oxygenation. In obstructive lung disease, for example, where the airways of the lung are of reduced calibre, aerobic capacity can be limited almost to resting values of oxygen uptake. Thus during exercise not only does the oxygen cost of respiration increase but the airway resistance effectively prevents adequate air flow to the alveoli. As a consequence the blood P_{O_2} remains very close to the value of mixed venous blood, about 40 mm Hg, and the oxygen needed for aerobic glycolysis or metabolism of free fatty acids is not available. Other respiratory disorders, such as those which reduce lung volumes or which interfere with respiratory muscles, may reduce the volume of air breathed and so impair the amount of blood which can be adequately oxygenated.

Changes such as these may reduce the maximum amount of air which can be exchanged in the lungs in a given time: i.e. there is reduced maximum breathing capacity (MBC) or maximum voluntary ventilation (MVV). Typically, a healthy adult would have a MBC of between 150 and 200 l/min. In disease this can be reduced to 10% of this value or less. Since the lungs only reduce the oxygen concentration of inspired air by about 3% (from 21% to say 18%), it follows that an MBC of only 15–20 litres represents a maximum uptake of about 0.66

Fig. 2.9. Oxyhaemoglobin dissociation curve of normal blood compared with that of blood with reduced oxygen-carrying capacity (e.g. during anaemia).

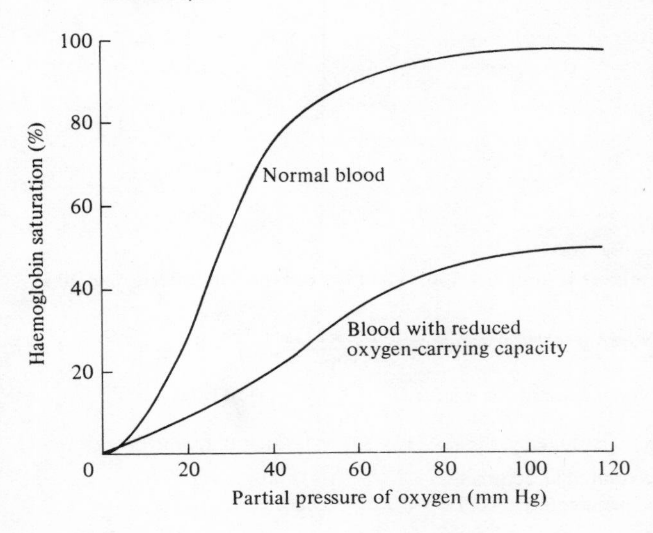

litres/min of oxygen.* This is quite inadequate to provide the oxygen necessary for most activities. For a full discussion of impaired lung function see Cotes (1979).

Physical restrictions on oxygenation of the blood

There are a number of physical (environmental) factors which serve to reduce blood oxygenation and hence limit energy production:

Hypoxia. The total pressure of the atmosphere is inversely related to altitude (see Fig. 2.10). Since the partial pressure of oxygen in the air, and hence in the lungs and blood, is directly related to total pressure, it follows that the driving force for diffusion in the lungs is reduced at altitude. Thus, in acute exposure to high altitude, blood oxygen saturation is reduced and the capacity to do work is severely limited. Few effects are felt below about 3500 m, but above this aerobic capacity is increasingly reduced until at about 6000 m the unacclimatised subject loses consciousness even when at rest. Acclimatisation induces physiological changes which counteract some of the effects seen during acute exposure to altitude.

Increased gas density. The total pressure underwater is a function of depth of immersion. Thus the density of the gas breathed is related to depth: air at a pressure equivalent to a depth of 10 m is twice as dense as at the surface, three

Fig. 2.10. Relationship between altitude and atmospheric pressure.

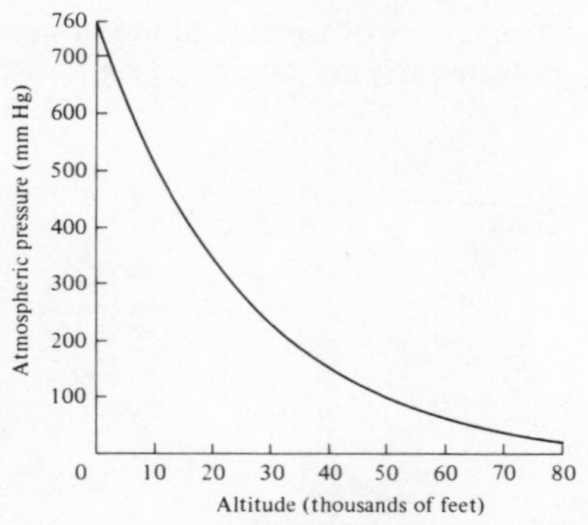

* If normal minute volume is 7.5 litres and the oxygen concentration of air is 21%,

volume of oxygen inspired per minute is $\frac{7.5 \times 21}{100} = 1.575$ litres

volume of oxygen expired per minute is $\frac{7.5 \times 18}{100} = 1.350$ litres

volume of oxygen uptake per minute is $1.575 - 1.350 = 0.225$ litres

For a minute volume of 20 litres,
maximum oxygen uptake = $20/7.5 \times 0.225 = 0.66$ litres

times as dense at 20 m, etc. The density of the gas influences the maximum volume of gas which can be breathed in a given time (MBC) according to the equation:

$$\text{MBC} \propto 1/\sqrt{(\text{density of gas breathed})}$$

It follows, that if the restriction imposed by the gas density is such that a particular activity demands a ventilation at or near to MBC, then the duration for which the activity can be sustained is limited by ventilation factors. If the density of the inspired air rises, the frictional resistance to gas flow rises, the work of breathing is increased and soon exceeds the value at which normal blood oxygen concentrations can be maintained. The aerobic work capacity is accordingly limited.

Increased resistance to breathing. This can occur, for example, when breathing apparatus or respirators are used. Increased breathing resistance usually reduces minute volume because of the additional work of breathing imposed by the apparatus. Thus during exercise oxygenation of the blood is reduced and aerobic capacity is lowered.

Maximum oxygen uptake

It has been shown above that a number of physiological and physical factors may limit the production of energy. These principally are of two types: cardiovascular limitations and respiratory limitations. The former are by far the most common cause of limited exercise capacity in healthy subjects at sea level. In diseased subjects or in subjects exercising in hypo- or hyperbaric environments (i.e. where air pressure is lower or higher than normal), respiratory factors may be the limiting feature. Whatever the cause, the common manifestation of these limitations is that exercise (aerobic) capacity reaches plateau values. In these circumstances the oxygen uptake of the subject is maximal ($\dot{V}_{O_2\ max}$).

There is a complex literature on the subject of $\dot{V}_{O_2\ max}$ (see, for example, Astrand and Rodahl, 1977), mostly defining the limits of $\dot{V}_{O_2\ max}$ in various populations. It is apparent that the maximum capacity to do work is dependent upon factors such as age, body size, gender, degree of physical fitness and possibly genetic factors (Fig. 2.11). Measurement of $\dot{V}_{O_2\ max}$ is important because it gives a basis on which the fitness of a subject may be judged; and may set the duration for which given individuals may work at certain rates. For instance, it is possible to work at 30% of one's $\dot{V}_{O_2\ max}$ indefinitely, whereas endurance at 70% of $\dot{V}_{O_2\ max}$ is about 1 hour. Thus in assessing either the fitness of an individual for a given job, or the appropriateness of a job for an individual, the $\dot{V}_{O_2\ max}$ plays a vital role.

Assessment is normally made by prediction of $\dot{V}_{O_2\ max}$ from submaximal workloads on a treadmill or cycle ergometer. The details of the various methods are discussed by Astrand and Rodahl (1977), and the subject is discussed at length by Shephard (1978). Values are usually related to body weight to minimise inter-subject differences. Typical values are: for most individuals, less than 40 ml/min •kg; physically fit individuals may exceed 75 ml/min •kg.

Efficient use of energy

Even though the body is often capable of producing energy for muscular exercise far in excess of its current demand, the obvious tight limitations on energy supply dictate that energy is used most efficiently wherever possible. It follows, therefore, that any method which reduces the oxygen demand to an individual of a given task, or which produces a certain result with the minimum of energy expenditure is obtaining the most efficient use of substrates for energy production. A number of ways of increasing the efficiency of energy expenditure are possible:

Fitness

Fig. 2.12 shows the effect of fitness on $\dot{V}_{O_2\,max}$. The extremely fit athletes had very high values while the 'unfit' non-athletes began with very low values (less than 40 ml/min •kg) which increased as training continued. The implications of these findings are that those individuals with low values of $\dot{V}_{O_2\,max}$ would find it taxing to carry out exercise demanding an oxygen uptake of say 2.0 l/min because they would be working at about 70% of $\dot{V}_{O_2\,max}$ and this level of activity can be sustained only for relatively short periods.* In contrast, a fit athlete of the same weight with a $\dot{V}_{O_2\,max}$ of 80 ml/min •kg would

Fig. 2.11. Maximum oxygen uptake in relation to age and gender.

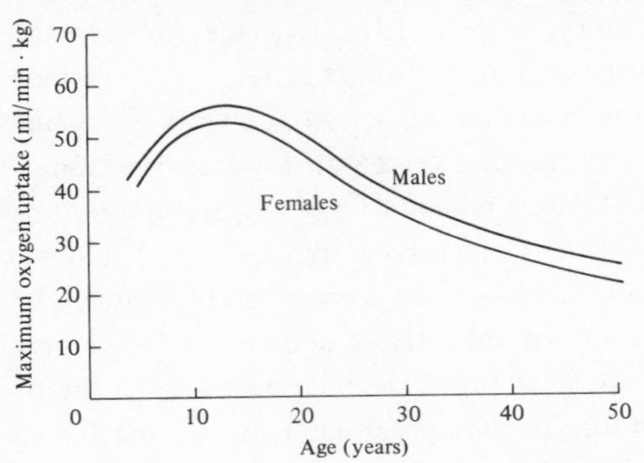

* If a 75-kg man has a $\dot{V}_{O_2}$ max of 40 ml/min • kg, then this is equivalent to 75 × 40 ml/min = 3.0 l/min.

be working at only 2.0/[(75 × 80)/1000] % i.e. about 35% of his $\dot{V}_{O_2}$ max, a level which he could maintain indefinitely.

Training

Training of the cardiovascular system or of specific muscle groups may be used to increase the efficient use of energy. The best form of training depends upon the type of physical activity for which the training is undertaken. For example, short-duration activities demanding strength (such as the throwing events in athletics, or heavy lifting tasks in an industrial situation) are best improved by training of specific muscle groups. Fitness for longer duration activities (most track events at athletics, or continuous moderate industrial work situations) is best improved by cardiovascular training. Such training results in a lower resting heart rate, a lower heart rate when undertaking a given physical activity, and higher maximal heart rate. The overall effect is to improve the cardiovascular response to work and so lead to the more efficient supply of oxygen for aerobic metabolism.

Continuous and intermittent work

Astrand *et al.* (1960) showed that a given amount of work could be accomplished far more easily using an intermittent work rate rather than a continuous one. They showed that whereas an individual could only maintain a work rate of 350 watts for nine minutes when working continuously, the same subject was easily able to complete 1 hour at the same rate of work when

Fig. 2.12. Maximum oxygen uptake in relation to physical training.

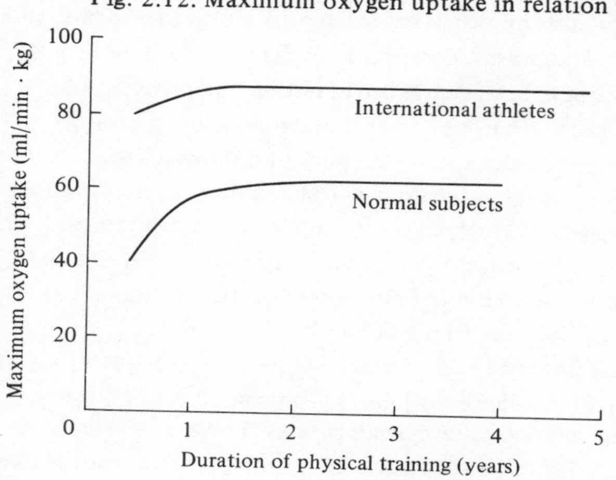

undertaking a 30-second work, 30-second rest schedule. Such finds give important clues about how such heavy work schedules are best undertaken.

Conclusion

We have seen, then, that in order to carry out any activity (even to sleep) requires the provision of energy. This is obtained from metabolism of the ingested food. By catabolic processes which need oxygen, energy is released in the shape of particular 'high-energy' molecules such as ATP. These molecules are used mainly by muscles to carry out all the daily activities of which man is capable. There are many limitations upon the production of energy, some physiological, others physical. Whatever the limitations the energy production reaches maximal values dependent upon age, gender and degree of physical fitness. These values of maximal oxygen uptake can be improved by certain manoeuvres all of which tend to maximise the efficiency with which the energy is used.

2.2 Techniques available to measure factors related to energy production

We have seen that energy production is based upon a number of interrelated factors. In order to understand fully the nature of these problems it is essential that they are amenable to quantification. A wide variety of methods is available to carry this out and these will be outlined below. Details are best understood by examination of the references cited.

Measurement of dietary components

The chemical composition of any diet or food can be measured. In the first instance it is necessary to measure the total carbohydrate, fat and protein content; it may, however, also be necessary to identify individual sugars, triglycerides, amino acids or vitamins. A description of detailed methodology is beyond the scope of this book; however, in broad outline the procedure for food analysis is as follows. Total carbohydrate is determined by hydrolysis of polysaccharides to monosaccharides and the amount of monosaccharide may be determined by any suitable chemical technique. The determination of total fat is extremely difficult because the analysis is susceptible to the method used. The most reliable method seems to be acid hydrolysis. Protein content is assessed following determination of the total nitrogen content of the food sample. For an introduction to methodology see Paul and Southgate (1978).

The other important feature of the diet which warrants careful quantification is the energy value (content of calories or joules). This can be determined extremely accurately by the method of bomb calorimetry, in which a sample of the food is completely burned in a closed container. From the temperature rise in a known mass of water surrounding the food sample, the calorific equivalent of the sample can be determined. As an example of the type of information contained in food analysis tables, Table 2.4 gives a few sample values drawn from Paul and Southgate (1978). Tables are available giving analysis by brand name, by ethnic origin of foodstuffs and for a great number of individual chemical constituents.

Measurement of energy expenditure

For detailed methodology and discussion of the principles involved in the measurement of energy expenditure and physical fitness see Astrand and Rodahl (1977). Green (1976) gives an extremely lucid account of the basic methods of measuring gas volumes and concentrations, and Cotes (1979) thoroughly discusses the principles of measuring all aspects of lung function. The following discussion outlines some of the more common methods which may be used to measure energy expenditure. Basic methods are of two types: direct or indirect calorimetry. In the former, the energy expenditure of a subject completely enclosed in a controlled and monitored environment is measured by changes in the heat content of water. In the latter methods, energy expenditure is estimated from measurements of oxygen uptake.

Direct calorimetry

For a detailed description of this technique see Atwater and Benedict (1905). A subject is confined in a closed, well-insulated box. Air is passed through the box at a controlled rate to maintain a normal atmosphere and water is passed around the outside of the chamber to remove the heat produced by the subject. If allowance is made for the heat content of everything entering and leaving the chamber (food, excreta, air, etc.) the change in the heat content of the water reflects the energy expended by the subject. Such a simple description belies the incredibly complex nature of performing experiments in such a device.

Table 2.4. *Simple analysis of some common foodstuffs*

Food	Carbohydrate (g/100 g)	Fat (g/100 g)	Protein (g/100 g)	Energy value (kcal/100 g)
Wholemeal bread	41.8	2.7	8.8	216
Salted butter	Trace	82.0	0.4	740
White sugar	105[a]	0	Trace	394
Tea (infusion)	Trace	Trace	0.1	<1
Fresh cows' milk	4.7	3.8	3.3	65

[a] Apparent anomaly arises because of the method of determination.

Modern developments in this field have been described by Benzinger *et al.* (1958), Blaxter, Brockway and Boyne (1972), and Webb, Annis and Troutman (1972).

Indirect calorimetry

Respiration chamber. The subject sits in a chamber which is ventilated at a known fixed rate and the oxygen and carbon dioxide concentrations of the ingoing and outgoing air are measured. When corrections are made for the changes in pressure and/or temperature of the gases, the concentration differences and volume flow reflect the oxygen uptake of the subject. This is basically a modification of the method of Atwater and Benedict (1905). It is simpler than the technique of direct calorimetry but has the disadvantage that energy expenditure is obtained only by an additional step which involves many assumptions (Weir, 1949). A development of this method was described by Ashworth and Wolff (1969). It involves the use of an enclosed, ventilated hood which is placed over the head of the subject and sealed at the neck. This is equivalent to having a head-sized respiration chamber and avoids the necessity of having the subject confined in a chamber.

Douglas bag techniques. Most twentieth-century laboratory studies of exercise have used the technique first described by Douglas (1911). Expired air is collected in a suitable closed bag. Its volume is measured with a wet-gas or dry-gas meter, or a Tissot spirometer; its oxygen and carbon dioxide concentrations are measured with a Lloyd–Haldane (or similar) apparatus or mass spectrometer; finally calculation of the difference between the amounts of oxygen taken in and given out (volume × concentration) gives the oxygen uptake.

Portable instruments. Most field studies rely upon the availability of instruments which are carried by the subject whilst undertaking the activity under investigation. A variety of these devices is available. For instance, in the meter (KM meter) described by Kofranyi and Michaelis (1940), expired volume is measured mechanically from a mouthpiece or facemask and a sample of expired air is collected for later analysis. Fletcher and Wolff (1954) developed an instrument, the integrating motor pneumotachograph, or IMP, which works in a manner similar to the device of Kofranyi and Michaelis but is battery powered. This gives it a lower resistance to respiration, which is a highly desirable feature at high rates of respiration.

In recent years further developments have been described. The Oxylog (Humphrey and Wolff, 1977) is a battery-powered device which measures both volume and oxygen concentration and gives a continual readout of oxygen uptake. The MISER (Eley *et al.*, 1978) is similar in concept to the KM meter and IMP, but relies on updated technology. The volume of the inspired air is measured by a mechanical device and a vacuum container collects a sample of expired air. Subsequent laboratory analysis of this sample allows oxygen uptake to be calculated.

Indirect methods of estimating oxygen uptake. Sometimes it is impracticable for the subject to wear any instruments to measure oxygen uptake and so it must be estimated by indirect methods. An example of a semi-quantitative method of doing this is the activity diary method described by Passmore, Thomson and Warnock (1952). For as long as possible each 24 hours, the subject or an observer keeps a detailed diary of activities. These activities are classified according to standard semantic descriptions such as light sitting, light standing. The total time spent in each category is obtained and an estimate of the total oxygen uptake made based upon values of oxygen uptake assigned to each of the semantic descriptions. Such methods find use in very prolonged activities or those activities where interference with the subject is unacceptable.

Another method of estimating oxygen uptake is to make use of the good, linear relationship between $\dot{V}_{O_2}$ and heart rate. Heart rate can often be monitored continuously for prolonged periods using either telemetry of the electrocardiograph (ECG) signal, or recording of the signal on miniature tape-recorders. Warnold and Lenner (1977) have reviewed the application of this method for clinical use.

Durnin and Edwards (1955) described a further indirect method of estimating oxygen uptake. This is based upon the essentially linear relationship between oxygen uptake and minute volume (the volume of air inspired in 1 minute), a relationship which holds true over a wide range of energy expenditure. Liddell (1963), and Datta and Ramanathan (1969), report application of the principle in field studies. They conclude that the method is suitable for use in circumstances in which other techniques may be unsuitable or impracticable.

The main disadvantage of these indirect methods of estimating oxygen uptake is that they rely upon *average* correlation between the variable being monitored and oxygen uptake. There can be very large intra-subject and inter-subject variability in these relationships. In addition, factors such as training or acclimatisation can alter the relationships and so worsen an already tenuous accuracy. A further disadvantage of such methods is that they are two steps removed from the variable which is being estimated, rather than the single step of the oxygen uptake methods.

Units of measurement

Very many units of measurement are employed in the literature of oxygen uptake and energy expenditure. This chapter has used those units most commonly found in the literature. Table 2.5 sets out the inter-relationships of a number of units.

Body size

Units of oxygen uptake are often related to body weight or surface area. Surface area is predicted from the equation of Du Bois:

$$\text{surface area } (m^2) = \text{weight (kg)}^{0.425} \times \text{height } (m)^{0.725} \times 0.00718$$

Details of the measurement of weight, height and other anthropometric variables are discussed by Harris (1978).

2.3 Examples of the application of the principles discussed

Effect of diet on physical performance

As we have seen, the source of substrate to supply the energy needed for all activity is the metabolisable fraction of the diet. If the amount of energy available from the diet is less than that expended (that is if there is a negative energy balance) then it is possible that there could be a progressive limitation of work capacity. This problem has been examined extensively in relation to prolonged calorie restriction or to starvation. A comprehensive review of the early findings in this field is given by Keys *et al*. (1950). However, examination of some important papers will serve to show how such a problem may be amenable to experimental investigation. It is assumed that the problem is related to calorie deficiency and not to mineral, vitamin or other deficiencies which may lead to a pathological condition.

Henschel, Taylor and Keys (1954) described an experiment in which there was acute exposure of subjects to a period of starvation (no calorie intake) after pre-training them to condition them to hard exercise (treadmill walking). The aim of the experiment was to ascertain the effect of such conditions on the exercise capacity of the subjects. Both aerobic and anaerobic work capacity were tested. Aerobic capacity was measured by direct measurement of the oxygen

Table 2.5. *Inter-relationships between units employed in exercise physiology*

1 kilocalorie (kcal)	=	1000 calories (cal)
1 calorie	=	4.186 joules (J)
1 joule/second	=	1 watt (W)
1 watt	=	6.12 kilopond-metres/minute (kpm/min)
746 W	=	1 horse-power (h.p.)

uptake when the angle of the treadmill was changed (increasing grade) but it was kept at a constant speed. An alternative approach would have been to infer $\dot{V}_{O_2}$ max from indirect measurement: measure oxygen uptake at several submaximal workloads and estimate $\dot{V}_{O_2}$ max from the relationship so obtained. Whilst this is a convenient and rapid method it does suffer from the disadvantage of being based upon an assumption of doubtful validity in the conditions of this experiment. Thus the choice of a direct measurement of $\dot{V}_{O_2}$ max gave a value limited only by experimental error (likely to be about 5%). In practice, the procedure was carried out by having subjects expire into a Tissot spirometer (to measure gas volume), with gas concentrations being measured by the Haldane volumetric analysis. Anaerobic work capacity was tested by measuring the blood concentration of lactate and pyruvate following a 75-second very hard exercise period on the treadmill.

The effect of 4 days of starvation on $\dot{V}_{O_2}$ max and anaerobic work capacity is shown in Table 2.6. It is apparent from these results that whilst in absolute terms $\dot{V}_{O_2}$ max fell, it did so only in proportion to the loss of body weight. The $\dot{V}_{O_2}$ max per kilogram of body weight remained unchanged, as did the ventilation needed to sustain this level of oxygen uptake. The anaerobic work capacity was, however, impaired.

The findings contrast with those obtained when studying the effects of prolonged calorie restriction (Keys *et al.*, 1950). Using similar techniques to those described above, these authors found that following 24 weeks of calorie deficiency there was a 25% decrease in $\dot{V}_{O_2}$ max and a very marked deterioration in the ability to carry out anaerobic work, although working pulse rates showed very small changes. It is not appropriate to discuss here the implications of these findings, or those reported by the authors cited for the rate of recovery from the

Table 2.6. *Effect of 4 days of starvation on the exercise capacity of 12 men (mean value ± 1 standard deviation)*

	Control	After 4 days of starvation
$\dot{V}_{O_2}$ max (ml/min)	3.46 ± 0.19	3.19 ± 0.23
$\dot{V}_{O_2}$ max (ml/min·kg)	49.5 ± 2.6	49.5 ± 3.2
Minute volume (l/min)	85.5 ± 10.6	86.5 ± 9.2
Lactate concentration (mg/100 ml blood):		
At rest	6.2 ± 1.8	14.2 ± 4.4
Following exercise	28.0 ± 10.4	36.6 ± 8.7
Heart rate (beats/min) during work	129 ± 9	151 ± 12

experimental circumstances. It is appropriate, however, to point out the essentially simple basis of the methodology involved. By applying standardised techniques in a logically progressive series of experiments, the mechanisms underlying the human response to an important and practically significant stress can be better understood.

Oxygen cost of daily activities

Passmore and Durnin (1955) give a detailed discussion of the energy expenditure of a wide variety of work and leisure activities. Almost inevitably, the methodology involved in gathering such information is based upon indirect calorimetry (i.e. measurement of the oxygen uptake). It would be impossible, for instance, to measure the energy expenditure of swimming, driving or lumberjacking in a human calorimeter, whereas measurement of the oxygen cost of these activities is relatively easy. The disadvantage of this indirect calorimetry is that conversion to kilocalories of energy expenditure is dependent upon the respiratory quotient (RQ) at the time. Since this can only be measured in a calorimeter* it follows that a value for RQ must be assumed in the conversion equations. However, the error involved in such assumptions is small. Given an oxygen uptake of 1.0 l/min and a correct RQ of 0.83, an assumed RQ of 0.7 would give a value 3% too low and an assumed RQ of 1.0 a value 4% too high. It follows, then, that a reasonably accurate assessment of energy expenditure can be obtained from indirect methods. An example of the application of such procedures to an outdoor activity will serve to exemplify the general approach.

Walking across rough terrain is more 'difficult' than walking on a smooth, hard surface. It is desirable in a number of circumstances to know the energy cost of such activities. Pandolf, Haisman and Goldman (1976) described the oxygen cost of walking on snow of various depths. The rationale for this particular experiment was to provide coefficients for a general equation for the prediction of the energy expenditure of walking at any speed across any terrain. In these circumstances, it is obviously essential that the oxygen cost is measured directly and not inferred from measurements of heart rate or minute volume. Two approaches are possible. One is to collect expired air in Douglas bags attached to the subjects. Alternatively, a portable instrument may be used. If a bag of gas is collected the volume and composition must be measured in the laboratory (which may be inaccessible). The portable instruments, however,

* The value $\dot{V}_{CO_2}/\dot{V}_{O_2}$ obtained when carrying out oxygen uptake studies is correctly termed the respiratory exchange ratio (R). This is not the same as RQ because RQ assumes a steady state of body oxygen and carbon dioxide stores. R can be elevated above 1 when hyperventilation is present and carbon dioxide is being lost from the body.

measure volume as the gas sample is collected (and the Oxylog will even measure oxygen concentration simultaneously). If a laboratory is close at hand, and if the activity is to be of relatively short duration, it is probably desirable to use a bag collection technique because the volume of gas can then be measured extremely accurately in the laboratory (a gas meter or Tissot spirometer being the most convenient way). Any mechanical device which is used on a portable instrument to measure volume must be carefully calibrated before and after use, as such devices are prone to damage and to calibration changes in field conditions. On the other hand, when the activity is to be prolonged, say a 4-hour march, then the constant collection of bags of expired air becomes logistically inconvenient and portable instruments are highly desirable.

In the experiment of Pandolf *et al.* (1976) the oxygen cost of walking across snow of various depths was measured using a Max Planck respirometer (KM meter). Expired air passes through the device, which measures the volume, and a pump is activated to take a small sample of the gas which is stored in a container. Over a period of time the gas in the sample bag reflects the composition of the mixed expired air. This can be determined in the laboratory using any appropriate technique. In this particular case the oxygen concentration was measured using a paramagnetic oxygen analyser whose accuracy was checked using a micro-scholander volumetric method. The accuracy of the paramagnetic oxygen analyser is good and so there can be an accurate analysis of the oxygen concentration of the mixed expired gas. It is then possible both to calculate the oxygen uptake of the activity and to estimate the energy expenditure involved, by using the equation of Weir (1949). The latter is a way of estimating energy expenditure from the measured concentration of oxygen in, and the volume of, the mixed expired air. For most purposes it is sufficiently accurate and is an extremely useful method if the concentration of carbon dioxide cannot be measured, so preventing the calculation of oxygen uptake by other methods.

By examining movement over a variety of snow depths, Pandolf *et al.* (1976) were able to produce the results shown in Fig. 2.13. From such data it is possible not only to give an estimate of the energy expenditure required for walking in the snow, but also to derive the coefficients needed for the prediction equations. Such equations themselves are very useful both as the basis of models of human responses in various conditions and, at the immediately practical level, for allowing accurate planning of the calorie requirements of expeditions etc.

Indirect estimates of workload

In some circumstances it is impracticable to use methods of measuring oxygen uptake which rely on delicate mechanical equipment, or which involve wearing mouthpieces. For example, in most very cold conditions the valves of valve-boxes tend to freeze up and mechanical devices are prone to stick. It is desirable, therefore, that in such circumstances a method of estimating energy expenditure is used which does not employ mechanical equipment. One such method is to use indirect estimates of oxygen uptake from some other variable such as heart rate.

Rodahl *et al.* (1974), and Rodahl and Vokac (1977), have described the use of such a technique in the estimation of the workload of trawler fishermen. The basic method consists of making a continuous recording on a miniature tape-recorder of the subject's ECG signal. This is analysed by computer in the laboratory to give the heart rate over any desired period. The measured heart rate is then converted to an equivalent oxygen uptake graphically, using data which have previously been obtained in the laboratory. This method relies on the linear relationship between oxygen uptake and heart rate over a wide range of values of oxygen uptake. The main disadvantage with the method is that the relation-

Fig. 2.13. Energy expenditure when walking on snow as a function of the depression of the feet. Observations of: A, Heinonen *et al.* (1959); B, Ramaswamy *et al.* (1966); C–F, Pandolf *et al.* (1976). Pandolf *et al.*'s observations expressed in terms of nude body weight without (C) and with (D) an estimated correction for total oxygen debt, and in terms of total clothed body weight without (E) and with (F) an estimated correction for total oxygen debt. (After Pandolf *et al.*, 1976.)

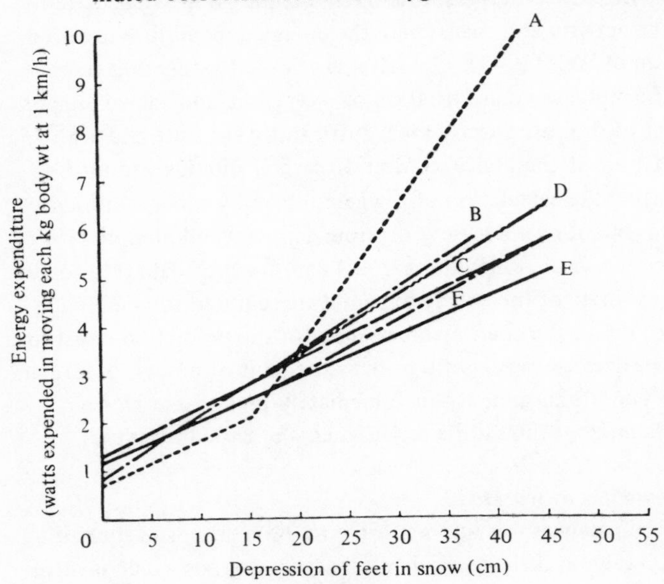

ship may be upset by a variety of factors unrelated to the activity in question. For example, the heart rate may be artificially elevated by emotion or depressed by drugs, or the relationship may change as the subjects become physically fit. Furthermore, the exact nature of the oxygen uptake/heart rate relationship is dependent upon the activity being undertaken, different values being obtained, for example, for arm or leg cycling and treadmill running. It will also be different if a significant portion of the work is isometric (i.e. involving pushing or pulling with the muscles contracted but the limbs not moving).

However, in spite of such limitations Rodahl *et al.* (1974) have shown results using this method to be very reproducible. Accuracy may be improved by making an activity diary to cover the entire period for which the recordings are undertaken. The heart rate of discrete periods covering discrete activities can then be assigned with great confidence. The limitations outlined above can also be minimised by recording oxygen uptake and heart rate *simultaneously* whenever possible. An oxygen uptake/heart rate relationship can then be established for that particular activity. Estimated values of oxygen uptake are therefore likely to be more accurate. Obviously, such a method cannot be used in many circumstances (otherwise oxygen uptake would be measured directly and there would be no need for the heart rate recordings), and this may make a laboratory simulation of the task, by means of which the oxygen uptake/heart rate relationship may be calibrated, an attractive alternative.

Using recordings of ECG together with activity diaries for the same period, Rodahl *et al.* (1974) and Rodahl and Vokac (1977) obtained the results shown in Fig. 2.14. It was possible to ascribe the mean values of heart rate to equivalent values of oxygen uptake, based upon laboratory studies. The authors found that the mean oxygen uptake during the periods of work in a 24-hour observation period was about 1 l/min, giving an estimated energy expenditure of about 3900 kcal. It is emphasised that this was in the work periods. To get the full 24-hour energy expenditure, the energy expenditure of the off-duty periods must be added.

It will be apparent that whilst the precision of this value may be questioned, the fact that it is reproducible suggests that it may be a reliable value. The attraction of the overall method is obvious. It would be difficult to carry out direct measurements of oxygen uptake in such experimental conditions. The miniature tape-recorders used in this study are small, light, robust and reliable and thus the field measurements are relatively simple. However, analysis of the tape-recordings is complex and time-consuming and requires access to computing facilities. Thus in considering a method suitable for field use it is essential to consider not only the problem of gathering the data, but also that of their subsequent analysis.

2.4 Conclusion

These examples show that the choice of a particular technique to measure the energy cost of daily activities depends both upon the activities themselves and the conditions in which they are carried out. Another factor which influences the choice is the precision required of the answer, which in turn depends upon the original reason for carrying out the measurements. The main reason for applying ergonomic principles to any work or leisure activity is to ensure that the activity is carried out with maximum efficiency. This may mean, for example, reducing the fatigue or stress of an industrial task so as to increase output (Vogt *et al.*, 1977), or measuring the performance of racing cyclists so as to improve their chances of winning races (Whitt, 1971). In both of these examples the fundamental purpose of understanding the metabolic responses of the subjects was to improve their performance. It is obviously important, there-

Fig. 2.14. Heart rates recorded during various activities of Norwegian trawler fishermen in the first (14.00–02.00 hours: open circles) and in the second (02.00–14.00 hours: filled triangles) twelve hours of investigation. (*a*) Handling trawl; (*b*) bleeding the fish; (*c*) washing the fish; (*d*) resting; (*e*) sleeping; (*f*) whole 12-hour period. (After Rodahl and Vokac, 1977.)

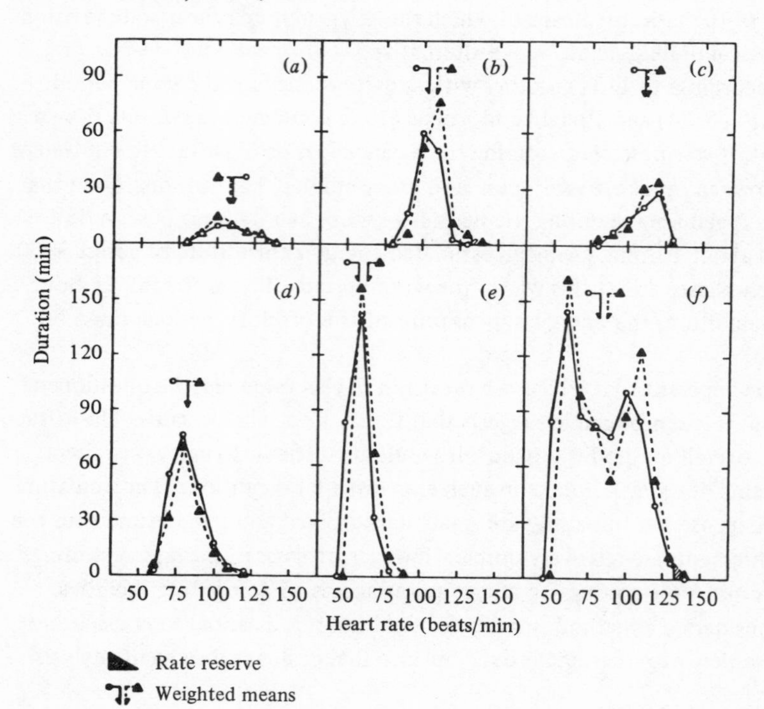

fore, that the measured metabolic responses do, in fact, reflect the degree of stress imposed upon the subjects. This may not be the case in many occupational conditions because the responses to the stress may be modified by such factors as emotion, drugs, time of day, age and sex of the subject, differences between individuals, posture, etc.

Vogt *et al.* (1977), for example, assessed the stress of blast-furnace workers partly by their heart rates. This is a measure used extensively to 'quantify' the strain resulting from environmental stresses, the line of reasoning being that the heart rate is related to oxygen uptake which, in turn, is a measure of the amount of energy output (and consequently is an estimate of the amount of energy input needed to sustain the activity). But as Vogt *et al.* (1973) have shown, heart rate itself varies with the thermal stress of the environment. So in the case of the blast-furnace workers a measurement of heart rate alone is not necessarily a measure of the oxygen cost of the activity. Even if it were, the measurements may not be valid in all circumstances. For example, men and women respond differently to thermal stress (Burse, 1979), so that any conclusion about the stress of a particular working task may not be valid for women if the original test subjects were men (and vice versa).

In a study of female nurses Fordham *et al.* (1978) showed that heart rate did not correlate with simultaneously measured oxygen uptake, so that if heart rate alone had been used to predict the energy cost of that job the conclusions would have been erroneous. The authors suggested that in this particular case intervening factors such as emotion and circadian rhythms confused any possible correlation between the measures.

In view of all these difficulties it is reasonable to question why any measurements of the physiological cost of activities are ever attempted. One answer is that it is not necessary to know in every case the exact level of physiological cost. For example, the value of any changes made to the working environment (e.g. reduction in noise levels, improving working posture, or relieving thermal stress) can be assessed by a simple comparison of the physiological cost of the work before and after the changes, a lower level of oxygen uptake or heart rate after the changes implying a lower level of stress. In these circumstances, then, the precise level of oxygen uptake is not important: what matters is the reduction in physiological cost brought about by the changes in the environment.

In some cases, however, it may be extremely important to know the *exact* level of oxygen uptake for particular activities. This is true, for example, in any situation where self-contained breathing equipment is used (e.g. in work underwater, in fire-fighting, in military aircraft, or in spacecraft). In these cases the duration of a mission may be limited solely by the finite supply of oxygen available to the worker. For this reason the quantification studies such as those of

Duncan, Gardner and Barnard (1979) and Bell and Wright (1979) are important.

A further reason for making measurements of the metabolic cost of activities is that whilst trends within a small group of subjects may not be at all significant, trends within larger groups or whole populations may be invaluable in assisting the design of the work place or the living area. This problem of individual differences is illustrated in the relationship between heart rate and oxygen uptake, and is very obvious in Fig. 10-5 of Astrand and Rodahl (1977). The same figure also shows clearly the trend which is the much-quoted classical relationship. It is this trend which enabled Rodahl and Vokac (1977) to evaluate the work stress of trawler fishermen; and it is the same trend which assisted the prediction of the limits to the manual handling of heavy loads (Ayoub *et al*., 1980), an area of active research to minimise occupational stress.

We have seen then that the functioning of the body requires a supply of chemical energy. This energy is derived from the metabolism of food, a process which leads to the formation of carbon dioxide, water and some complex metabolites. The metabolism of food ultimately requires oxygen, so that an indirect measure of the amount of energy being used in a task is the amount of oxygen being taken into the body. We have seen that it is not always possible to measure oxygen uptake directly, so various indirect methods have been devised from which estimates of oxygen uptake can be made. The most commonly used indirect method is based on the linear relationship between oxygen uptake and heart rate or minute volume. However, as we have also seen, there are very many problems with these indirect methods. The linear relationship is not the same for men and women; it is not the same for fit and unfit individuals; it is not the same for the same individual at different times of the day; it will be altered by drugs, working posture, environmental conditions, emotion and very many other factors. In spite of these difficulties, though, the relationship can still be used to evaluate the interaction between man and his work and play: it is the optimisation of this interaction which is the aim of ergonomics.

References

Ashworth, A. and Wolff, H.S. (1969) A simple method for measuring calorie expenditure during sleep. *Pflügers Archiv,* **306**, 191–94.

Astrand, I., Astrand, P.-O., Christensen, E.H. and Hedman, R. (1960) Intermittent muscular work. *Acta Physiologica Scandinavica,* **48**, 448–53.

Astrand, P.-O. and Rodahl, K. (1977) *Textbook of Work Physiology: Physiological Bases of Exercise,* 2nd edn. New York: McGraw-Hill.

Atwater, W.O. and Benedict, F.C. (1905) *A Respiration Calorimeter with Applicances for the Direct Determination of Oxygen.* Publication 42, Carnegie Institution of Washington. Washington, DC: Carnegie Institution.

Ayoub, M.M., Mital, A., Asfour, S.S. and Bethea, N.J. (1980) Review, evaluation and comparison of models for predicting lifting capacity. *Human Factors*, **22**, 257–69.

Becker, W.M. (1977) *Energy and the Living Cell.* Philadelphia: J.B. Lippincott.
Bell, D.G. and Wright, G.R. (1979) Energy expenditure and work stress of divers performing a variety of underwater work tasks. *Ergonomics*, **22**, 345–56.
Benzinger, T.H., Huebscher, R.G., Minard, D. and Kitzinger, C. (1958) Human calorimetry by means of the gradient principle. *Journal of Applied Physiology*, **12**, Supplement 1, S1–S28.
Blaxter, K.L., Brockway, J.M. and Boyne, A.W. (1972) A new method for estimating the heat production of animals. *Quarterly Journal of Experimental Physiology*, **57**, 60–72.
Burse, R.L. (1979) Sex differences in human thermoregulatory responses to heat and cold stress. *Human Factors*, **21**, 687–99.
Cotes, J.E. (1979) *Lung Function: Assessment and Application in Medicine*, 4th edn. Oxford: Blackwell Scientific.
Datta, S.R. and Ramanathan, N.L. (1969) Energy expenditure in work predicted from heart rate and pulmonary ventilation. *Journal of Applied Physiology*, **26**, 297–302.
Daws, T.A., Consolazio, C.F., Hilty, S.L., Johnson, H.L., Krzywicki, H.J., Nelson, R.A. and Witt, N.F. (1972) Evaluation of cardiopulmonary function and work performance in man during caloric restriction. *Journal of Applied Physiology*, **33**, 211–17.
Douglas, C.G. (1911) A method for determining the total respiratory exchange in man. *Journal of Physiology (London)*, **42**, 17P–18P.
Duncan, H.W., Gardner, G.W. and Barnard, R.J. (1979) Physiological responses of men working in fire fighting equipment in the heat. *Ergonomics*, **22**, 521–7.
Durnin, J.V.G.A. and Edwards, R.G. (1955) Pulmonary ventilation as an index of energy expenditure. *Quarterly Journal of Experimental Physiology*, **40**, 370–7.
Eley, C., Goldsmith, R., Layman, D., Tan, G.L.E., Walker, E. and Wright, B.M. (1978) A respirometer for use in the field for the measurement of oxygen consumption. 'The Miser', a Miniature Indicating and Sampling Electronic Respirometer. *Ergonomics*, **21**, 253–64.
Fletcher, J.G. & Wolff, H.S. (1954) A lightweight integrating motor pneumotachograph (imp) with constant low resistance. *Journal of Physiology (London)*, **123**, 67P–69P.
Fordham, M., Appenteng, K., Goldsmith, R. and O'Brien, C. (1978) The cost of work in medical nursing. *Ergonomics*, **21**, 331–42.
Garrow, J.S. (1978) *Energy Balance and Obesity in Man*, 2nd edn. Amsterdam: Elsevier/North-Holland.
Green, J.H. (1976) *An Introduction to Human Physiology*, 4th edn, pp. 63–77. Oxford: Oxford University Press.
Harper, H.A., Rodwell, V.W. and Mayes, P.A. (1977) *Review of Physiological Chemistry*, 16th edn. Los Altos: Lange Medical Publications.
Harris, A. (1978) *Human Measurement.* London: Heinemann Educational.
Heinonen, A.O., Karvonen, M.J. and Ruosteenoja, R. (1959) The energy expenditure of walking on snow at various depths. *Ergonomics*, **2**, 389–93.
Henschel, A., Taylor, H. and Keys, A. (1954) Performance capacity in acute starvation with hard work. *Journal of Applied Physiology*, **6**, 624–33.
Humphrey, S.J.E. and Wolff, H.S. (1977) The Oxylog. *Journal of Physiology (London)*, **267**, 12P.
Keys, A., Brozek, A., Henschel, A., Mickelsen, O. and Taylor, H.L. (1950) *The Biology of Human Starvation.* Minneapolis: University of Minnesota Press.
Kofranyi, E. and Michaelis, H.F. (1940) Ein tragbarer Apparat zur Bestimmung des Gasstoffwechsels. *Arbeitsphysiologie*, **11**, 148–50.
Lehninger, A.L. (1971) *Bioenergetics: The Molecular Basis of Biological Energy Transformations*, 2nd edn. Menlo Park: W.A. Benjamin.
Liddell, F.D.K. (1963) Estimation of energy expenditure from expired air. *Journal of Applied Physiology*, **18**, 25–9.

Passmore, R., Thomson, J.G. and Warnock, G.M. (1952) A balance sheet of the estimation of energy intake and energy expenditure as measured by indirect calorimetry using the Kofranyi Michaelis calorimeter. *British Journal of Nutrition*, **6**, 253–64.

Passmore, R. and Durnin, J.V.G.A. (1955) Human energy expenditure. *Physiological Reviews*, **35**, 801–40.

Pandolf, K.B., Haisman, M.F. and Goldman, R.F. (1976) Metabolic energy expenditure and terrain coefficients for walking on snow. *Ergonomics*, **19**, 683–90.

Paul, A.A. and Southgate, D.A.T. (1978) *McCance and Widdowson's The Composition of Foods.* London: HMSO.

Ramaswamy, S.S., Dua, G.L., Raizada, V.K., Dimri, G.P., Viswanathan, K.R., Madhaviah, J. and Srivastava, T.N. (1966) Effect of looseness of snow on energy expenditure in marching on snow-covered ground. *Journal of Applied Physiology,* **21**, 1747–9.

Rodahl, K. and Vokac, Z. (1977) Work stress in Norwegian trawler fishermen. *Ergonomics*, **20**, 633–42.

Rodahl, K., Vokac, Z., Fugelli, P., Vaage, O. and Maehlum, S. (1974). Circulatory strain, estimated energy output and catecholamine excretion in Norwegian coastal fishermen. *Ergonomics*, **17**, 585–602.

Shephard, R.J. (1978) *Human Physiological Work Capacity. IBP Synthesis Series 15.* Cambridge: Cambridge University Press.

Vogt, J.J., Foehr, R., Kuntzinger, E., Seywert, L., Libert, J.P., Candas, V. and van Peteghem, Th. (1977) Improvement of the working conditions at blast-furnaces. *Ergonomics*, **20**, 167–80.

Vogt, J.J., Meyer-Schwertz, M. Th., Metz, B. and Foehr, R. (1973) Motor, thermal and sensory factors in heart rate variation: a methodology for indirect estimation of intermittent muscular work and environmental heat loads. *Ergonomics*, **16**, 45–60.

Warnold, I. and Lenner, R.A. (1977) Evaluation of the heart rate method to determine the daily energy expenditure in disease. A study on juvenile diabetics. *American Journal of Clinical Nutrition*, **30**, 304–15.

Webb, P.A. (1973) Work, heat and oxygen cost. In *Bioastronautics Data Book*, 2nd edn, J.F. Parker and V.R. West, chapt. 18. *Special Publication 3006,* National Aeronautics and Space Administration. Washington, DC: NASA.

Webb, P.A., Annis, J.F. and Troutman, S.J. (1972) Human calorimetry with a water-cooled garment. *Journal of Applied Physiology*, **32**, 412–18.

Weir, J.B. de V. (1949) New methods for calculating metabolic rate with special reference to protein metabolism. *Journal of Physiology (London)*, **109**, 1–9.

Whitt, F.R. (1971) A note on the estimation of the energy expenditure of sporting cyclists. *Ergonomics*, **14**, 419–24.

3 BIOMECHANICS

D. Grieve and S. Pheasant

3.1 Introduction

The role of biomechanics in ergonomics

Human biomechanics deals with the mechanical aspects of the body. Ergonomic applications are principally concerned with the linked body segments and their musculature, and the mechanical interfaces of man with his environment. A knowledge of these is appropriate if man is to be accommodated in the workplace, if foot and hand controls are to be convenient in operation, if tasks are to be within his strength to perform and if stresses on particular regions of the body, which might cause fatigue or even injury under more extreme conditions, are to be avoided. Biomechanics has no sharp boundaries and blends with anatomy, physiology and industrial medicine.

Posture, movement, muscle activity and forces upon and within the body comprise a quartet of phenomena with which the biomechanist is usually concerned. The work physiologist shares this interest but it is possible to make distinctions. The physiologist arms himself with measurements of heart rate, oxygen consumption, body temperature and other quantities which reflect his concern with integrative phenomena such as energy expenditure, cardiovascular response and homeostatic mechanisms. He is less likely to be concerned with stresses within the musculo-skeletal system and his techniques rarely permit consideration of isolated events (as distinct from repetitive activities) which the intact body may perform in brief periods of a few seconds. Activities which create stresses upon localised regions may pass unnoticed as far as cardiovascular, respiratory or thermal monitoring is concerned. Indeed, to apply the more common physiological techniques in a valid manner to these activities would suggest that the wrong ergonomic questions were being asked.

The biomechanist, using, for example, force transducers, cine camera and an electromyograph, can make instantaneous records of man operating in the field of gravity and can examine his posture and movements in detail. He also has techniques for proceeding logically from the monitored interfaces of man with the environment to a consideration of the implications of his observations in terms of anatomical structures. In short, human physiology and biomechanics complement each other and both deserve the attention of the 'compleat ergonomist'.

Terminology

The student of ergonomics cannot expect to master anatomical structure in detail, but neither can he ignore it. Most anatomical texts are intended as a background to medicine. The ergonomist will find more than sufficient anatomical description of the body in Basmajian (1970), and Basmajian (1967) is recommended for its functionally oriented discussion of the musculature.

Biomechanics makes use of anatomical terminology; it is a technical jargon which is especially useful for the stereotyped assessment of the sick on couches, but not ideal for describing the worker at his bench. Other approaches to descriptions of posture and movement will be encountered in the text, each with its advantages and disadvantages. However, anatomical nomenclature is universal and, even though it may prove cumbersome, has the merit of being unambiguous when properly used.

Terms of position

Consider a person standing upright, feet together and arms by the sides, with palms forward. This reference posture, the anatomical position, is bilaterally symmetrical about a so-called *mid-sagittal plane;* planes parallel to it are also called *sagittal.* Vertical planes perpendicular to the sagittal are termed *coronal* planes, and *horizontal* (or *transverse*) planes are mutually perpendicular to the sagittal and coronal planes. Care is required to define planes of reference when the body is far removed from the anatomical position.

Terms of *relative* position are *medial* and *lateral,* meaning nearer to or further from the mid-sagittal plane; *superior* and *inferior,* meaning nearer to or further from the top of the body; and *anterior (ventral)* and *posterior (dorsal),* meaning in front of or behind another structure. Nearer to and further from the body surface are termed respectively *superficial* and *deep,* and nearer to or further from the trunk are called *proximal* and *distal.*

There are few terms which refer to absolute position. Structures in the mid-sagittal plane are termed *median.* Structures on the back are *dorsal* and on the front of the body are *ventral* (cf. their use above as relative terms). We stand on the *plantar* surface of the feet and the upper surface is the *dorsum* of the foot. The back of the hand is its dorsum, on the opposite face to the *palmar* surface.

Terms of movement

Movements, for simplicity, will usually be considered as taking place from the anatomical position. The descriptions that follow are adequate but not complete; anatomical texts should be consulted for discussion of detailed joint movements in the context of structure.

(i) *Trunk*

Sagittal plane: bend forward = *flexion*

bend backwards = *extension*

Coronal plane: bend sideways = *lateral flexion*

Transverse plane: rotate about the long axis = *axial rotation.*

NB: These terms apply to head on neck, to vertebrae on each other and to the trunk as a whole.

(ii) *Shoulder girdle*

Raise shoulders = *elevation*

Lower shoulders = *depression*

Draw forward = *protraction*

Draw backward = *retraction*

Rotate scapula on rib cage as in arm-raising = *outward rotation*

NB: The shoulder as a *region* should be distinguished from the shoulder *joint*, i.e. gleno-humeral joint.

(iii) *Shoulder and hip joints*

Sagittal plane: raise arm or thigh forward = *flexion* (opp. = *extension*)

Coronal plane: raise arm or thigh to side = *abduction* (opp. = *adduction*)

Transverse plane: rotate arm or leg along its long axis = *lateral* or *outward rotation* (opp. = *medial* or *inward rotation*).

NB: Shoulder and hip joints are ball and socket types, having three degrees of rotational freedom.

(iv) *Elbow and knee joints*

Sagittal plane: bend from the fully straightened position = *flexion* (opp. = *extension*)

The elbow joint is a hinge type with one degree of rotational freedom. The knee joint approximates to a hinge with one degree of rotational freedom but its structure permits a limited second degree of rotational freedom when it is flexed.

(v) *Wrist joint* (combined with carpal joints)

Sagittal plane: bend palm upwards = *flexion* (opp. = *extension*)

Coronal plane: move hand away from trunk (to side of radius bone) = *abduction* or *radial deviation* (opp. = *adduction* or *ulnar deviation*)

(vi) *Forearm*

The radius bone rotates around the ulna. With the hands on a table, rotation of the forearm to a palm-up position is *supination* and from there to a palm-down position is *pronation.*

(vii) *Ankle*

Sagittal plane: raise dorsum of foot = *extension* (opp. = *flexion*)
It is less confusing, at the ankle, to refer to raising the dorsum as *dorsiflexion* and the opposite movement as *plantarflexion.*

(viii) *Sub-talar joints*

For many purposes the ankle region may be considered as a universal joint. In fact, the movement which brings the sole of the foot to face medially (inwards) takes place between bones below the ankle joint proper and is called *inversion* (opp. = *eversion*).

(ix) *Digits*

The thumb is the first digit and lies in a plane perpendicular to the others. M/P and I/P refer to metacarpo-phalangeal and inter-phalangeal respectively.
Movement of thumb across palm = *flexion* of M/P and I/P joints (opp. = *extension*)
Movement of thumb away from palm (perpendicular to it) = *abduction* (opp. = *adduction*)
Movement of thumb to oppose pulp to pulp of the other digits = *opposition*
Curling the fingers as when making a fist involves flexion of the M/P and I/P joints. The hand is considered to have an axis defined by the 3rd metacarpal and digit. Movement of digits 2, 4 and 5 away from the axis in the plane of the palm (at M/P joints) is *abduction* (opp. = *adduction*). Movement of all digits in plane perpendicular to palm = *flexion* and *extension.* The 3rd M/P joint is said to flex, extend and also to exhibit *radial* and *ulnar deviation* (cf. wrist).

Conventions for joint and limb angles

There is no universal convention governing the measurement of the posture of a limb segment in space or the (joint) angle between adjacent segments. Conventions used in this chapter to describe various limb and joint angles are summarised graphically in Fig. 3.1. Because of the lack of standardisation, care must be taken when comparing and collating information about joint ranges and posture from different sources.

Mechanical principles applied to the human body

The ergonomist is expected to judge whether human speed, force, accuracy and endurance are sufficient for the performance of a particular task in a given environment. He should appreciate the ways in which various working methods may be fatiguing or hazardous for the worker. Analyses will sometimes be confined to the interfaces between the body and its environment; for example, anthropometric data may be used to estimate the suitability of a proposed workplace for the operatives and whether they will have the strength to perform the

Fig. 3.1. Conventions for joint and limb angles that have been used to describe postures and movements in the sagittal plane: (1) arm angle, (2) forearm angle (3) elbow angle, (4) hip angle, (5) thigh angle, (6) leg, or shank, angle, (7) knee angle, (8) foot angle, (9) ankle angle.

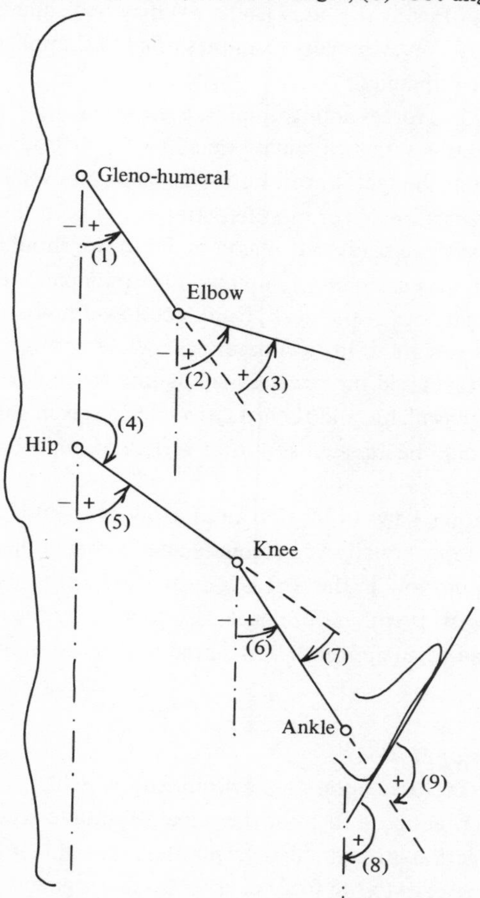

tasks. Other studies will consider the implications of the tasks for strains set up within the body; a more extended range of techniques is required for this purpose.

In the sections which follow, topics such as anthropometry, muscle posture and heavy work are discussed. They are not truly separable and we shall begin by setting out a framework for discussion which contains the minimum essential mechanical knowledge. The student should not construe biomechanics as the quantitative application of complex equations of motion, even though that is necessary in some research problems. It is more important that the *qualitative* implications of Newton's Laws of Motion, the principles of levers and of living in a gravitational field be grasped; correct qualitative deductions can then be made from observation, even if we cannot resort to measurement. A qualitative introduction to the mechanics of using the body (in sport) may be found in Dyson (1977), and Le Veau (1977) gives many examples of clinical application of calculations based upon the principles.

Every segment of the body has forces acting upon it, even when completely at rest. Apart from the continuous action of gravity on all parts, the body may interact with the environment at the feet, knees, buttocks, back, elbows, hands and elsewhere. The use of muscles creates or modifies stresses between the body segments and across interfaces. If we study one region in depth, we should not forget that *any* use of the body has mechanical implications for *all* parts. Raising one arm, for example, affects stresses in the neck, trunk and lower limbs and at foot–floor interfaces. The stresses are of little consequence while raising an unloaded limb, but if a weight is raised the stresses may be intense and even hazardous. Options are usually available and their relative merits, given some biomechanical knowledge, should be assessed and, if of sufficient importance, be amenable to measurement.

We shall now discuss Newton's Laws of Motion and introduce the use of the free-body diagram. The laws apply equally to inanimate and animate objects, but the words 'rest' and 'uniform motion' in the first law mean different things to the engineer and to the biologist. It is therefore necessary to amplify the laws to show how the distinctions lead to an approach best suited to discussion of the living body.

Newton's Third Law of Motion

When a person exerts a force upon the environment, which is measurable in principle and often in practice, a force of the same magnitude acts upon the person in the opposite direction through the same point on the man–environment interface. The law also applies to the force exerted by one segment of the body upon an adjacent segment. If a torque (turning moment, couple) is exerted upon the environment, an equal and opposite torque acts upon the person. We may therefore replace the environment conceptually by a set of forces and torques and thus free the body, or the part of the body which concerns us, for discussion. Fig. 3.2 shows how the person in the environment may be separated from it in terms of a free-body diagram and how a segment of the body is separated from the remainder by the same technique.

Newton's First Law of Motion

Newton's First Law of Motion states that a body will remain in a state of uniform motion or at rest, i.e. will conserve its momentum, unless acted upon by a net force. The motion is detected by measuring it relative to the surroundings, e.g. the workplace or laboratory. Momentum, a vector quantity, is the sum of the products of all the masses comprising the body and their velocities in the direction we choose to consider. We are principally concerned with the human body, which is not rigid and redistributes its mass with each change of posture. We may only demonstrate a state of uniform translational motion or rest in the Newtonian sense by computing the movement of the centre of gravity. It is important to remember that the centre of gravity is not fixed relative to any anatomical landmarks; procedures exist for estimating its position (see the

Fig. 3.2. Diagram illustrating how, according to Newton's Third Law of Motion, a body, or body part, may be conceptually isolated from its surroundings by replacing the latter by sets of forces and torques which act across the interfaces.

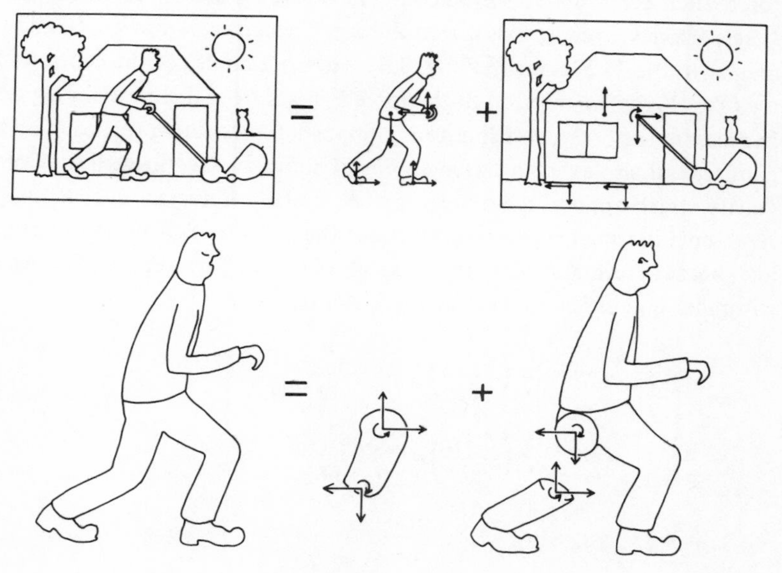

section on Newtonian anthropometry, p. 115). Conversely, we know that if no net force is acting on the body we may deduce that the body is in a state of uniform motion. If wind resistance is discounted, uniform horizontal motion may be deduced for a long-jumper while he is in the air but uniform vertical motion does not exist because a net force of his weight is acting.

A man in repose is not truly in a state of Newtonian rest, because he is continually making slight adjustments to his posture. Examples will be given later in which true equilibrium is assumed because the advantages of simple statements sometimes outweigh the disadvantages of dealing with an approximation (see section on epidemiology of injuries due to handling, p. 174).

A corollary to the First Law of Motion is that the body will conserve its angular momentum unless acted upon by a net torque. The angular momentum of a non-rigid body will require definition, and we will discuss this in the following section.

Newton's Second Law of Motion

Newton's Second Law of Motion tells us what the *instantaneous* state of affairs is if the net forces and torques on the body are *not* zero. The body (or body segment) will change its linear momentum at a rate proportional to the *net* force that is acting. A corollary is that the angular momentum about any chosen axis will change at a rate proportional to the *net* turning moment applied about that axis.

Consider the motion of a body mass, in one ($X-Y$) plane, under the action of forces such as F whose resolved components are F_x and F_y at distances l_y and l_x respectively from an axis drawn through the centre of gravity, G, and perpendicular to the $X-Y$ plane (see Fig. 3.3). The coordinates of the centre of gravity are (X_G, Y_G). The reaction of the force F arises from the inertial drag of all the particles in the body, which not only opposes forces acting vertically and horizontally but also exerts a turning moment about an axis through the centre of gravity which opposes the moments of F_x and F_y acting about that axis. If A represents the angular momentum about the axis through the centre of gravity and, because several forces may be acting, the symbol Σ represents the sum of all similar quantities, we have at any instant:

$$\Sigma F_x = m \cdot \frac{\mathrm{d}}{\mathrm{d}t}\left(\frac{\mathrm{d}X_G}{\mathrm{d}t}\right) = m \cdot \frac{\mathrm{d}^2 X_G}{\mathrm{d}t^2} \tag{3.1}$$

$$\Sigma F_y = m \cdot \frac{\mathrm{d}}{\mathrm{d}t}\left(\frac{\mathrm{d}Y_G}{\mathrm{d}t}\right) = m \cdot \frac{\mathrm{d}^2 Y_G}{\mathrm{d}t^2} \tag{3.2}$$

$$\Sigma (F_x \cdot l_y + F_y \cdot l_x) = \frac{\mathrm{d}A}{\mathrm{d}t} \tag{3.3}$$

It is now appropriate to consider the angular momentum, A. Each particle, mass δm, of which the body is comprised, may be considered at some distance, r, from the axis through the centre of gravity of the body. The vector between the centre of gravity and the particle rotates about the axis with angular velocity ω (radians per second). The particle has a speed of $r\omega$ perpendicular to the position vector and linear momentum of $\delta m \cdot r\omega$ due to the rotation. The angular momentum of the particle is the product of the linear momentum (perpendicular to the radius vector) and the distance from the axis through the centre of gravity. For all p particles in the body, we have the total angular momentum given by:

$$A = \sum_{1}^{p} \delta m \cdot r^2 \omega \tag{3.4}$$

If the body is *non-rigid* and undergoing complex changes of form, ω is different for each particle. At the other extreme, all particles in a *rigid* body rotate with the same value of ω and we have

$$A = \omega \sum \delta m \cdot r^2 \tag{3.5}$$

It is customary to assume that the human body consists of rigid segments, linked together. For the purpose of applying Newton's Laws to the segments, the positions of the centres of gravity of segments relative to their anatomical

Fig. 3.3. Quantities referred to in the text which are required when calculating the influence of a force F upon the motion of a body in one $(X-Y)$ plane.

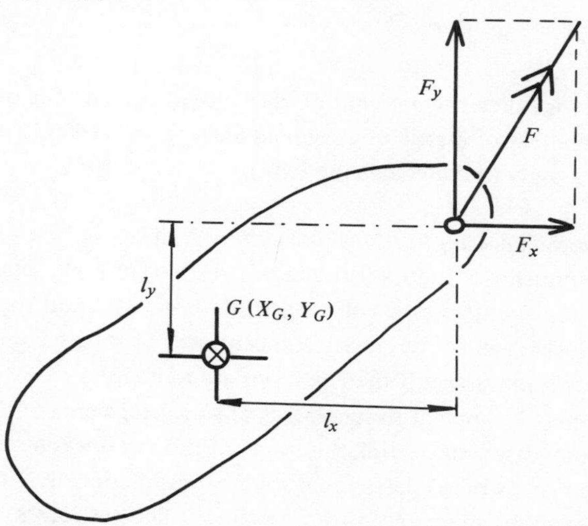

landmarks, the masses of the segments are quantities known as radii of gyration (k) are to be found tabulated in the literature (see section on Newtonian anthropometry, p. 115). By definition, k is given by

$$k^2 = \frac{\Sigma\, \delta m \cdot r^2}{\Sigma\, \vartheta\, m} \qquad (3.6)$$

and, therefore, for a rigid segment of mass m

$$A = m\, k^2\, \omega \qquad (3.7)$$

Now suppose that equations 3.1, 3.2 and 3.3 apply to the motion of the whole human body in the X–Y plane. The intact body may be altering its posture; we require a more useful equation than 3.4, but cannot apply equation 3.5.

We obtain the value of A by calculating the angular moments of the segments about axes through their own centres of gravity and add further quantities which are the products of the segment masses, the square of the distances R of the centres of gravity from the centre of gravity of the whole body, and the angular velocities, ω_1, of the segments about the centre of gravity of the whole. We then have, for n segments:

$$\underset{\text{whole body}}{A} = \sum_{1}^{n} m\, k^2\, \omega + \sum_{1}^{n} m\, R^2\, \omega_1 \qquad (3.8)$$

It can be seen that if the whole body is made rigid ω and ω_1 are identical. In that case:

$$\underset{\text{whole body}}{A} = \omega \sum_{1}^{n} m\, (k^2 + R^2) \qquad (3.9)$$

The quantity $m\,k^2$ is called the moment of inertia of a segment; the quantity $m\,(k^2 + R^2)$ is the moment of inertia of a segment about a parallel axis, distance R from its centre of gravity (parallel-axis theorem).

Some biomechanical implications of Newton's Laws

1. Masses, moments of inertia (or radii of gyration) of body segments and the locations of their centres of gravity constitute important anthropometric data. The ability to locate centres of gravity and centroids of forces acting upon the body at interfaces is important if the equations are to be used.

2. The equations may be applied to successive, adjacent segments of the body by the use of free-body diagrams, to link the forces at the interfaces between man and his environment (which relates to characteristics of controls, floor surfaces and demands of a job) and forces within the body. The net forces, and

torques at joints, calculated by this means provide a starting point for the consideration of demands upon particular anatomical structures.

3. Equations 3.1, 3.2 and 3.3 are statements of instantaneous conditions. The equations may also be expressed in the form of definite integrals over a period of time t_1 to t_2.

$$\int_{t_1}^{t_2} \Sigma F_y \cdot \mathrm{d}t = m \left| \frac{\mathrm{d}\, Y_G}{\mathrm{d}t} \right| \tag{3.10}$$

$$\int_{t_1}^{t_2} \Sigma F_x \cdot \mathrm{d}t = m \left| \frac{\mathrm{d}\, X_G}{\mathrm{d}t} \right| \tag{3.11}$$

$$\int_{t_1}^{t_2} \Sigma \left(F_x \cdot l_y + F_y \cdot l_x \right) \mathrm{d}t = |A| \tag{3.12}$$

Newton's concept of rest or uniform motion is that of the engineer, whereas the ergonomist, observing persons standing still or moving at a steady speed, may consider the biological concepts of rest or uniform motion, i.e. in which no appreciable net change of kinematic state occurs over a period of time, as more relevant.

Equations 3.10, 3.11 and 3.12 are very suitable for the discussion of posture and locomotion. If steady states exist (in the biological sense) the right-hand sides of the equations are set to zero; the time interval must be chosen suitably, e.g. over a complete cycle of events in the case of steady locomotion. The application of *statics* to postural analysis should be clearly understood to be a simplification.

4. The equations refer to *net* forces on the body or its segments. Observed postures and movements are not therefore the result of a *unique* set of forces and torques. If implications for particular anatomical structures, e.g. bones, ligaments or muscles, are sought, supplementary evidence is required. Electromyography, despite its many drawbacks, is a valued tool in the search for further evidence.

5. The choice of particular axes (X and Y) for a planar analysis, or X, Y and Z for a three-dimensional analysis, is arbitrary providing that the axes are mutually perpendicular. Vertical and horizontal axes often yield simpler equations, but an axis between the centre of pressure at the hands and feet has a special significance in the analysis of manual exertion. In the analysis of swinging

implements, the plane of swing deserves more attention than vertical or horizontal planes.

3.2 Human muscle function

Introduction

Skeletal (or striated) muscles are the organs responsible for initiating or restraining the relative movements of skeletal members. This results from the muscle's capacity to exert a modulated tension between the two bony regions to which it is attached (known to anatomists as its origin and its insertion). Only at the extremes of joint movement are other structures (e.g. taut ligaments or opposing body surfaces) capable of exerting restraining forces.

Skeletal muscle tissue accounts for approximately 40% of the total mass of the body. It can exist in either of two conditions, which are known as the 'resting' and 'active' states. In the resting state muscle behaves like a non-linear spring with negligible viscous damping. Stiffness is minimal in the lower part of its range, but increases as the muscle is stretched towards its maximum length in the body. The tension in the resting state arises because of strain in the strands of connective tissue which surround the muscle cells and bind the anatomical muscle together. 'Energy expenditure' (over and above resting level) as measured by work physiologists is determined by the proportion of the muscle mass which is in its active state (averaged over a period of time).

Voluntary or reflex neural signals from the central nervous system cause the muscle to enter the active state in which tension, in excess of that which is due to passive strain, is exerted between the skeletal attachments. In the active state a complex sequence of metabolic reactions is occurring, by virtue of which chemical energy (ultimately derived from the combustion of foodstuffs) is transformed into mechanical energy (either potential or kinetic) and heat. (The reader wishing to review the biochemistry and thermodynamics of muscular activity should consult Carlson and Wilkie, 1974.) The associated potential charges of the membranes of the muscle cells may be detected as the electromyogram (see section on electrical correlates of muscle activity, p.93). A delay of variable duration (in the order of 100 milliseconds) occurs between the initiation of the active state and its mechanical consequences.

Each thread-like individual nerve cell (motor neurone) branches along its path from the spinal cord to the muscle, mainly within the muscle substance, to eventually supply many muscle cells. The nerve cell and the muscle cells that it supplies are a functional group known as a motor unit.

A single nerve impulse gives rise to an active state which is rapid in onset and decays completely in less than 0.1 second. Impulses in nerve fibres may range from nil to about 50 per second, each muscle may contain several thousand motor units and several muscles act about each joint in the body, so that the possibilities for the control of posture and locomotion are virtually infinite. The intensity of the active state depends on the number of neural impulses received in unit time, whereas the tension, the work done and the efficiency with which this work is achieved depend on a variety of factors discussed later in this section. Any purposive muscular action is controlled by a complex hierarchy of feedback loops which depend upon sense organs located within the muscle itself as well as elsewhere in the body.

The precision with which muscular tension may be modulated depends upon the number of muscle fibres within each motor unit (the motor unit providing the quantum of muscular activity). This number ranges in man from 1934 in the gastrocnemius (a calf muscle which is adapted for the postural and locomotor control of the whole body) to only 9 for the lateral rectus muscle (which controls the position of the eye within its socket) (Feinstein *et al.*, 1955).

Physiologists usually refer to the switching on of a muscle into its active state as a muscle contraction. It is important to realise that an active muscle may remain at the same length (isometric contraction), may shorten (concentric contraction) or may lengthen (eccentric contraction). The mechanical outcome of an active state will depend on whether the tension which the muscle exerts is equal to, greater than or less than the opposing force against which it is acting. In concentric action the muscle may be compared to a motor, in eccentric action to a brake and in isometric action to a guy rope.

It is most unusual for a single anatomical muscle to be solely responsible for a given movement. Movements generally demand the coordinated activation of several muscles which are said to act as synergists. Hence, although the deltoid is the *prime mover* in abduction at the shoulder joint, the deeper 'rotator cuff' muscles (see any anatomical text) are required as *synergists* to maintain the position of the head of the humerus in its socket on the scapula, and other muscles (e.g. trapezius, serratus anterior) are required to act as *muscles of fixation,* to control the position of the scapula and clavicle with respect to the trunk. In the slow lowering of the arm (adduction of the shoulder) the prime mover is gravity, and the deltoid muscle is called into play as an *antagonist* which acts as a brake and controls the rate of descent. (In rapid or ballistic actions antagonistic muscles are generally completely relaxed.)

The mechanics of muscle

The mechanical properties of muscle are frequently described or analysed in terms of a three-component analogue of the type shown in Fig. 3.4. The contractile component, c.c., is the 'motor', which may exist in an 'on' or 'off' condition as described above. It is coupled to its skeletal attachments through an undamped elastic element (the series compliance, s.c.). A further undamped elastic element (the parallel compliance, p.c.) also runs between the same points of attachment. These elements are abstractions, in the sense that they may be clearly defined in physiological experiment but cannot be located in specific anatomical structures.

The properties of these three components together determine the length-tension relationship of the resting and fully active muscle. The remaining characteristics of the muscle are specifically associated with the contractile component.

The length–tension curve

Let us consider a muscle which has been removed from the body but maintained in a viable condition (a common procedure in physiological laboratories). By measuring the tension required to maintain the muscle (under isometric conditions) at lengths greater than its resting length (L_0) a *passive length-tension curve* may be produced (Fig. 3.5). It must be emphasised that the term 'resting length' refers to *isolated* muscle and has no direct significance for the function of muscles within the body. If this procedure is repeated on a muscle which has been electrically stimulated into its fully active state a second length–tension relationship is obtained, the tension at any length now being the sum of the elastic resistance and the contractile tension. This relationship is known as the *total length–tension curve.* Subtracting the passive tension from the total tension at any given length allows the plotting of an *active length–tension curve,* which represents the capacity of the contractile mechanism alone to exert tension (Fig. 3.5). This relationship was first described by Blix in 1892; more recent descriptions are to be found in Gordon, Huxley and Julian (1966) and Rack and Westbury (1969).

Fig. 3.4. Three-component analogue of muscle: c.c., contractile component; s.c., series compliance; p.c., parallel compliance.

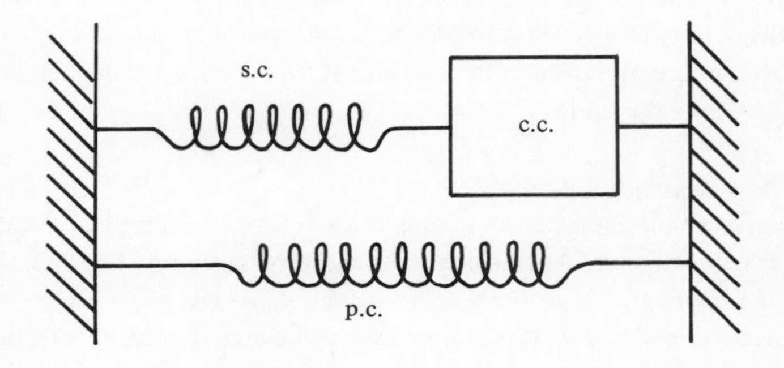

The angle–torque curve

Although the length–tension relationship of isolated muscle is well documented, it is not immediately relevant to the intact organism in which muscles work between skeletal attachments which articulate at joints of complex geometry. The leverage of individual muscles in various portions of their ranges cannot be determined accurately, and the only functional quantity which can be measured empirically is the net torque which they exert about a joint. For practical purposes, therefore, the length–tension curve must be replaced by an angle–torque curve in which the maximum torque which an individual may exert about a joint under isometric conditions is plotted against the angle of that joint. (This curve therefore relates voluntary strength to posture.)

Fig. 3.6 shows the angle–torque curve of the human ankle joint in which plantarflexor torque is plotted against ankle angle (see Fig. 3.1). The subjects' feet were strapped on to an instrumental pedal and at any given angle, passive torque was recorded when electromyographic silence (see p. 93) prevailed in both the calf muscles and the anterior tibial muscles. A second torque reading was taken in a maximum voluntary exertion, and the active torque plotted is the difference between the two. Further experimental details are to be found in Pheasant (1977). Gastrocnemius (a calf muscle) arises above the knee; its length, and therefore the contribution it makes to both active and passive torque about the ankle joint, is dependent upon knee posture (see Fig. 3.6).

Fig. 3.7 shows the torque which a sample of subjects could exert during

Fig. 3.5. The length–tension curve.

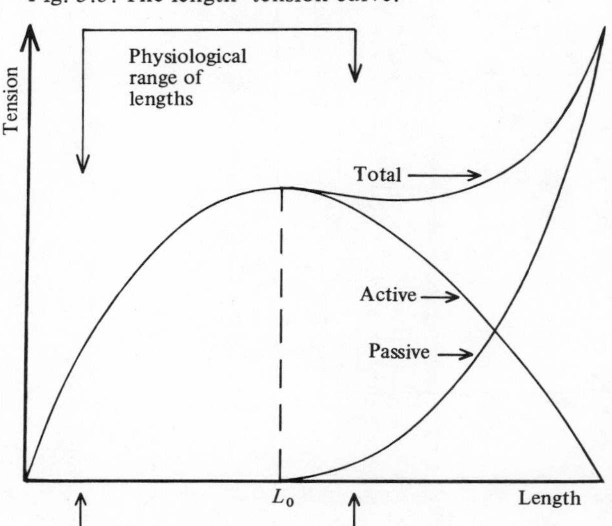

isometric pronation and supination of the forearm whilst gripping a stirrup-type handle. In this case it is not possible to distinguish the passive components of such torques as the human hand forms too 'slack' a linkage between musculoskeletal system and measuring instrument.

The above examples suggest that muscles *in situ* are capable of exerting their greatest tensions in those postures in which they are at their greatest length. (The extreme right-hand portions of the length–tension curve are unphysiological in that bony and ligamentous restraints prevent their employment in the living person.)

Smidt (1973) published a detailed analysis of the angle–torque curves of flexion and extension of the knee joint which is in agreement with the above assertion, as are the data of Murray and Sepic (1968) for abduction and adduction of the hip. The strength of elbow flexion is greatest in the middle portion of the range and is substantially reduced in either extreme posture (Doss and Karpovitch, 1965). Fick (1911), however, showed that all of the elbow flexor

Fig. 3.6. The angle–torque relationship for plantarflexion. Torque is plotted as a function of ankle angle for four different knee angles: 0° (open circles), 45° (filled circles), 90° (crosses), 135° (upright crosses). Filled squares represent all knee angles. Mean of nine subjects.

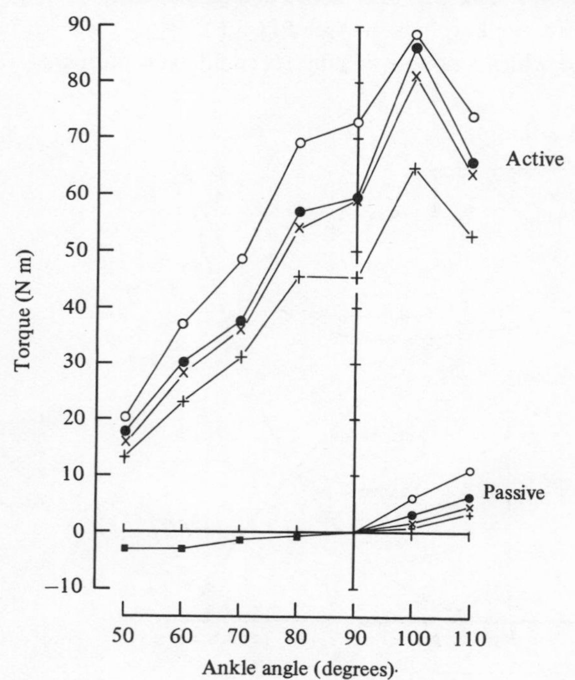

muscles have a very much greater leverage in positions of flexion than they do in extension where their intrinsic strength is presumably greatest. By induction, therefore, the principle of 'greatest length = greatest strength' is a general one for all muscle groups. (The authors know of no exceptions to this rule.)

Fig. 3.7. Angle–torque curves for pronation and supination. Torque is plotted against forearm angle (0° is prone, 180° is supine; supinator torques are positive, pronator torques negative). The data for each subject were normalised by expressing each item as a percentage of the mean strength of that subject in all positions. Best fit curves were derived by polynomial regression (± 1 r.m.s. residual).

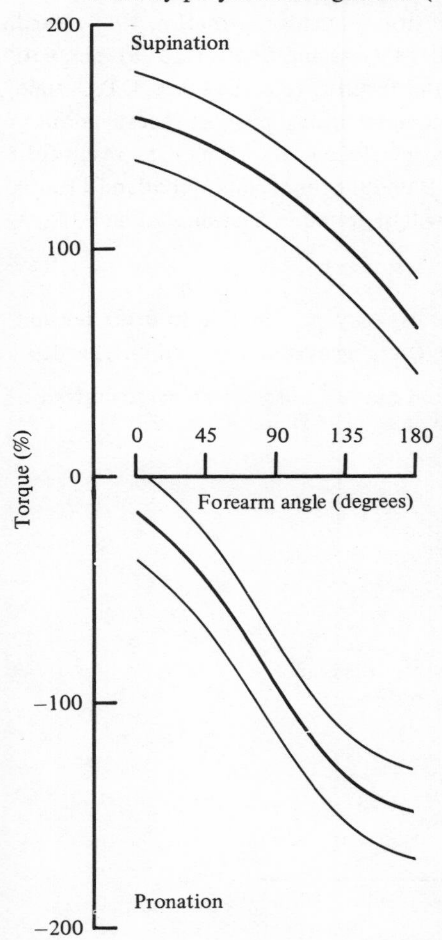

Strength

Angle–torque curves describe the strength of single articular elements of the musculo-skeletal system. Muscular actions of interest in ergonomics (such as the operation of heavily loaded control devices) commonly require the integrated exertion of many muscle groups. (Thrusting on a pedal, for example, requires plantarflexion of the ankle, extension of the knee, extension of the hip and the stabilisation of the pelvis and trunk with respect to the seat.) The maximum force which can be exerted in a complex action is determined by the weakest link in the kinetic chain concerned. (Although this assertion has not been proved experimentally, it is difficult to conceive how things might be otherwise.) Several attempts have been made to utilise sets of angle–torque curves in predictive models of more complex activities. The most important of these have been developed at the University of Michigan (Chaffin, 1969; Martin and Chaffin, 1972; Park and Chaffin, 1974; Garg and Chaffin, 1975). Simulation of this kind requires substantial computer facilities (storage space, C.P.U. time), and at present it remains necessary to conduct special-purpose investigations of human strength in the operation of any specific control device. The results of many such studies are summarised in Damon, Stoudt and McFarland (1966). Further references to human strength will be found in Sections 3.3 and 3.8.

Dynamic properties of muscle

The force–velocity curve. The capacity of a muscle to exert tension whilst changing in length (or to change length against an imposed load) is des-

Fig. 3.8. Force–velocity curve of human muscle. Data were derived from Wilkie (1950), Asmussen *et al.* (1965) and Komi (1973).

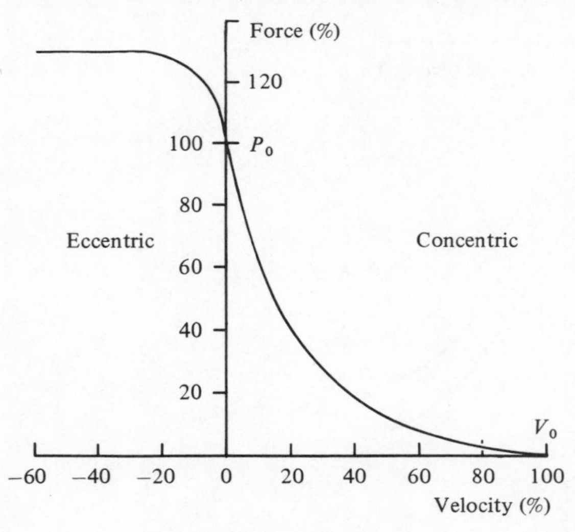

cribed in a relationship known as the force–velocity curve (Fig. 3.8). A variety of equations have been fitted to the concentric portion of this curve (i.e. that portion in which the velocity of contraction is positive). The best known is the equation of Hill (1938):

$$(P + a)\ (V + b) = (P_0 + a)b = \text{a constant}$$

where P is the tension in the muscle,
P_0 is the isometric tension,
V is the velocity of shortening, and
a and b are constants.

Hill's equation was based originally on experiments using isolated amphibian muscle, but Wilkie (1950) demonstrated that it holds well for voluntary actions of elbow flexion provided that corrections are made for the inertia of the forearm and the hand.

The condition P_0 can be equivalent to any point on the length–tension curve (see above), and in the practical measurement of the force–velocity curve all readings must be taken as the muscle passes through a constant part of its range of length. This is because Gordon *et al.* (1966) showed that while for all lengths greater than the resting length (L_0) the maximum velocity of shortening (V_0) remained constant, at lengths less than L_0, V_0 decreased with P_0. Since the physiological range of human muscle does not extend greatly above L_0 the latter condition is more realistic.

Hill's equation does not, however, hold for the eccentric condition in which the muscle is actively resisting stretch imposed by an external force (i.e. the muscle is acting as a brake rather than as a motor). The tensions which a muscle may exert in eccentric action are substantially greater than isometric (P_0). Asmussen, Hansen and Lammert (1965) studied a horizontal pulling task and registered forces as high as 140% of the isometric maximum. Grieve and Arnott (1970) studied axial rotation of the trunk and by indirect means calculated an equivalent torque–angular velocity curve for this complex multi-link activity. Their results indicated the existence of torques of 200% of isometric maximum during the eccentric phase of this activity.

The power output (rate of work) of the muscle is given by the product of the force and velocity ($P\,V$). This is zero at both P_0 and V_0, and passes through a maximum at a point somewhere between them. (In Wilkie's experiments maximum power output was given with a loading of approximately 30% isometric maximum.) Fig. 3.9 shows the relationship between velocity and power output. It will be noted that in eccentric conditions power output becomes negative and increases rapidly with velocity. In this condition the muscle is absorbing energy from external sources. In extreme cases, muscle fibres or their tendinous attachments may be ruptured. In less extreme cases, eccentric contractions are extremely effective in providing the 'overload' stimulus necessary for both muscle training and its associated soreness (Komi and Buskirk, 1972).

History dependence. Impulsive or forceful muscular activities are performed most effectively when preceded by a 'back-swing' or 'wind-up' movement in which the active muscle is stretched by its antagonist or by an external force. (Kicking, punching, striking a blow with a hammer, jumping from a standing position, the golf swing, and the tennis service all exhibit this sequence of events.) Grieve, Pheasant and Cavanagh (1978) have shown that the calf muscles also demonstrate this same eccentric preceding concentric pattern of activity during the walking cycle.

Asmussen and Sørensen (1971) showed that the work done in the early stages of a concentric contraction is greater if that contraction is immediately preceded by an eccentric exertion than if it is preceded by an isometric exertion, which in turn is greater than the work done if the contraction is initiated from rest. Cavagna, Dusman and Margaria (1968) calculated that the quantity of work gained by pre-stretching a muscle in this way is greater than can be stored by the deformation of the series compliance, and it has even been suggested (Hill and

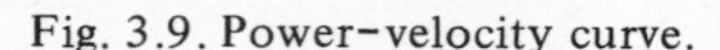

Fig. 3.9. Power–velocity curve.

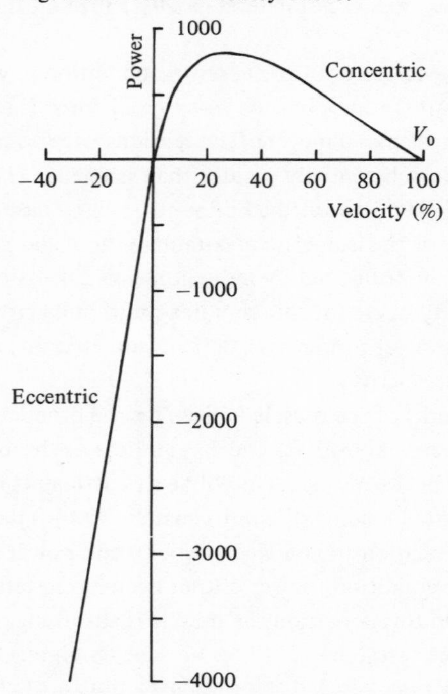

Howarth, 1969) that the work absorbed in an eccentric exertion may actually reverse the chemical processes by which potential energy is liberated in muscle function. Edman, Elzinga and Noble (1978) have recently published a very detailed account of the ways in which pre-stretch modifies both the length–tension and force–velocity curves of isolated muscle fibres. Their results point the way to experiments which need to be performed on human subjects in the future.

Although the precise mechanism is uncertain, it is clear that the action of muscle is history-dependent in a way that cannot be predicted from the length–tension and force–velocity characteristics alone, or from the simple models of muscle currently in use. The widespread occurrence of pre-stretch in muscular action suggests that it is a mechanism of very great importance.

Fatigue

'Fatigue is a term used to cover all those determinable changes in the expression of an activity which can be traced to the continuing exercise of that activity under normal operational conditions, and which can be shown to lead, either immediately or after delay, to deterioration in the expression of that activity, or, more simply, to results within the activity that are not wanted' (Bartlett, 1953). The present discussion is concerned solely with fatigue in activities of forceful muscular exertion, and will concentrate on measurements of performance rather than on the subjective, psycho-physiological or cardio-vascular phenomena which may be concomitant with such exertions.

The underlying mechanisms of muscular fatigue remain unclear in spite of extensive research. Some authors choose to locate the phenomenon in the muscle itself, whereas others emphasise changes which take place in the control nervous system. Some experiments indicate that performance decrement results from the depletion of stores of chemical 'fuel' within the muscle whereas others point to the disturbance of ionic balance or to the accumulation of metabolic waste products. Rival theories are reviewed in Astrand and Rodahl (1970).

Any one of these factors can become critical under certain circumstances. Pressure increases within the active muscle may partially or wholly occlude blood flow. Hence, sustained isometric actions, which effectively isolate the muscle from its blood supply, are much more rapidly fatiguing than intermittent or dynamic actions which 'pump' the blood through the active tissue. Chemical changes within the muscle (particularly the accumulation of acidic metabolic waste products) stimulate sensory nerve endings in the vicinity. The resulting neural activity is perceived by the subject as pain, and when this approaches an intolerable intensity output is reduced. Pain tolerance is, of course, greatly influenced by motivation. Learning to tolerate this pain and to employ the muscles closer to their true physiological limits is an important part of athletic training.

The skeletal muscles of quadrupedal mammals may be divided, on the basis of certain morphological and biochemical criteria, into two (or perhaps three) types: 'fast' muscles which fatigue rapidly and 'slow' muscles which fatigue more slowly (see Close, 1972, for a review of this complex subject). Each anatomical muscle in man contains motor units with a variety of contractile properties. Evidence is now accumulating which demonstrates that the proportions in which the fibre types are present in a given muscle in any particular individual may have important functional consequences, especially for athletic performance (Hulten *et al*. 1975; Thorstensson, Grimby and Karlsson, 1976; Thorstensson and Karlsson, 1976; Komi *et al*., 1976).

In spite of the complexity of the above factors, certain tests of muscular endurance yield remarkably consistent results. Fig. 3.10 shows a set of load–endurance curves in which the limit time, or maximum voluntary duration, of a continuous static exertion is plotted as a function of the strength of that exertion. The force is expressed as a percentage of the isometric maximum measured in the unfatigued state. This relationship has been determined for 'arm, leg and

Fig. 3.10. Load–endurance curves of human muscle groups: 1, 'arm, leg and trunk muscles' (Rohmert, 1960); 2, upper limb pulling task (Caldwell, 1963); 3, 'biceps brachii, triceps brachii, middle finger flexor, quadriceps femoris' (Monod and Scherrer, 1965); 4, trunk extensors (Jørgensen, 1970).

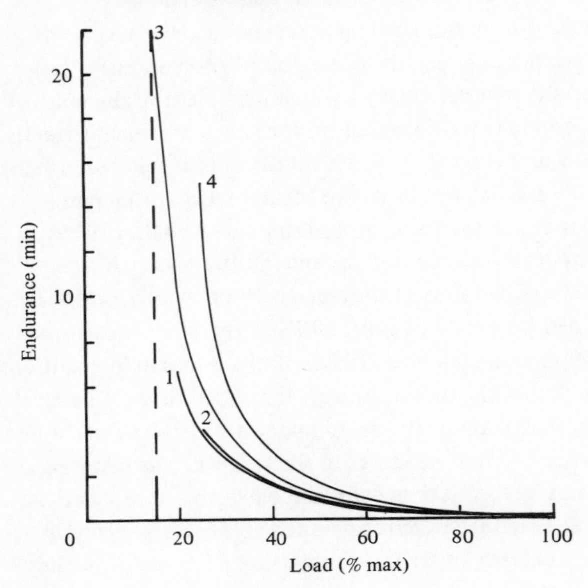

trunk muscles' by Rohmert (1960), for an upper limb pulling action by Caldwell (1963) and for biceps brachii, triceps brachii, the middle finger flexor and quadriceps femoris by Monod and Scherrer (1965). The results of these three studies are very similar. The limit time tends to infinity at a force of 15–20% maximum strength and tends to zero at 100% maximum strength. (It has been shown that vascular occlusion commences at forces of around 20% isometric maximum.) Jørgensen (1970) plotted a somewhat different load–endurance curve for the trunk extensors (see Fig. 3.10). The high level of agreement between the above studies is perhaps surprising since the different muscle groups were presumably acting in different parts of their tension–length relationships.

Since strength is itself a function of posture (see sections on the angle–torque curve and strength, pp. 85–91), it might be deduced that a given absolute force may be sustained for a longer time in a 'strong' posture than it can in a 'weak' one. Caldwell (1964) demonstrated experimentally that this deduction is correct – a finding which has considerable importance for workspace design and working posture.

Monod and Scherrer (1965), Pottier *et al.* (1969) and Monod (1972) have extended the above concepts to deal with intermittent static efforts and with dynamic activities.

Although these laboratory studies have elucidated many important aspects of prolonged muscle function, great circumspection must be urged in extrapolating the results to 'real-world' situations in which an intermittent, variable, submaximal workload is spread over a working shift, five days a week, 48 weeks a year.

Electrical correlates of muscle activity

Visible bulging and palpable hardening of the muscles are not reliable indicators of their use. The adoption of posture, execution of movement and the exertion of force imply the use of muscles but do not imply a unique set of patterns of muscle activity. Active muscle contributes to energy consumption, but the contributions do not reveal themselves until some time afterwards.

Rapidly fluctuating differences of potential within the tissues and across the skin are created when muscles are activated by signals from their nerves; a record of the potential changes (ranging from a few microvolts to a few millivolts) between two electrodes placed on the skin surface or embedded in the tissues is called an electromyogram (EMG). Inspection of an EMG record can be likened to viewing the acknowledgement of instructions received by the muscle motor through a window with a restricted view. The ergonomist's interest in the musculature (as distinct from the final performance) lies in whether muscles are being used economically and skilfully, and whether avoidable or unacceptable local stress and fatigue are present. Such decisions require value judgements to which the EMG lends support. The EMG is valuable as a qualitative tool, but must be treated with caution when used quantitatively.

The EMG reflects local muscular activity in the region of the recording electrodes, provided that some anatomical knowledge guided the choice of muscular regions probably involved in a performance. It is a common and serious mistake to equate mechanical with electrical activity. The former is reflected in the EMG as in a distorting mirror whose shape may alter considerably as conditions change.

Registration and quantification of the EMG are achieved in many ways (Grieve, 1975). Full-wave rectification followed by smoothing or integration is most common, but voltage excursions and turning points in the waveform, and power-spectral analysis have all been used for purposes of description. The intriguing feature is that if the EMG is recorded from part of the musculature which is clearly involved in a movement or an exertion, a close relationship will be found between any (of several) measures of the EMG and any measure of the performance providing that the experimental conditions are rigidly controlled. Thus the EMG will relate to the force exerted if the posture does not alter, and to the amplitude or speed of movement if the loading remains the same (Bouisset, Lestienne and Maton, 1976). Unfortunately, the introduction of another variable, including shifting the electrodes, fatigue, change of limb temperature, change of subject and even subtle changes in the use of muscle groups (consider the many ways of sitting and pedalling on a bicycle) is likely to alter the quantitative relationship. The mechanical interpretation of an EMG is therefore often problematical and ambiguous, as will now be illustrated.

Effect of posture on EMG–torque relationships

Suppose that a person applies a clockwise torque to a handle with his right hand (see Fig. 3.11). Reference to an anatomy text shows that biceps brachii is partly responsible for the transmitted torque and that electrodes on the anterior aspect of the arm will overlie the muscle. If the rectified and smoothed EMG is plotted against the torque as the subject slowly increases his exertion to maximum and then slowly relaxes, while the handle remains vertical, a relationship is obtained similar to that shown in curve B of Fig. 3.11. If the handle is rotated through 90° and the experiment repeated with the hand in a fully supine (curve A) or a fully prone (curve C) position, dramatically different relationships are obtained.

The maximum torque is always, and the maximum intensity of EMG is usually, altered by the posture of the hand. We would expect the former because of the length–tension characteristics of muscles responsible for supination. The reasons for the latter are probably associated with the need to send more frequent signals to a muscle in its shortened condition than when its length is greater, in order to maintain full activity (Grieve and Pheasant, 1976). If only one maximum were altered by posture, some mechanical interpretation of the EMG would be possible; changes in both render the mechanical interpretation totally impossible without a knowledge of the posture. The phenomenon may be observed with other muscle groups in other regions of the body and the ergonomist who is contemplating making comparisons of working methods based upon quantative EMG ('more EMG = worse') is urged to consider the implications of the simple experiment described.

Multiple roles of muscles within a group

In the second example, we demonstrate the effect of a more subtle change of performance upon an EMG–force relationship. The electrodes are placed as before on the skin overlying biceps. Deep to biceps is another muscle,

Fig. 3.11. Rectified and smoothed electromyograms (EMG) from biceps brachii of a subject standing with his arm vertical and forearm horizontal, while exerting a torque of supination on a T-bar handle. The relationship between EMG and torque is very sensitive to the posture of the forearm. A, forearm fully supine; B, forearm mid-prone; C, forearm fully prone. The subject increased, then decreased the torque steadily to counts of ten. Surface electrodes and an $X-Y$ chart recorder were used.

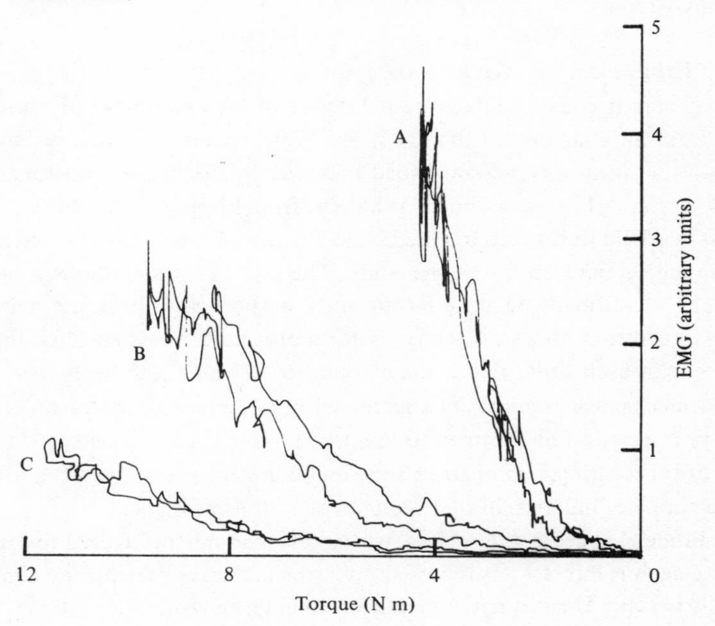

brachialis, which also is a flexor of the elbow, but further from the electrodes. Other muscles such as brachioradialis and pronator teres are also elbow flexors but are so far from the electrodes that little of their activity would be detected. The electrical signals are coming from muscles which contribute to elbow flexion but the signals do not reliably indicate the flexor torque about the elbow except under the most carefully controlled conditions.

In Fig. 3.12, a person is depicted supporting a weight from the hand. The relationship between the weight and the EMG is very difficult depending on whether the weight is applied symmetrically or on the medial or lateral borders of the hand. No change in posture is involved and the EMG relationships can only be interpreted if it is appreciated how the different loadings on the hand require different degrees of supinator effort in addition to the constant flexor demand at the elbow. If EMG from the biceps area was to be quantified as part of a comparative assessment of levers or hand tools, it would be important to appreciate that subtle shifts of loading in the grip between the fingers and the thenar and hypothenar eminences could influence the EMG as much as could changes in the loading at the elbow. Several simultaneous channels of EMG recording from various regions of the arm and forearm may be valuable in such instances in order to understand qualitatively how the muscle groups are employed, but the complexity of such studies militates against their use in the field. The foregoing examples are useful illustrations of the pitfalls of quantitative electromyography.

Fatigue and the electromyogram

There is considerable current interest in the assessment of muscular fatigue from the character of the EMG. The EMG associated with a sustained contraction against a submaximal load increases in amplitude with time. The example in Fig. 3.13 was obtained, as before, from biceps brachii. In this case a weight was held in the outstretched hand (forearm horizontal, arm vertical) until the subject gave up from exhaustion. The increase in amplitude is mainly due to the recruitment of more motor units as those already in use become weaker. In addition, there is a tendency for motor units to synchronise their activities with each other under these conditions (which also leads to a pronounced mechanical tremor). The increased involvement of motor units as fatigue progresses is not confined to the prime movers. Lundervold (1951), in studies of typewriting, demonstrated the increasing involvement of the forearm, arm and shoulder muscles, in that order, as the subject fatigued.

Amplitude changes in themselves, with a test load, might be used to assess local fatigue, but interest has focussed upon the changes of frequency content which also occur. The ultimate objective is to determine whether fatigue

Fig. 3.12. Relationships between rectified and smoothed surface electromyograms (EMG) biceps brachii, and weights held in the hand. The load was supported with the forearm pronated and the load symmetrical (A) or asymmetrical (B), then with forearm supinated and load symmetrical (C) or asymmetrical (D). The torque about the elbow in the sagittal plane was the same in each case but the asymmetrical loading required an additional torque of supination in the coronal plane. Mean ± S.D. of EMG levels are shown after normalisation of each subject's results (five subjects) for held masses of 4 and 8 kg.

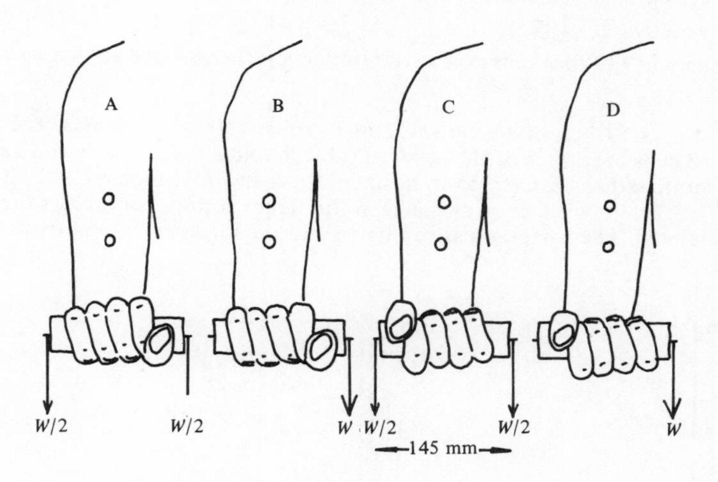

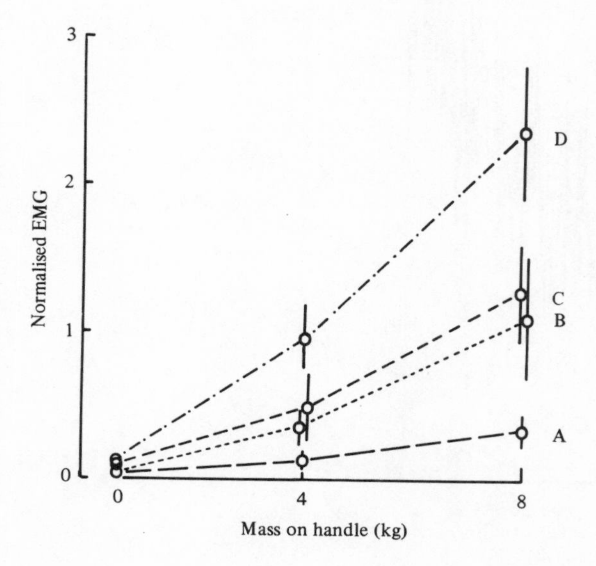

(objectively defined in terms of EMG) is present, from samples recorded during the course of the subject's work without pause for the application of a special test. The EMG power-spectrum of a fatigued muscle has a greater content in the lower frequencies and diminished power in the upper frequencies compared with that of fresh muscle.

An introduction to the EMG phenomena in fatigue may be found in Lippold, Redfearn and Vuco (1960), De Vries (1968), Mortimer, Magnusson and Petersen (1970), Lindstrom *et al.* (1970, 1977), Petrofsky *et al.* (1975) and Broman (1977). Industrial application of the phenomena are described by Chaffin (1973) and Ortengren *et al.* (1975).

Petrofsky (1979) has shown that the frequency-changes due to increased limb

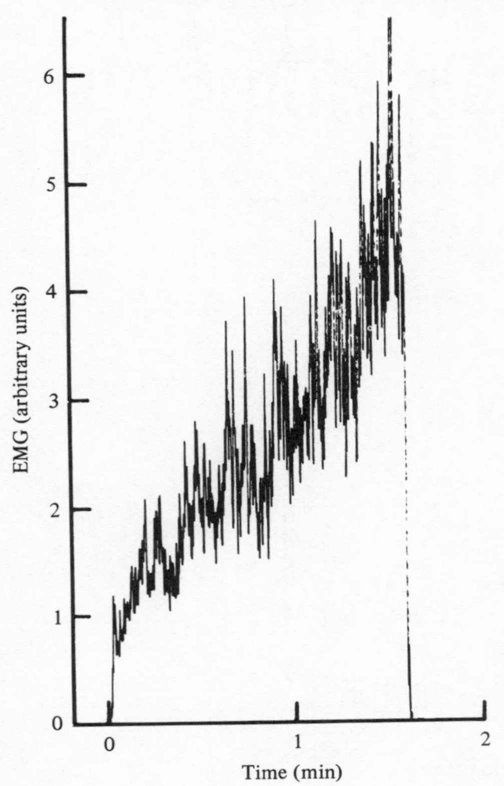

Fig. 3.13. Rectified and smoothed surface electromyograms (EMG) of biceps brachii, recorded while a subject held a 9-kg mass in the hand until exhaustion (forearm horizontal). A five-fold increase in EMG occurred, without any change in the flexor torque required at the elbow. The forearm was supinated and the load was symmetrical.

Fig. 3.14. Surface electromyograms (EMG) from the thenar eminence during unloaded movements of opposition and reposition. The hand was gloved and immersed in water whose temperature was lowered to ice-point, raised 50 °C and returned to room temperature. The records were taken in sequence during the half-hour experiment at the same gain when a skin thermocouple indicated the temperatures shown. The galvanometer recorder had a limited frequency response (below 100 Hz). The amplitude and timing of the movements, and the posture were not altered.

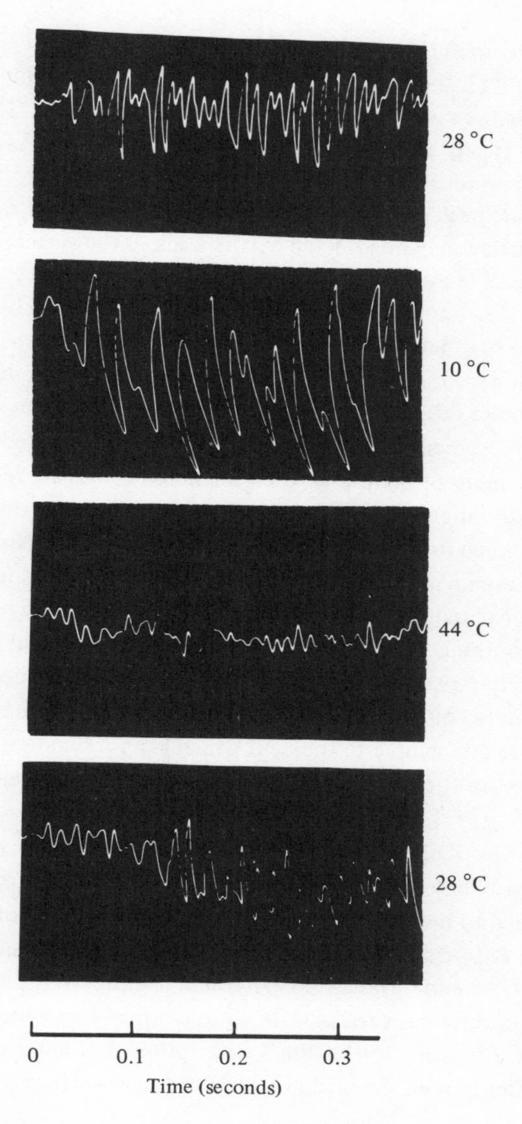

temperature (which may arise in dynamic exercise) are in the opposite direction to the fatigue-changes which may be observed in a sustained contraction. Fig. 3.14 shows the dramatic changes in the quality of an EMG of the unfatigued muscles of the thumb during a movement of opposition that occurred when the rubber-gloved hand was slowy cooled and warmed in a water bath.

As in the examples of the previous sections, fatigue phenomena in EMG require very careful interpretation.

Electromyography under dynamic conditions

By loose analogy with the lightning which precedes a clap of thunder, events in the EMG precede the mechanical events associated with them. This is of no consequence when steady exertions without movement are discussed. During a human muscular response to a single action potential, the delay between the action potential and the peak of recorded mechanical force is in the range of 20–50 milliseconds. The delay may be affected by the slack in the system and the posture (see p. 85 re elastic element and active length–tension curve), and also by the fraction of the slow and fast fibre population in use. The delay between the onset of EMG and the first detectable onset of mechanical force is much shorter (but is a matter of definition because the measurement requires decisions concerning the signal to noise ratio and the sensitivity of the instruments). In the most general case, when movement is occurring, the delays are influenced by the existing direction of movement of the joints when the nerve signals are received, since rise in tension is faster under eccentric than under concentric conditions.

No meaning can be attached to the instantaneous voltage of an EMG. Numeric descriptions of temporal events in the EMG range from sequential measurement of the amplitude of rectified and smoothed signals (which is distorted by the time constant of the smoothing circuit), to stepwise descriptions based upon true integration over discrete periods. In practice, sample periods cannot be much shorter than 50 milliseconds because the individual signal fluctuations cause large variations from one sample to the next which cannot be interpreted and are in marked contrast to the relatively smooth changes of posture and force that are occurring.

Quantitative descriptions of the time-course of dynamic EMG have rarely been used in ergonomic studies and are more commonly found in electromyographic kinesiology applied to medicine or sport. A variety of methods of description will be found in Battye and Joseph (1966), Milner, Basmajian and Quanbury (1971) and Grieve and Cavanagh (1973). One might expect published work concerning common activities such as walking to contain data which are capable of confirmation. Confirmation of EMG description is qualitatively possible, but humans differ in their detailed anatomy and have sufficient options available to their locomotor systems that quantitative descriptions never repeat themselves. This is even true of the same subject performing two consecutive cycles of steady walking (Grieve and Cavanagh, 1974), which is relevant to the application of EMG for control purposes in prosthetics. Less stereotyped activities that would arise in various jobs do not justify detailed numerical description of their temporal patterns although many more qualitative descriptions than are available would be welcome. Examples of the art of qualitative assessment of dynamic EMG will be found in Lundervold (1951), Person (1958), Basmajian (1967) and Tichauer (1978).

Quantification over long periods, without concern for the detailed temporal pattern, is of some use in ergonomics. Harding and Sen (1969) obtained a global figure for the combined integration of EMG signals from several parts of the body, thus providing a technique which indicated the level of energy expenditure with less restriction on the subject than the more accurate respirometer would impose. Grieve and Rennie (1967) used a device (ANDIG) which counted the number of 1/5 second periods in which EMG levels about 50 μV occurred, displaying the result in the field of view of a time-lapse camera which photographed the seated subject. In 24-hour experiments the ANDIG provided information that could otherwise be found by tedious measurement by ruler of EMG bursts on a very slow paper record.

Concepts of efficiency

Our present knowledge of the human musculo-skeletal system provides some insight into the factors which can influence performance but does not yet permit us to predict conditions under which maximum power output can be obtained, or minimal effort required in the performance of a task. In the unusual conditions of a fully active muscle group acting with constant torque about a joint, we might predict the particular speed and posture in which optimal power transfer would be expected. A real movement from rest might achieve the optimum briefly, but the initiation and eventual halting of the movement constitute appreciable fractions of the manoeuvre and make detailed prediction impossibly complex.

Physiological concepts of efficiency can be applied to a succession of repetitive movements, because under those conditions – e.g. man-powered flight, or rhythmic lifting – both the useful work and the net metabolic cost can be determined. In contrast the cost of a single, possibly unidirectional movement cannot be monitored metabolically; neither would the cost be of much importance. In such cases the effficacy of the manoeuvre must be judged from some other standpoint, probably related to its purpose. The feature or quantity that we wish to optimise is a matter of choice in the particular circumstances (see Barnes'

(1968) principles of motion economy). We might, for example, consider the force required in a movement as something to be traded against the time, or bulk of active muscle required to complete it. If a very small amount of external work is required, some movements (out of a choice) may demand the disproportionate expenditure of energy in moving the body parts. It is up to the task designer to choose the features that he wishes to optimise.

Whatever features the task designer incorporates into the work, providing that no physiological limits are reached, a validation study would reveal substantial variations in the techniques adopted by individuals. A corollary of this is that observation of people who are not given the opportunity to familiarise themselves with the task can be grossly misleading. Person (1958) studied the acquisition of skill in metal-filing and used electromyography to form a qualitative judgement about the efficiency of the apprentices during training. The initial efforts were not only mechanically uneven, but were performed with marked co-contraction of opposing muscle groups; as skill developed, co-contraction (of the arm muscles) was no longer observed. Co-contraction is not necessarily undesirable, and may be expected if a task requires the steadying of an object against forces in any direction. On the other hand, if a task were redesigned to reduce the need for stabilisation, the likelihood of local fatigue would be reduced.

An insight into the way in which alternative concepts of efficiency may be formulated is to analyse mechanically the task-demands and consider the options available to the worker which would still produce satisfactory performances. The following is a simple example in which the task is to raise a 10-kg mass through exactly 1m. Suppose two persons, A and B, exerted themselves for 1 second in each case; timing is important if we assume the task to be an element in a paced situation. An infinite number of temporal patterns of force will achieve the desired result but each will have different implications for the body. The model is a simple one in which postulated patterns of vertical force are applied to a load initially resting on the floor. The subsequent velocities, rates at which work is done on the load and the trajectories are computed by iterative application of Newton's Second Law of Motion.

Fig. 3.15 shows the force patterns of A and B. Worker A develops force faster than B and relaxes more slowly from the peak output. The trajectories of the loads are very similar and reach the required maximum height of 1 m at 1.1 seconds after commencement. It is clear that very careful observation would be required in real life to distinguish the two performances. If peak force development was used as a criterion of efficiency, A's performance is better (16%), thanks to his faster attack and slower decay of effort. If performance were judged on how fast the men work on the load, the difference (25%) is even more marked in A's favour. Although blood flow and recovery from a *single* performance would be of little consequence, the frequent repetition of the manoeuvre might tilt the judgement in B's favour because the fraction of the cycle in which circulatory interference could be expected is less. Such a model is useful for suggesting ways of judging performance, but tells us nothing about the ability of the person to achieve it. It would be biologically ridiculous, for example, to postulate a force which rose to peak immediately at time zero, or fell to zero from peak instantaneously. Another important omission is the fact that the body parts must also rise with the load, and that the manner in which they do so will considerably influence the demands (see Section 3.8).

If we consider vertical lifting strengths at various heights above the ground (as was done with a group of students for Fig. 3.16), it is apparent that strength is markedly greater at about 0.75 m above the ground than at other heights. Individual differences in performance are considerable (in an untrained group). The stick figures in Fig. 3.16, constructed from visual observations of the students

Fig. 3.15. Model of lifting a 10-kg mass through 1 m from the floor, by two methods A and B. The temporal patterns of vertical force in the upper chart were postulated. The rate of working on the load (middle) and resulting trajectories (lower chart) were computed by applying Newton's Second Law of Motion to the motion of the load.

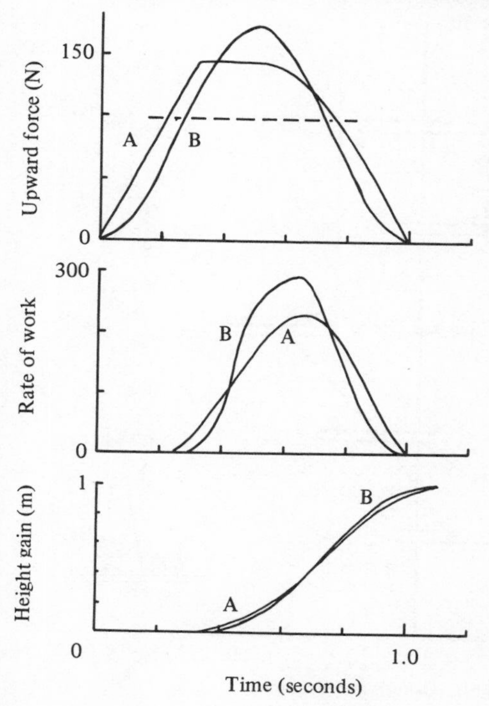

by sighting of landmarks against vertical and horizontal scales, strongly suggest that appreciable fractions of the variance in observations of strength arise from differences of technique (see PSD analysis on p. 176).

In the dynamic situation of competitive weightlifting, where the maximum possible load has to be raised successfully, the performer must reach a compromise between his strength at various heights above the ground, the rapidity with which he can control the activity of his motor units, and the speeds with which the weight and body parts are moving at each height since they detract from the force that can be generated. No biomechanical model yet has the subtlety to examine such manoeuvres quantitatively but it can provide a framework for considering the essential elements.

Fig. 3.16. Maximum vertical lifting strengths at five heights above the floor in freely chosen symmetrical postures (five male and five female students). Forces are expressed as percentages of mean performance at the five heights. The postures of ankle (A), knee (K), hip (H), trunk line and elbow (E), relative to the hand grip (HG) in the sagittal plane, as observed visually against measuring scales for two good and two poor performances, strongly suggest that the variance in the results arises partly from the technique adopted.

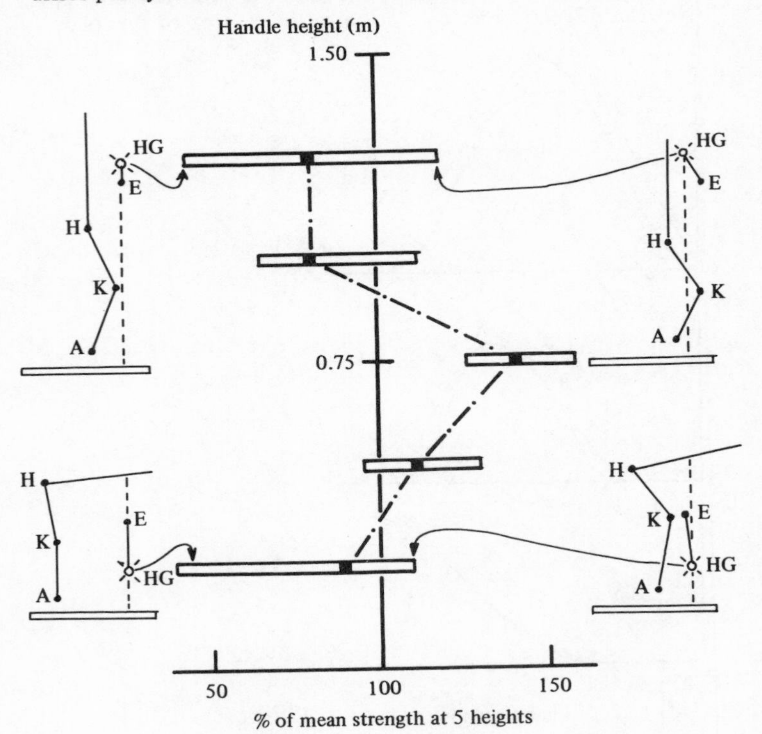

etry

anthropometry' is given to the collection, analysis and tabu-
data concerning the physical attributes of human beings.
ata of interest to the ergonomist can be divided into three

nthropometry (often called 'static anthropometry') deals with
nsions of the human being, e.g. stature, and the lengths, breadths,
umferences of body structures. Body weight is included in this

ional anthropometry (often called 'dynamic anthropometry') deals
und measurements such as reach and working clearances which can-
n be predicted from sets of structural data. The angular ranges of joints
a strengths of various body actions (see section on strength, p. 88)
m included in this category, but more complex measures of human
pe such as reaction time, power output and proficiency in skilled tasks
are excluded.

(ian anthropometry is a term we shall use to describe the parti-
cular subset of structural and functional data which is required in order to apply Newton's Laws of Motion (see the section on mechanical principles applied to the human body, p. 75) to the analysis of human activity. These data describe the segmental parameters of the 'linked body man' (see p. 137).

The statistical description of data

Although a few examples exist of situations in which the dimensions of single individuals are of importance for design purposes (bespoke tailoring, astronautics), in the vast majority of cases the ergonomist or designer is concerned with the target population, which must be described in statistical terms. Many adequate textbooks of statistics are available; Chapanis (1959) and Roebuck *et al*. (1975) include introductions to the subject which are useful for the student of ergonomics. A few points need to be emphasised here.

The mathematical description of the 'normal' or 'Gaussian' distribution was initially developed to describe random errors in the measurement of physical phenomena. Empirical observation indicates that many biological (e.g. anthropometric) variables are approximately normally distributed within any given population. The theoretical justification for this finding is complex and obscure. In cases where the variable is normally distributed, the population may be completely described in terms of its mean ($\bar{x}$) and its standard deviation (s), and percentile values may be calculated as shown in Fig. 3.17(a) (n% of the population exhibit values less than or equal to the nth percentile).

Fig. 3.17(*a*). The probability density function of the normal distribution. The function plotted is given by the equation

$$\phi(x) = \frac{1}{\sqrt{(2\pi)}} \cdot e^{-\frac{1}{2}x^2}$$

which describes a normal distribution with a mean of zero and a standard deviation of unit. In general, 1st percentile = $\bar{x} - 2.326\ s$; 5th percentile = $\bar{x} - 1.645\ s$; 95th percentile = $\bar{x} + 1.645\ s$; 99th percentile = $\bar{x} + 2.326\ s$.

(*b*). The cumulative probability curve of the normal distribution (normal ogive). The curve plotted describes the percentage of members of a normally distributed population exhibiting values of the measured variable less than x.

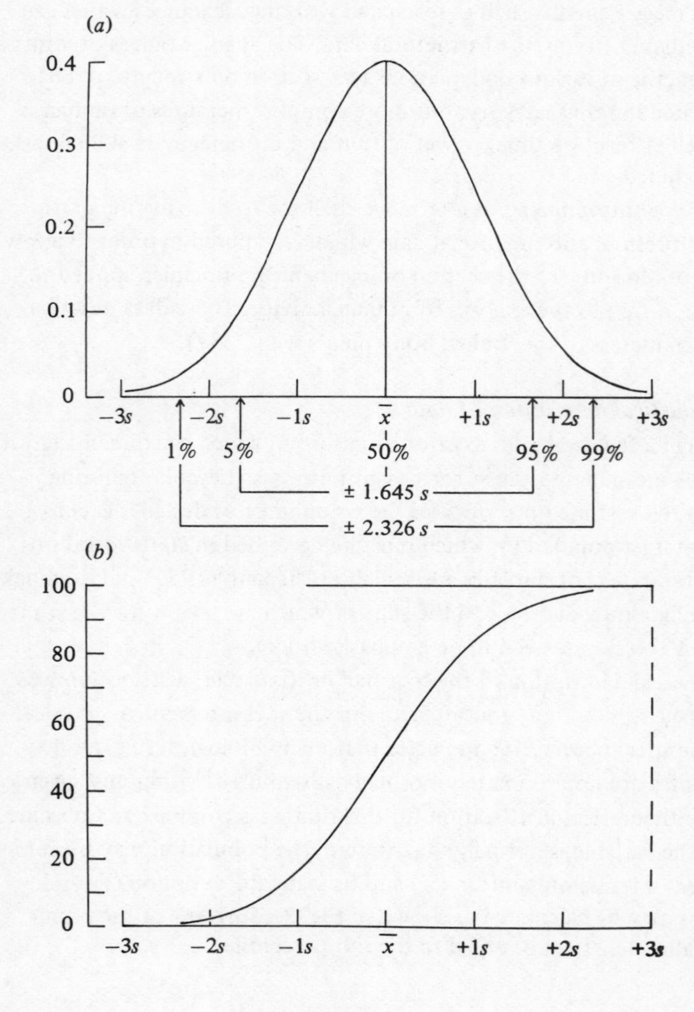

The familiar smooth bell-shaped curve of the normal distribution is properly termed a 'probability density function', and the ordinate expresses the relative likelihood of an individual drawn at random from the population exhibiting a certain value of the measured variable. An alternative way of plotting such data is as a cumulative probability curve (Fig. 3.17*b*) from which the percentile values can be read directly.

Anthropometric data, however, are not necessarily normally distributed in any given population. (This is a subject which has been little investigated as the sample sizes required to define accurately populations in the general, non-normal case are larger than most investigators are prepared to tackle.) Asymmetric distributions are said to be skewed: either positively (exhibiting a 'tail' of large values) or negatively (vice versa). Body weight exhibits a modest positive skew in most populations (Damon *et al.*, 1966), and for functional measures such as strength, positive skew may become very pronounced. Other human performance measures which exhibit pronounced skew include reaction time (Chapanis, 1959), maximum oxygen uptake (Hermansen, 1974), industrial aptitude test scores (Tiffin and McCormick, 1966) and the examination results of medical students (authors' observations).

The coefficient of variation (c.v. = (S.D./ mean) × 100%) is a valuable non-dimensional indication of the variability of a measure. Table 3.1 shows the ranges of coefficients of variation exhibited by the various data sets listed in Damon *et al.* (1966). All of these are based on United States Air Force or samples specifically chosen to represent them. It may be seen that functional measures have an inherently greater variation than structural ones (although the inordinately high c.v. values for some strength measures reflect the extreme skew of the distribution).

Population characteristics

The above statistical methods represent ways of succinctly summarising numerical data concerning a given population. The section which follows is con-

Table 3.1. *Variation in United States Air Force aircrew*

Data set	Coefficient of variation (%)
Stature and body lengths	3 - 5
Other dimensions (incl. reach)	3 - 11
Joint ranges	7 - 28
Weight	13
Strengths	13 - 85

cerned with the biological and social variables which define a population with respect to anthropometric measures.

Sex

It is a matter of everyday experience that men are, on average, taller and heavier than women. Due to the strong correlation which exists between most pairs of anthropometric variables, men will demonstrate proportionally greater values in most bodily dimensions. There are also sex differences in bodily proportions (see Table 3.2).

It is commonly reported that women are, on average, two-thirds as strong as men. Hettinger (1961) lists sources showing that this ratio ranges from 55% for the elbow flexors and extensors, to 80% for the hip flexors and extensors, the knee flexors and the muscles of mastication. He very reasonably concludes that the differences are most marked in predominantly 'male' activities and that they are associated with differences in muscle bulk rather than with differences in the intrinsic contractile properties of the muscle tissue. Sinelnikoff and Grigorowitsch (1931) reported that the joint ranges of women exceed those of men at all joints except the knee.

Age

The relationship between any human characteristic and age may be studied either by cross-sectional studies (in which 'cohorts' of individuals of different ages are studied at the same time) or by longitudinal studies in which individuals are followed over the years. The results of cross-sectional studies reflect the interaction of the ageing process *per se,* long-term secular trends within the population and the effects of differential patterns of survival.

Tanner's longitudinal study of British children (Tanner, 1962) showed that males reach their maximum stature by 20 years and females by 17 years. Changes in bodily proportion (e.g. width of shoulders or hips) continue into the mid-twenties. Bainbridge (1972) conducted a cross-sectional study of British school-children aged 3 to 18 years and published a data set appropriate for the design of school furniture. Snyder *et al.* (1977) published an exhaustive account of the anthropometry of American children aged 2 weeks to 18 years. Body weight and its related variables have a tendency to increase steadily throughout life in response to a complex set of environmental variables.

It is well established that the average stature of people in Western Europe and North America is increasing: Tanner (1962) estimated that this was occurring at the rate of approximately 1 cm per decade in the average height at maturity. It is not clear whether these long-term trends are hereditary or environmental in origin. These trends are reflected in the results of the cross-sectional studies of

Table 3.2. *Anthropometric predictions for British civilians (all dimensions in centimetres)*

	E_1	E_2	Percentiles 1st	5th	50th	95th	99th	S.D.
					Males			
Stature			159	164	175	186	191	6.7
Eye height	0.936	1.017	148	153	164	175	180	6.8
Cervical height (C7)	0.855	0.950	135	139	150	160	165	6.4
Acromial height	0.815	0.932	128	132	143	153	157	6.2
Elbow height	0.622	0.742	97	101	109	117	121	5.0
Perineal height	0.474	0.683	72	75	83	91	94	4.6
Knuckle height	0.439	0.605	67	70	77	84	86	4.0
Fingertip height	0.369	0.541	55	57	63	69	72	3.6
Elbow–fingertip length	0.271	0.339	42	44	47	51	53	2.2
Elbow–wrist length	0.163	0.222	25	26	29	31	32	1.4
Sitting height (erect)	0.523	0.522	83	86	92	97	100	3.5
Sitting height (slumped)	0.502	0.619	78	81	88	95	98	4.1
Eye height (sitting)	0.460	0.537	72	75	81	87	89	3.6
Acromial height (sitting)	0.345	0.492	53	55	60	66	68	3.3
Elbow rest height (sitting)	0.137	0.399	18	20	24	28	30	2.6
Thigh clearance	0.087	0.228	12	13	15	18	19	1.5
Knee height (sitting)	0.312	0.410	48	50	55	59	61	2.7
Popliteal height (sitting)	0.254	0.354	38	40	44	48	49	2.3
Buttock–knee length	0.341	0.436	53	55	60	65	67	2.9
Buttock–popliteal length	0.280	0.429	42	44	49	54	56	2.8
Bi-deltoid breadth	0.264	0.379	40	42	46	50	52	2.5
Elbow–elbow breadth	0.259	0.507	37	40	45	51	53	3.4
Hip breadth (sitting)	0.206	0.356	30	32	36	40	42	2.3
Foot length	0.153	0.194	24	25	27	29	30	1.3
Heel–ball length	0.112	0.153	17	18	20	21	22	1.0
Foot breadth	0.058	0.089	9	9	10	11	12	0.5
Ankle height (lat. malleolus)	0.042	0.097	6	6	7	8	9	0.6

(*Continued on next page*)

Table 3.2 (*contd*)

	E_1	E_2	Percentiles 1st	5th	50th	95th	99th	S.D.
				Females				
Stature			148	152	163	173	178	6.4
Eye height	0.928	1.081	136	140	151	162	166	6.5
Cervical height (C7)	0.855	0.950	128	129	139	149	153	6.1
Acromial height	0.815	0.992	118	122	132	143	147	6.3
Elbow height	0.611	0.762	88	91	99	107	111	4.9
Perineal height	0.473	0.699	66	69	77	84	87	4.5
Knuckle height	0.435	0.596	62	64	71	77	80	3.8
Fingertip height	0.369	0.541	52	54	60	66	68	3.4
Elbow–fingertip length	0.273	0.340	39	41	44	48	49	2.1
Elbow–wrist length	0.163	0.222	23	24	26	29	30	1.4
Sitting height (erect)	0.526	0.550	77	80	85	91	94	3.5
Sitting height (slumped)	0.514	0.630	74	77	84	90	93	4.0
Eye height (sitting)	0.459	0.512	70	69	75	80	82	3.3
Acromial height (sitting)	0.351	0.513	49	52	57	62	65	3.3
Elbow rest height (sitting)	0.141	0.388	17	19	23	27	29	2.5
Thigh clearance	0.086	0.331	9	10	14	17	19	2.1
Knee height (sitting)	0.309	0.391	44	46	50	54	56	2.5
Popliteal height (sitting)	0.258	0.337	37	38	42	46	47	2.1
Buttock–knee length	0.355	0.484	50	53	58	63	65	3.1
Buttock–popliteal length	0.296	0.444	43	44	49	53	55	2.6
Bi-deltoid breadth	0.258	0.423	36	37	42	46	48	2.7
Elbow–elbow breadth	0.240	0.864	26	30	39	48	52	5.5
Hip breadth	0.232	0.544	30	32	38	43	46	3.5
Foot length	0.148	0.189	21	22	24	26	27	1.2
Heel–ball length	0.107	0.141	15	16	17	19	20	0.9
Foot breadth	0.056	0.092	8	8	9	10	10	0.5
Ankle height (lat. malleolus)	0.042	0.097	5	6	7	8	8	0.6

various samples of the British population, shown in Fig. 3.18. (Such data sets also reflect the fact that people shrink in old age, due to degenerative changes of the vertebral column.)

Asmussen and Heebøll-Nielson (1962), in a cross-sectional study, measured the strengths of 25 different muscular actions in subjects ranging from 15 to 65 years in age. The strength of the hands and upper limbs reached a plateau at 20 years and began to decline slowly from 40 onwards, whereas the trunk and lower limb muscles had a peak of shorter duration at around 30. The ages of peak and decline were slightly earlier in women than in men.

Fig. 3.18. Stature as a function of age in cross-sectional studies of British populations. Data are derived from Cathcart *et al.* (1935), Kemsley (1950), Khosla and Lowe (1968), Montegriffo (1968) and Thompson *et al.* (1973).

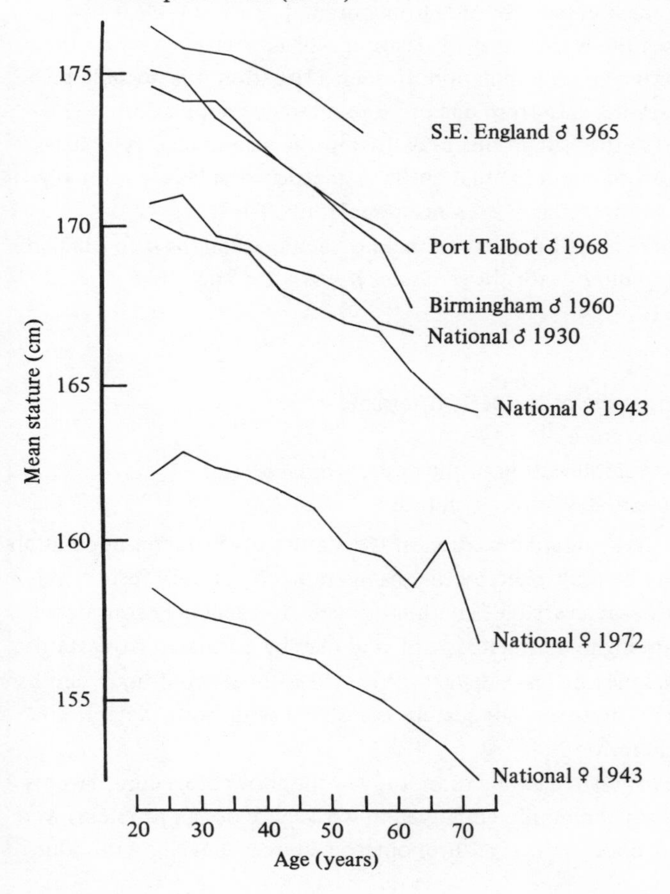

Racial, regional and social factors

It is well known that different racial and ethnic groups exhibit pronounced differences in stature and physique. Smaller differences also exist on a social and geographic basis. The last major surveys in Great Britain were conducted during the Second World War. Clements and Pickett (1957) in a 1941 survey found appreciable differences between the regions, and differences between two social classes in any one region. It is impossible to predict how subsequent social trends and patterns of human migration may have affected these differences.

An interesting example of occupational differences in physique is the study of Morris, Heady and Raffle (1956) who showed that London bus-drivers were fatter at any given age than London bus-conductors.

Anthropometric estimates for unknown populations

In view of the complex set of factors outlined above it is clear that in the majority of cases the ergonomist or designer will not have access to the appropriate data for his target population. In such a situation it is possible to make estimates by scaling data from one or more reference populations. (This procedure assumes (*a*) that subgroups drawn from the same racial type have similar bodily proportions, and (*b*) that anthropometric variables are normally distributed in both the target and reference populations.)

The most expedient and direct way of making such predictions is to calculate two scaling ratios, E_1 and E_2, for the *reference population,* such that

$$E_1 = \bar{A} / \bar{S}$$

$$E_2 = S_A / S_S$$

where $\bar{A}$ is the mean value of the desired variable,

$\bar{S}$ is the mean stature,

S_A is the standard deviation of the desired variable, and

S_S is the standard deviation of stature.

If the mean and the standard deviation of the stature of the target population are known, they may be multiplied by the scaling ratios E_1 and E_2 respectively and estimates of the desired variable may thus be made. It is generally considered that body lengths are highly correlated with, and may be estimated from stature. The above procedure may be less satisfactory for the estimation of body depths and girths, which are, however, adequately correlated with body weight, and may be estimated therefrom.

Pheasant (1980) conducted a validation study of the above procedure. Twenty-two bodily dimensions commonly employed in workspace design problems were chosen for study. A collection of anthropometric sources, dealing with adult

European and North American populations, were assembled. The sources were divided into two groups (A and B). Group A consisted of the following sources: Daniels *et al.* (1953), Hertzberg *et al.* (1954, 1963), Roberts (1960), Damon and Stoudt (1963), Kroemer (1964), Damon *et al.* (1966), Andrae *et al.* (1971), Garrett and Kennedy (1971), Ince, *et al.* (1973), Bolton *et al.* (1973), Thompson *et al.* (1973). For each of the 22 anthropometric variables, ratios E_1 and E_2 were derived for each of the anthropometric sources in which they appeared and mean values of E_1 and E_2 were then calculated. Group B sources comprised the following: British Infantrymen (Gooderson and Beebee, 1976), United States Air Force women (Clauser *et al.*, 1972), British male and female technical college students (DES, 1970), British male and female motorcar drivers (Haslegrave, 1979). The group B sources were treated as model target populations for which predictions were required. Using the stature data only from the six group B sources, and the ratios E_1 and E_2 from the group A sources, predictions of the 1st and 99th percentile of each anthropometric variable were obtained. Since these data were in fact available, predictions of the percentiles could be compared with the real values (error = predicted percentile − real percentile). A total of 136 comparisons were made in this way. The errors which were found demonstrated no systematic bias and were normally distributed with a mean of − 3 mm and a standard deviation of 13 mm. Ninety three per cent of the errors fell within the range of ± 25 mm. Most practical ergonomists would agree that anthropometric data of this accuracy are adequate for most workspace design applications. No detailed anthropometric data exist for the adult civilian population of Great Britain (with the exception of the relatively small number of variables measured by Thompson, Barden and Kirk, 1973, or by Haslegrave, 1979). Table 3.2 is an attempt to estimate such data by the process described above. The stature inputs were obtained by combining the figures of Montegriffo (1968), Thompson *et al.* (1973) and Haslegrave (1979). The values of E_1 and E_2 used were the means of the group A sources listed above. The reader is invited to make his or her own estimates of other target populations using the E_1 and E_2 values provided.

Predictions of this nature are easy to criticise and difficult to improve upon. They must of course be less reliable than the results of a detailed survey of a single target population. Since comprehensive anthropometric surveys are both expensive and time-consuming, a trade off between required accuracy and economic expediency will determine the procedure adopted in any individual case.

Functional anthropometry

As the application of anthropometry to engineering design problems developed, it became apparent that the measurement of body dimensions in formal standing and sitting postures needed to be supplemented with data concerning body mobility, reach and the space requirements of the moving person.

The simplest forms of such data are tabulations of the ranges of motion of individual bodily articulations. The definitive study of joint ranges remains that of Dempster (1955) as elaborated by Barter, Emmanuel and Truett (1957).

An alternative, more versatile approach is the construction of a kinetosphere or workspace envelope. Such a presentation defines the limits of the region which may be reached by a person under given conditions. It usually takes the form of a volume of space defined with reference to a datum such as the 'seat reference point' (SRP: defined as the point in the mid-sagittal plane of the subject where the squab of the seat meets its back). Workspace envelopes for seated

Fig. 3.19. Maximum single-handed reach (right hand, centre of grip), from a central toe-line reference with feet together. Upper row: 45° oblique planes. Lower row: coronal and antero-posterior planes. Based on 35 men and women, executing reaches at 0, 0.5, 1.0, 1.5 and 2.0 m above the floor. All data expressed as a percentage of stature. The maximum vertical reach (above the right shoulder) was executed with heels on the ground, but all other reaches were performed freestyle with only the toe position constrained. Means ± 1 S.D. shown. Scale intervals equal 10% of stature.

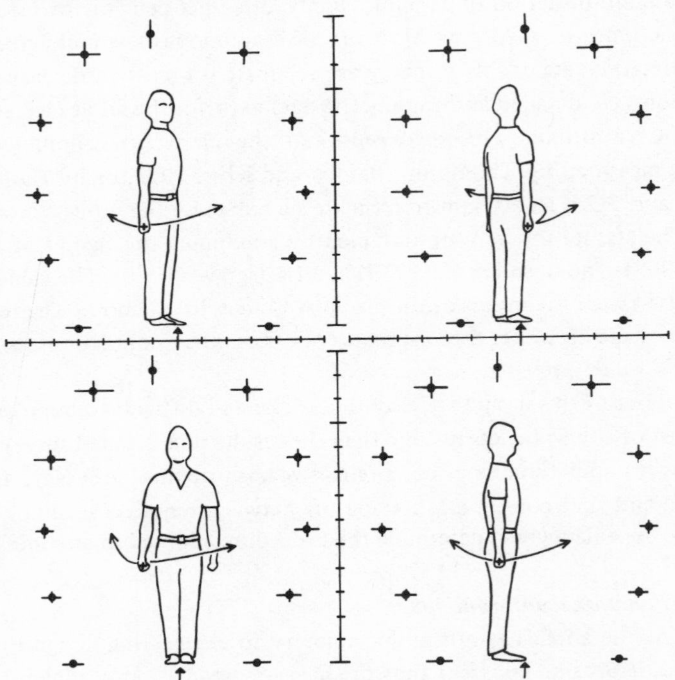

subjects have been described by Dempster (1955), Dempster, Gabel and Felts (1959) and Kennedy (1964).

The workspace envelope of the unsupported standing person is a somewhat elusive concept and remains to be investigated fully. Since standing reach is limited principally by balance, factors such as foot placement and loading in the hands are of paramount importance. Some examples of standing workspace envelopes are shown in Figs. 3.19 and 3.20.

Certain limitations must be borne in mind in the application of workspace envelopes of the type described above. The data cannot be readily applied to situations other than those in which the measurements were made. (The reach of a seated subject might, for example, be modified by the geometry of his seat.) Also, the extreme portions of either a workspace envelope or a joint range can only be used for transient actions because of the discomfort which rapidly develops if an extreme posture is maintained.

Newtonian anthropometry

'Newtonian' anthropometric data are required in those circumstances in which we wish to apply Newton's Laws of Motion to the human body. The complete solution to many problems in body mechanics would require a knowledge of the motions of every individual particle within the body. Clearly this is not a

Fig. 3.20. Maximum single-handed reach (10 men and women) with data from Fig. 3.19 (open circles) superimposed for comparison. Right: effect of placing a wall 300 mm behind the toe-line, which limits the capacity to reach forward below waist level. Left: maximum reach obtained by placing one foot forward to increase the foot-base. The larger envelope shows the reach relative to the trailing toe, the smaller envelope the reach relative to the leading toe. Scale intervals equal 10% of stature.

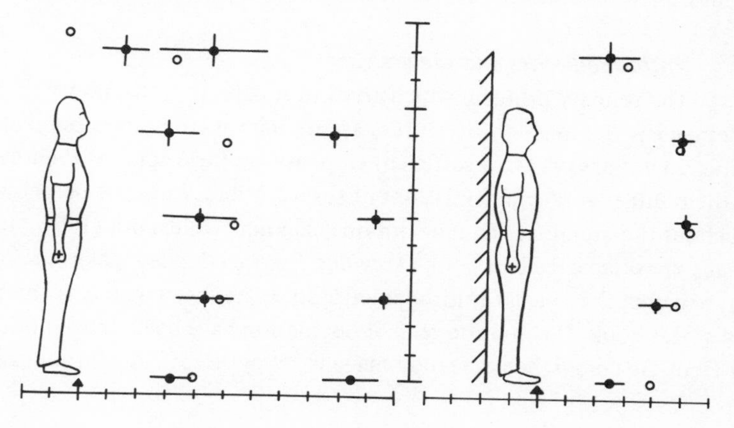

practical proposition, and a realistic approach is to divide the body into a minimal number of segments in such a way that the specification of their mechanical properties yields adequately accurate solutions to the problems under consideration.

Dempster (1955) published the definitive study of 'linked body man'. This model of man assumes that he may be divided into a finite number of rigid linkages, which articulate at joints which may be represented as simple pivots in any given plane. (Hence the thigh link extends from the hip joint to the knee joint, the leg link from the knee joint to the ankle joint, etc.) It is further assumed that these links have constant length, mass and moments of inertia and that their centres of gravity are precisely located with reference to the terminal articulations of the link.

The requisite data may be obtained by the systematic dismembering of cadavers. The cadavers are frozen and sectioned along defined planes and the parameters of the segments are defined by ordinary physical means which are described in the original source (Dempster, 1955). The number of specimens treated in this way has to date been regrettably small. Dempster's study from which the data of Table 3.3 were taken was in fact based on only eight specimens. The data in this table are all expressed in proportional terms. Insufficient evidence exists at present to estimate the variance in these proportions. Dempster's moments of inertia have been converted to radii of gyration for uniformity of presentation. Centres of gravity of the body segments may be held to lie upon the line joining the centres of rotation of their terminal articulations at the percentage distance specified from the proximal end.

The principal limitations of Newtonian anthropometry for practical problem-solving lie in its assumptions. Centres of rotation of joints (especially the shoulder) are not truly constant and segments such as the trunk are by no means rigid. Solutions therefore must be treated as first-order approximations only.

Anthropometrics and the designer

The primary problem which arises in supplying anthropometric data to the designer is the location of sources appropriate to the target population. Authors do not always place sufficient emphasis on the origins of their data, with the result that recommendations of suspect validity may be perpetuated throughout the literature. To take but one example, Goldsmith (1976), in his generally speaking excellent book *Designing for the Disabled* makes the following statements: 'No special anthropometric study has been made for the purposes of this book. Data on the general population have been derived principally from the comprehensive study made by Dreyfuss . . . As a group, ambulant

Table 3.3. *Proportional anthropometry of linkages*

	Length (% stature)	Weight (% body weight)	Location of centre of gravity (% link length from proximal articulation)	Radius of gyration	
				% link length about proximal articulation	% link length about distal articulation
Trunk + head	47.1 (hip to vertex)	58.4	39.6 (hip to vertex)	60.7 (about hip)	
Head + neck (vertex to C7)	14.5	8.1	56.7 (C7 to vertex)		
Arm	17.3	2.7	43.6	54.2	64.5
Forearm	15.5	1.6	43.0	52.6	64.7
Hand	4.0 (wrist to c. of g.)	0.6	–	58.7	–
Thigh	24.8	9.9	43.3	54.0	63.3
Leg	23.4	4.6	43.3	52.8	64.3
Foot	4.7 (ankle to c. of g.)	1.4	Midway between ankle joint centre and ball of foot at head of metatarsal III	69.0	–

Data from Dempster (1955).

disabled people are not in anthropometry terms distinguishable from able-bodied people'.

In the small print of Dreyfuss (1967) we read that 'The surveys from which our information is drawn have been based on fairly narrow statistical samples. Much of it, for example, is taken from studies of armed services personnel.' To what extent will ambulant disabled people in Great Britain be assisted by specifications based on American servicemen?

Once an appropriate data set has been located, it is still necessary for the designer to make certain value judgements concerning the percentiles of the target population which he must accommodate. It may be necessary to specify a minimum allowable dimension in which a high percentile value is relevant (e.g. body clearances), a maximum allowable dimension in which a low percentile value is relevant (e.g. reach and accessibility), or an adjustable range in which both must be specified (e.g. height and location of seats with respect to fixed surfaces).

In all the above situations the shape of the normal distribution (Fig. 3.17*a*) dictates that catering for extreme percentiles of the population yields rapidly diminishing returns in terms of the number of people accommodated by a unit increase in design constraint or range of adjustment (Fig. 3.21). The final specification of dimensions must be decided with reference to the expected conse-

Fig. 3.21. The percentage of members of a normally distributed population who will be accommodated by a given range of adjustment. (The latter is plotted in units of the standard deviation.)

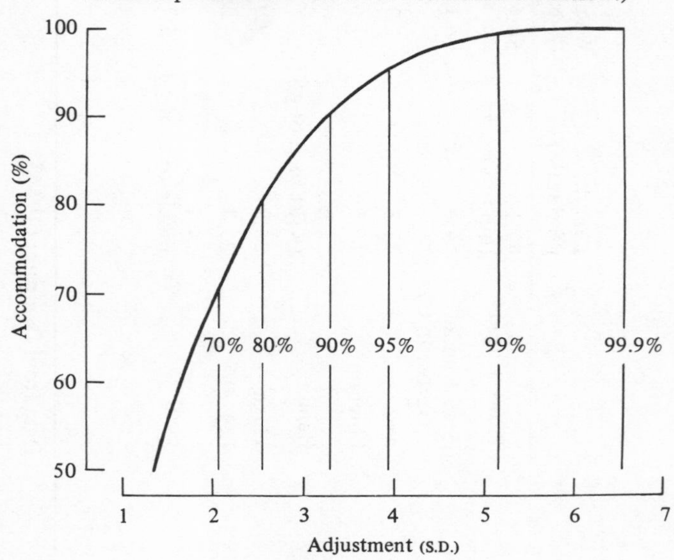

quences of a mismatch between the system and its operator or the product and its user.

There are certain important situations where it is necessary to design for extreme individuals (e.g. pregnant women, chair-bound invalids, travellers with unwieldy luggage) who represent a numerically insignificant proportion of the target population but for social reasons cannot be excluded from the system.

It is a relatively simple matter to describe a 99 percentile fighter pilot and to make judgements about cockpit design accordingly. To make similar statements concerning the passengers of a London bus is very much more difficult. The traditional concepts of percentiles are only meaningful for a population which can be isolated and investigated. As ergonomics expands from the design of military hardware into an involvement with broader ranges of human artefacts, more flexible ways of considering anthropometric data may be required.

3.4 Methodology

The reader interested in the full range of techniques used in biomechanics is advised to peruse the *Journal of Biomechanics* and the *Proceedings of the International Conferences in Biomechanics*. Further descriptions will be found in Miller and Nelson (1973), Grieve *et al.* (1975) and Winter (1979). The three most common types of measurement are those of position (and movement), of force, and of signals from muscles, each of which will be treated briefly.

Measurement of position

Cameras, film and lighting

Suppose that a photograph is to be taken for measurement of a person in a workplace. Familiarity with the camera is essential. It is recommended that ergonomists should be thoroughly familiar with the use of a Polaroid camera (e.g. Type 340), which is valued for its immediacy and performance in dim light, and, for photogrammetric purposes, with a single-lens-reflex (SLR) 35 mm camera of which there is a wide choice. If movement is of interest, familiarity with a basic 16 mm cine camera (e.g. Bolex H16) is desirable. Where budgets allow, instrumentation cameras (e.g. Automax or Cine-Pulse Locam) are available which operate with pin-registered accuracy from single shots, time lapse from 1 to 10–12 frames per second and cine to 128–500 frames per second.

There is scope for ingenuity in photography for biomechanical purposes. Movements may be recorded, using the bulb setting of a still camera, by means of strobe flash (not for potential epileptics!), by rotating a sector in front of the lens to give multiple exposures, or by pulsing lights (e.g. light-emitting diodes) mounted on the subject. Special clothing and body markers may be required to give a clean result. If extreme contrast between subject and background can be achieved, the image density obtained with a pinhole camera will indicate space occupancy over extended periods. When other measurements are made simultaneously with photography, links between other recorders and either the operation of the shutter or a display in the field of view deserve careful planning to simplify analysis.

The choice of film and its development will influence definition. Insufficient depth of focus and movement of the subject may also degrade the image. Both problems may be alleviated in stronger light by an increase in shutter speed and a decrease in aperture. At least 3 kilowatts of photoflood or quartz–halogen lighting should be available. There is no space to discuss features of cameras, lenses and films here; the reader is referred to Horder (1958), amongst a vast literature, for discussion and practical advice.

Measurement from photographs may, with care, yield three-dimensional information within about 1 cm in a space that can be swept out by a standing man, while 1-mm accuracy is virtually impossible to achieve. Measurements under field conditions, e.g. in a factory, usually involve compromise.

Parallax and the single camera

Let us take a simple example of a metre rule set up at 5 m from the centre of a lens, perpendicular to the optical axis, and another rule parallel to the first, but 20 cm further away (i.e. about half the width across the shoulders). The optical images on the film will be 4% different in length. A photograph provides a perspective view of a scene and the illusion of depth depends on familiarity with or knowledge of the arrangement of objects to be photographed. Since objects of very different sizes may subtend different angles at the lens, depending upon their placement relative to the camera, the three-dimensional determination of relative positions in space from the location of images on a two-dimensional photograph is only possible if one coordinate of each object is known independently. A person may occupy a volume 1–2 m across, and an unobstructed camera distance of even 5 m may be difficult to obtain under field conditions. Large errors in photogrammetry may arise under these conditions unless the most careful orientation of the optical axis and reliable information about distances from the camera have been obtained. Further discussion of parallax may be found in Miller and Nelson (1973).

The vital information about the third dimension may be obtained with a mirror system or a second camera, but the alignment of optical axes is always required if cameras are to be used for measurement rather than as 'note-takers'. Methods of achieving this will be discussed next.

Alignment of the optical axis

The optical axis of a camera must always be oriented in a known manner relative to the coordinates of the workspace. A convenient way of achieving this is to define axes with the beam of a low-power helium/neon gas laser. The laser beam may be set up to be truly horizontal in the workspace. A mirror mounted on the lens mounting flange or in the film plane with the lens removed or, best of all, with the lens in place and set to its smallest aperture enables the camera to be positioned so that the laser beam is reflected back along its own path. A very accurate location of the lens height and orientation of the optical axis is thereby achieved.

Stereophotogrammetry

It is assumed that camera and projector lenses are subject to neither pincushion nor barrel distortion. Nonlinear distortions necessitate complex mathematical procedures in analysis. The principles for using a camera for obtaining three-dimensional information from film are as follows:

1. Define X, Y and Z axes of the working area.
2. Align the *optical axis* of the camera with one (Z) axis.
3. Place calibration marks in the field of view at known X, Y and Z positions.
4. Either (*a*) know the Z coordinate of the objects of interest from non-photographic evidence; or (*b*) obtain a second view on the film, either (i) from a mirror system, noting that the optical axis for the reflected view has been rotated through an angle and must remain in the XY plane of the laboratory (or in the YZ plane if an overhead mirror is used), or (ii) using a second camera, ensuring that its optical axis is either in the XZ or YZ plane of the laboratory.
5. Apply the principles of geometry to interpret the relative positions of the images on the film(s).

Omission of any one of these steps will make the analysis impossible.

In the general case, suppose two views are obtained with camera(s) and projector free of geometric distortion and the optical axes of the two views coincide (*a*) with the Z axis and (*b*) in the XZ plane at an angle to the Z axis, respectively. The coordinates (M, N) of the points on the film are measured from an origin which is the image of a reference made visible in both views. The M coordinate corresponds to the horizontal X axis dimension and N to the vertical Y axis. Three measurements are required for each point, i.e. M_A, N_A and M_B.

The X, Y and Z coordinates are then computed using the determinants given below. Before they can be used, the 13 constants a to q must be found from measurements of M_A, N_A, M_B from the images of calibration markers which are at known coordinates in the workspace.

$$\frac{-X}{\begin{array}{ccc} 0 & a+b\cdot M_A & c+d\cdot M_A \\ 1 & e+f\cdot N_A & g+h\cdot N_A \\ 0 & j+k\cdot M_B & l+l\cdot M_B \end{array}} = \frac{Y}{\begin{array}{ccc} a+b\cdot M_A & c+d\cdot M_A & 1 \\ e+f\cdot N_A & g+h\cdot N_A & 0 \\ j+k\cdot M_B & l+l\cdot M_B & P+Q \end{array}}$$

$$\frac{-Z}{\begin{array}{ccc} c+d\cdot M_A & 1 & 0 \\ g+h\cdot N_A & 0 & 1 \\ l+l\cdot M_B & p+q\cdot M_B & 0 \end{array}} = \frac{1}{\begin{array}{ccc} a+b\cdot M_A & 0 & a+b\cdot M_A \\ e+f\cdot N_A & 1 & e+f\cdot N_A \\ j+k\cdot M_B & 0 & j+k\cdot M_B \end{array}}$$

If much work is to be done, the projected image should be measured with a digital trace reader (e.g. PCD Ltd photo-optical analyser or d-MAC pencil follower, or Vanguard analyser) and simple means of transfer of the digital information to a computer should be available.

Measurement of body contours

In special circumstances, e.g. the measurement of the contours of the exposed body which might be applied to the design of facemasks or back supports, a projector and graticule system can be used to project a coordinate system onto the skin surface or create a moiré fringe system which is then photographed. A comprehensive review of special techniques related to body form will be found in Herron (1972); the subject is developing rapidly (Burwell, 1978).

Alternatives to film

If posture and movement of a region of the body are of interest, it may be possible to attach a device to the body, e.g. an electrogoniometer (Elgon), which provides the information instantly. If movements are approximately confined to one plane, simple opto-electronic devices are available such as the polarised-light goniometer (Grieve, Leggett and Wetherstone, 1978*a*) for the on-line recording of angular movement of the limbs. Other devices exist such as CODA (Mitchelson, 1975) or SELSPOT (Lindholm, 1974) and closed-circuit television systems (Winter, Greenlaw and Hobson, 1972) in which on-line recording of position may be achieved. The latter three systems depend on advanced technology, large budgets and computer expertise which, for the moment, restricts their use to large research groups. A demand exists for reliable and cheap user-oriented systems for movement analysis which avoid the delays inherent in film analysis (which can be accurate and relatively cheap), especially for medical and sport applications. However, possibly because of the cumbersome techniques, the use of three-dimensional movement analysis in ergonomics has developed very slowly so far (see Section 3.7).

Measurement of force

A spring balance and a domestic weighing machine may be used for measurement of force, but their dynamic properties and lack of recorded output restrict their usefulness. Most force measurement is achieved by mechano-electrical transduction and depends upon the elastic deformation of a structure in proportion to the forces applied to it. The sensing element may be capacitative, resistive, inductive, photoelectric or piezo-electric; resistive and piezo-electric elements are most common.

It will be remembered (from Newton's Second Law of Motion) that force is proportional to acceleration, so that the deduction of forces acting in a dynamic situation can be achieved with accelerometers; these will not be discussed here.

Many aspects of force measurement at floor level (a common requirement in biomechanics) may be found in publications of the Force Platform Group of the International Society of Biomechanics, and of force transducers and accelerometers in general by reference to commercial literature (e.g. Statham, Pye Dynamics, Welwyn Electric, Kulite and Kistler). The design of suitable supporting structures requires reference to engineering texts and relevant information will be found in Timoshenko and Young (1956) and Case and Chilver (1971). Strain gauge techniques are discussed by Doebelin (1966).

Resistance strain gauges

The electrical resistance of a wire or foil changes when it is strained. If a resistance wire is stretched between two posts on a structure which move apart when force is applied to the structure, the wire is called an unbonded strain gauge. If the wire or strip of foil is attached intimately (but not electrically) to a surface so that its strain closely reflects that of the surface, the sensor is called a bonded strain gauge. The fractional resistance change is approximately twice the mechanical strain (gauge factor $\triangleq 2$) although factors of 200 may be achieved with semiconductor (or piezo-resistive) elements. The attractively high output of the latter is useful when monitoring structures which must be very stiff, but problems of temperature compensation and suitable circuits for dealing with large changes of resistance attend their use.

Unbonded gauges are found mainly in commercial force transducers whereas bonded foil gauges are usually chosen for laboratory construction, where the elastic support is to be specially designed, or forces within existing objects (such as a handle) are to be monitored. The gauge elements are never used singly because resistance is sensitive to temperature changes (ambient and due to current flowing in the gauge) as well as to strain. Both temperature compensation and two- or fourfold increased sensitivity to strain are achieved by the use of two or four gauges in half-wave or full-wave Wheatstone bridge circuits. The surface of the elastic structure to which the gauges are bonded should be designed to give modest strains of not more than 0.3% under full loading. The structures are frequently cantilevers but may be rings, diaphragms, rods and tubes. Electrically, resistance strain gauges are low-impedance devices (a few hundred ohms), typically giving full outputs of the order of 25 mV per volt applied to the bridge. They respond to static as well as dynamic strains.

Piezo-electric force transducers

The piezo-electric force transducer is usually a commercial device; the sensor is mechanically strong and therefore may be used as part of the load-bearing transmission without bonding to a stronger elastic support in the manner of the resistance strain gauge.

The transducer generates an electric charge in proportion to the mechanical strain; very-high-impedance- (or charge-) amplifiers are required to measure the charge without dissipating it rapidly and distorting the measurement. Leakages within leads, amplifier and the piezo material itself all conspire to make the transducer a device for dynamic rather than static use. Time constants of minutes are achieved in some devices, allowing near-static measurement to be made. The Kistler Company has particularly pioneered the development of quartz piezo-electric transducers for biomechanical application. Simultaneous measurement of forces in three axes is possible with composite transducers and four of these transducers, supporting a platform, are especially suitable for the measurement of forces and torques at the feet. Versatility is obtained from a combination of the transducers and the electronic circuitry, but a considerable budget is required to provide a fully instrumented system.

Equipment for electromyography

Equipment for electromyography may be constructed but is more likely to be commercial. The amplifier must have a high input impedance and, for most purposes, be a differential amplifier with high in-phase rejection. Readers are recommended to compare specifications of the major manufacturers, such as Medelec, S.E. Laboratories or Grass. Laboratory and factory conditions are usually sources of electrical and electromagnetic interference; advice on the reduction of unwanted interference is given in HMSO (1965). Techniques are further discussed in Grieve *et al.* (1975) and Basmajian (1967).

Surface electrodes are most likely to be employed in ergonomic studies. Skin impedance should be as low as possible and may be achieved by degreasing of the skin with an alcohol wipe, followed by shaving to remove excess hair, and light abrasion with a fine-grade emery paper. A rub with electrode jelly on a cotton wool swab and wiping dry completes preparation of the skin before attachment of the electrodes. Electrodes are commonly silver cups, coated with silver choride which (since skin contains chloride ions) is electrically reversible with respect to the body tissue. After attachment with double-sided adhesive rings, conducting jelly is introduced through a hole in the top of the electrode cup. If wire electrodes are inserted through the skin in order to reach deeply placed muscles (Basmajian, 1968; Jonsson, 1968), the strictest sterile precautions must be observed. Their use is not recommended for ergonomic studies except by medical personnel who enjoy adequate insurance cover.

3.5 Posture

The posture of a person may be defined as the relative spatial orientation of the members of his body. Human beings never stand or sit completely still and it is more correct, therefore, to think in terms of a mean posture being sustained over a period of time. A sustained mean posture requires an overall balance of forces between the body and its surroundings and between parts of the body. The most ubiquitous external force is gravity and this must be constantly opposed by muscular activity or by passive tensions in soft tissues (see section on the mechanics of muscle, p.83). Physiologists often refer to this muscle activity as postural or static work. This introduces semantic problems, because to the physicist or engineer no mechanical work is being done in static conditions, although the muscles demonstrably consume energy in order to maintain their active states. For the present purposes we shall refer to the mechanical loading imposed upon the body by virtue of its posture as 'postural stress'. The body's various responses to this stress we shall call 'postural strain'. Continuous tension in muscle (or other soft tissue) results in fatigue, discomfort and pain (see also the section on fatigue, p.91).

Under normal circumstances the physiological changes which underlie these experiences are reversible and the symptoms subside when the muscles are rested. The discomfort which results from a poor working posture may distract the subject from the task at hand, reduce his work output and predispose him to errors and accidents. Corlett and Manenica (1980) and Corlett (1981) describe a number of laboratory and field studies of the above phenomena.

The long-term consequences of postural stress are much less understood. It is common knowledge that an individual whose musculo-skeletal system has suffered injury or degenerative change (osteoarthrosis, spondylosis, etc.) will show a much lower tolerance of postural stress than an individual whose musculoskeletal system is 'fit'. There seems little doubt that postural strain aggravates existing pathological conditions of the body part concerned. Many authorities would accept that long-term postural stress is an important causative factor in a variety of chronic disorders of the musculo-skeletal system. There is, unfortunately, little direct epidemiological evidence for this belief. (Low back pain and neck pain have been widely investigated in the occupational context; their basic pathology remains obscure and interpretation of the clinical literature is confused by the absence of an agreed terminology for these syndromes and the proliferation of synonyms.) Wickström (1978) and Waris (1979) have written comprehensive reviews of the occupational aspects of back and neck disorders respectively.

Human beings vary greatly in their responses to postural stress; a cramped working posture or a poorly designed seat which is agonisingly uncomfortable for one person may be entirely tolerable for another. The psychological and subjective elements of postural pain and discomfort cannot be ignored. The present discussion will concentrate on the demands of given postures, expressed in terms of the mechanical loads they impose, rather than on the less readily identifiable consequences of these loads.

Standing posture

A man remains in equilibrium in the gravity field if the vertical line drawn through his centre of gravity falls within the boundaries of his base of support. This base is most extensive in the recumbent position, less extensive in the standing position, and least extensive during activities such as balancing on a tightrope, when the base has negligible dimensions in the coronal plane. (Activities in which the man also contacts the outside world through his hands will be discussed in a later section: see p. 139.)

The centroid of the supporting forces is known as the centre of pressure, and is vertically below the centre of gravity provided that no horizontal accelerations are occurring. If the centre of gravity passes outside the base of support, balance can be regained either by a ballistic righting reaction (in which rapid postural changes give the body momentum in the opposite direction) or by moving the base of support. Whitney (1962) showed that for sustained leaning actions with symmetrical foot placement, the anterior limit of the effective foot-base was in the region of the metatarsals, and the posterior limit just in front of the ankle joint. Other portions of the total length of the foot could only be deployed under dynamic conditions.

It is possible to imagine a standing posture in which the centres of gravity of each body segment are vertically aligned with each other and with their respective articulations. The net gravitational torque at any articulation would be zero. (Such a posture was referred to by the nineteenth-century German anatomists as *Die normal Stellung;* their term *normal* had the special connotation of perpendicular and did not mean common, usual or healthy.) The line of gravity through the centre of mass of the whole body is relevant only to postural maintenance at the foot–floor interface. At any other region, for example the hip, the relevant line of weight is that through the centroid of the masses above the hip. Many textbooks ignore this elementary fact.

Numerous investigators have reported periodic swaying motions in the relaxed, symmetrical, upright standing position (Hellebrandt, 1938; Smith, 1957; Thomas and Whitney, 1959). If a man stands thus upon a sufficiently sensitive force platform, oscillations in output may be recorded in the frequency bands associated with muscle tremor (8–12 Hz) and the heart beat (the 'ballistocardiogram'). These are superimposed on low-frequency changes in the centre of foot pressure (up to 1.13 cm in amplitude at a frequency of less than 0.4 Hz: Thomas and Whitney, 1959).

Electromyography rarely shows consistent activity in any given muscle group under the above conditions. It is more common to find intermittent alternating activity between the members of an antagonistic pair. (A possible exception is the calf muscles which are usually, though not invariably, active, as the weight of the superincumbent body segments usually falls in front of the ankle joint.) Basmajian (1967) has reviewed the extensive electromyographic literature on this subject; his discussion, however, is marred by an uncritical use of the concept of the 'line of weight' (see above).

Anti-gravity torques may also be supplied by passive tensions transmitted in soft tissues (muscle, connective tissue, ligaments, skin, etc.). Smith (1956, 1957) has discussed the role of such elements in the postural control of the knee and ankle joints, and Floyd and Silver (1955) with respect to the trunk. In general, passive restraints are only important in the extreme portions of the ranges of articulations.

The most realistic model of the control of the human standing posture is one in which intermittent ballistic corrections of alignment occur so as to keep the centre of foot pressure within safe limits well within the base of support. (In this respect human postural control probably resembles tracking behaviour as described by Craik, 1947.)

The analysis of postural stress

It is frequently necessary to investigate, and if possible quantify, the stresses to which an individual is exposed by virtue of his working postures. Many techniques have been used for this purpose, but few have been adequately validated. In the present authors' opinion, the simplest approach is the most satisfactory and this will be outlined first.

Let us now consider two modifications of the erect standing posture. Figs. 3.22 and 3.23 show a subject leaning forward and backward to the anterior and posterior limits of his foot-base respectively. He was photographed in a calibrated space (see p. 119) with his joint centres marked. The centres of gravity of each segment were established with reference to Table 3.3 and the locations of the centres of gravity of the whole body and of those parts superincumbent to the hip, knee and ankle joints respectively were calculated. The net anti-gravity counter-torque required for equilibrium at each articulation could therefore be calculated. (Hence, in Fig. 3.23 the required extensor torque at the hip is 0.2 W newton metres where W is the body weight in newtons.) Such calculations yield the *net torques* across any given plane of analysis, which cannot be attributed to specific anatomical structures transected by that plane. (A given net torque could be produced by an infinite variety of different patterns of muscular activity about a given articulation.)

In order to make further statements about the cost of a working posture, the investigator might choose to embark on any of a number of procedures. The classical measures of work physiology (such as oxygen uptake, heart rate, etc.)

Fig. 3.22. Subject leaning forward to the anterior limit of his foot-base. The postural stress at each articulation is expressed in terms of the proportion of body weight (W) which must be counterbalanced at each level, and the distance (in metres) from the articulation at which this weight acts. In the left-hand figure the cross indicates the centre of gravity (c. of g.) of the whole body, the filled circles the c. of g. of a segment, the open circles the axis of a joint, and the filled triangle the line of weight at the feet.

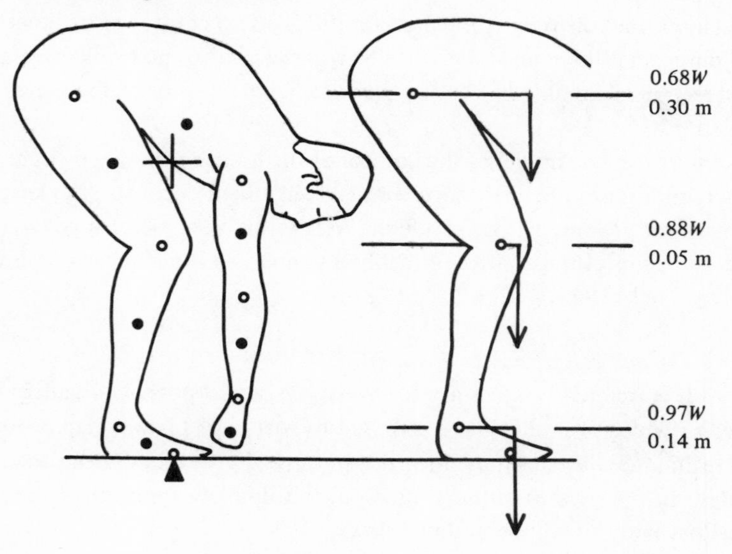

are not appropriate for this purpose. These techniques respond to the total demands placed on the body's musculature (and in the case of heart rate many other variables as well). In the postural case it is necessary to pinpoint small regions which are subjected to excessively high levels of stress against a background of relatively low-level activity. Traditional physiological techniques do not have the resolving power for this purpose.

Electromyography directly registers the activity of a given region of muscle, but the relationship between any quantitative measure of the EMG signal and the mechanical load on the muscle is not constant (see section on the effect of posture on EMG-torque relationship, p. 94). Since muscle length strongly influences this relationship, and is itself a function of posture, quantitative statements concerning the intensity of the EMG signal cannot readily be used to compare the mechanical loads imposed by different postures (Grieve and Pheasant,

Fig. 3.23. Subject leaning backward to the posterior limit of his footbase. For explanation of symbols see legend to Fig. 3.22.

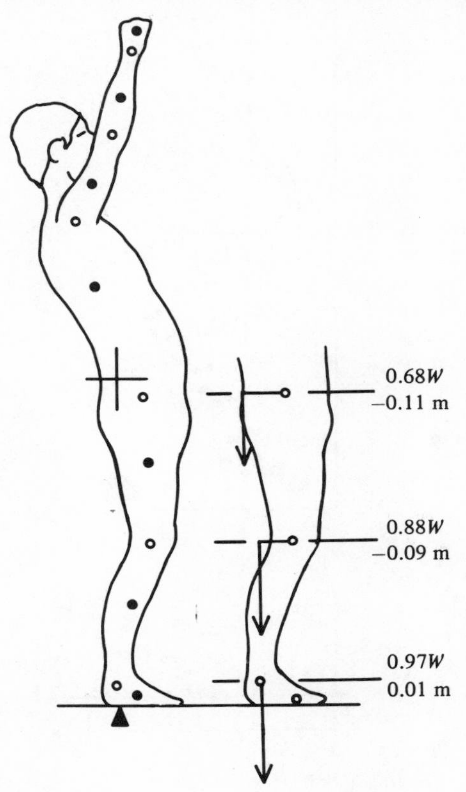

1976). To take just one of many possible examples: EMG from the erector spinae muscles has been used to investigate working surface height (Ward, 1971; Jonsson, 1974) and seating design (Andersson and Ortengren, 1974). Fig. 2.34 illustrates the way in which the EMG–torque relationship for erector spinae is

Fig. 3.24. Instantaneous readings of EMG from the erector spinae muscles plotted against extensor torque about the third lumbar intervertebral disc. Results for two subjects have been plotted. Each subject performed a graduated effort against a hand-held transducer bar, which was mounted at shoulder height (S) hip height (H) and knee height (K). For further details of the experimental method see Grieve and Pheasant (1976).

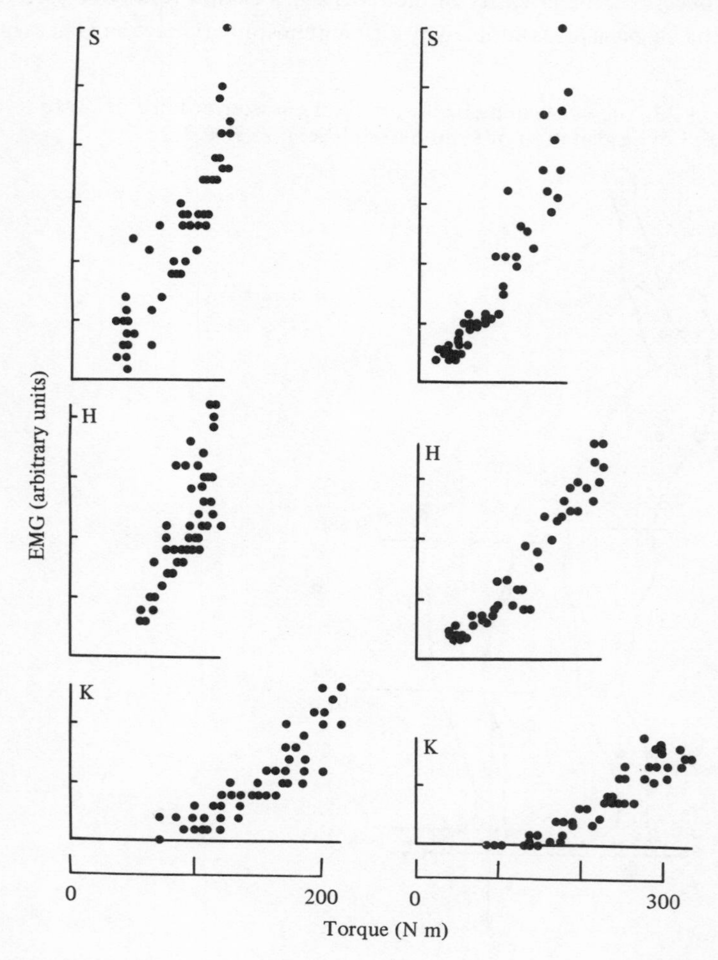

posturally dependent. The results of the above studies can only be interpreted in the light of these findings.

When a muscle fatigues, its EMG signal changes, not only in its intensity (see Fig. 3.13) but also in its power-spectrum, which exhibits a pronounced shift towards the lower frequencies. Several authors have employed this phenomenon in the investigation of the postural strain of prolonged work (Chaffin, 1973; Kadefors, Peterson and Herberts, 1976). It is not clear, however, whether the above changes reflect a peripheral (i.e. muscular) change or one related to the functioning of the central nervous system. In the absence of such evidence, their precise interpretation remains questionable. It should also be recognised that the power-spectrum of the EMG is dependent on muscle length (Sato, 1966) and hence, by inference, on posture.

Electromyography and other physiological measures, whilst they are of undeniable value in other situations, are only of peripheral interest in postural analysis. In this context there is no substitute for direct observation, preferably aided by carefully calibrated photographs (still, time-lapse, cine, or video-tape – whatever is considered appropriate). The interpretation of such observations does, however, demand a sound understanding of the functional anatomy of the human body.

It is appropriate to remind the reader that any such analysis will be greatly strengthened by systematically collected data concerning the subjectively expressed comforts/discomforts (or 'aches and pains') of the subjects or workers themselves. Effective techniques for the gathering of such data are described by Bennett (1963) and by Corlett and Bishop (1976).

Workspace design and working posture

The design of the workspace constrains the working posture by virtue of the spatial disposition of the man–machine interfaces. The postural options available to the worker are diminished as the number of contacts with the environment are increased. Due to the multi-linked structure of the body (see sections on Newtonian anthropometry, p. 115, and quantitative descriptions of movement, p. 155), constraints on hands and feet alone leave a wide choice available. When seated (particularly if a pedal must be used) the options are considerably reduced by the location of both ends of the lower limb. The contact with the environment need not be mechanical. A restricted angle of view, as in a car, or in a more extreme case, the constraints of an optical sight in a combat vehicle, leave few postural options available.

Fixed postures *per se* are to be avoided, as they are liable to lead to constant stress on muscles or other soft tissues, or to the disturbance of blood flow due to the compression of soft tissues. Maximum freedom for postural adjustment is a desirable goal in workspace design. The most significant types of continuous postural stress are as follows:

(i) The need to maintain the position of unsupported body members. (Various kinds of 'stooping' may place an unacceptable strain on the muscles of the neck or back; working with the hands away from the body stresses the muscles of the shoulder girdle.)

(ii) The need to maintain the body in asymmetrical or twisted positions.

(iii) The use of joints in extreme portions of their total ranges which therefore have an *inherent cost* in terms of passive tensions in soft tissues (see section on the mechanics of muscle, p. 83). (The postures of two adjacent articulations may frequently interact due to the presence of two-joint muscles: e.g. the hamstrings flex the knee and extend the hip; postures of hip flexion are therefore more tolerable when the knee is flexed also.)

The first step in laying out a workspace is, of course, the production of scale drawings made with reference to tables of anthropometric data of the target population. This should at least ensure that the operator will physically fit into the space, be able to reach the controls and be able to see the visual displays. It is then necessary to consider whether the range of operators can adopt satisfactory postures. Many designers find that a two-dimensional manikin, made to scale from cardboard or Perspex, and articulated to resemble the linkages of the human body, is an aid to thought in this process. Templates for the production of such manikins may be found in Dempster (1955), Dreyfuss (1967) and A.J. (1963). It must be remembered that manikins represent persons of a specified stature and average bodily proportions for that stature. The portion of the variability in a given bodily dimension which is not associated with stature is ignored in manikin design.

We shall now discuss the four man–machine interfaces mentioned above in terms of the influences which they have on working posture.

Visual requirements

The location of visual information ('displays') can be specified in terms of distance from the eyes and direction of gaze. (Sophisticated information can only be acquired within a solid angle of approximately 5° about the line of central fixation, i.e. the area of 'foveal' vision.) The greatest distance at which a display can be placed is determined by visual acuity (i.e. the capacity to resolve small details), which in turn is determined by many factors including figure–ground contrast and the overall illumination of the visual field.

As an object is brought closer to the face, the eyeballs must be rotated in their sockets by the extrinsic muscles of the eye in order to achieve 'convergence' of the image on to the foveal region of each eye. Fig. 3.25 shows the angle of convergence demanded by different viewing distances. The limit of this process occurs at approximately 8 cm from the eyes. The effective near-point of vision is, however, determined by the accommodation capacity of the lens of the eye (i.e. its capacity to change its focal length) and this is on average 12 cm at the age of 25 years, receding to 18 cm by the age of 40 years and changing very rapidly thereafter. (This change, due to the stiffening of the lens tissue, is known as presbyopia. When the near-point recedes to a distance which, by reason of visual acuity is too great for the performance of everyday tasks, spectacles are required.) A detailed description of these mechanism will be found in any textbook of human physiology (e.g. Ruch and Patton, 1965).

As both convergence and accommodation are muscular mechanisms, continuous 'close work' leads to fatigue. This occurs, for example, when exacting visual tasks must be performed in poor illumination and the subject voluntarily shortens the distance between eye and workpiece in order to achieve adequate resolution. Experience suggests that for comfortable viewing, objects should be

Fig. 3.25. The angle of convergence (d) required when viewing an object at different distances (d) from the eyes. The inter-ocular distance (o) is assumed to be 6 cm.

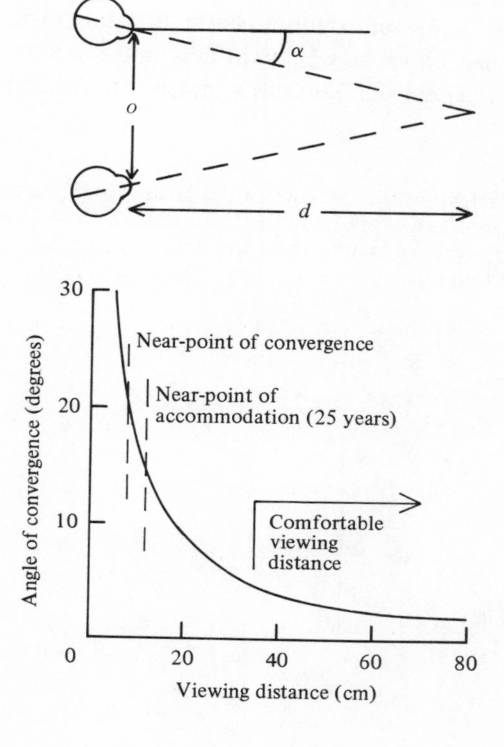

placed at least 50 cm from the eyes. (Properly prescribed spectacles will make this suitable for presbyopic individuals.) A typical preferred viewing distance for instrument panels is 60–70 cm. It is also important that major sources of visual information should be the same distances from the eye, since frequent changes of focal length can be problematic, especially for the older worker.

The direction of gaze is shifted by the movements of the eyes in the orbits and by movements of the head. Taylor (1973) states that the eyes may be elevated by 48° and depressed by 66° without head movements. Weston (1953), however, concluded that any raising of the eyes was rapidly fatiguing and also demonstrated that there was a self-imposed limit of 24°–27° of ocular depression, beyond which any greater downward angulation of gaze was achieved by neck flexion. Such an angulation of the head places a continuous postural stress on the neck extensor muscles (Fig. 3.26). Chaffin (1973) demonstrated that the time taken for the neck extensor muscles to reach a certain electromyographic criterion of fatigue decreased with the angle of head inclination. Hünting, Grandjean and Maeda (1980), in a survey of accounting machine operators, showed that the incidence of pain and stiffness in the neck muscles increased with the angle of head inclination. The above research supports the traditional ergonomic rule that primary visual displays should be in a zone no more than 30° below eye level (see Fig. 3.28). An extensive clinical literature (Travell, 1967; Dalessio, 1980) describes the way in which 'stiffness' in the neck muscles may lead to referred pain in the head and face: this is probably the cause of

Fig. 3.26. Postural stress on the neck when the head is inclined forward. Torques (*b*), expressed as a proportion of the product of body weight (*W*) and stature (*S*), were calculated from the data in Table 3.3. C7, seventh cervical vertebra.

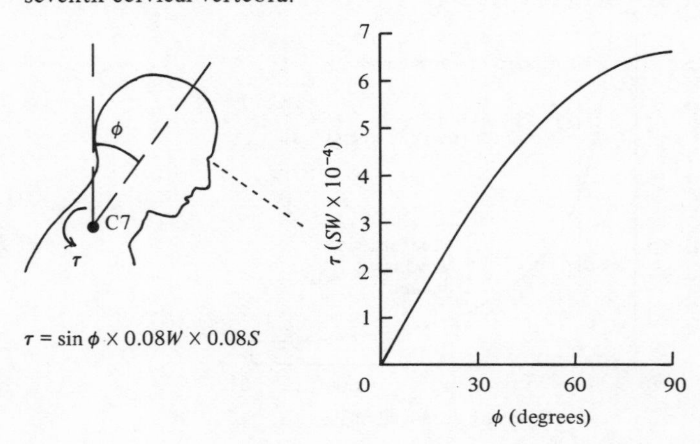

many of the so called 'eyestrain' symptoms which are so prevalent amongst the operators of the visual display terminals of computers.

The hands

The hands may, in principle, be located anywhere within the workspace envelope (see p. 114). If the upper limbs are unsupported, remote regions of the workspace envelope will impose greater postural stress on the trunk and shoulder girdle than central regions. Hünting *et al.* (1980) showed an increased incidence of pain and stiffness in the shoulder region with increased forward reaching in the workplace. Manipulative tasks which are located beyond the arc described by the hands of the fully erect standing person (see Fig. 3.27) demand forward

Fig. 3.27. The manual workspace of a standing person. The workspace envelope is based on the present authors' unpublished measurements; the arc of rotation is taken from the data of Table 3.3; the preferred working height is based on the empirical findings of Ellis (1951), Konz (1967) and Ward and Kirk (1968).

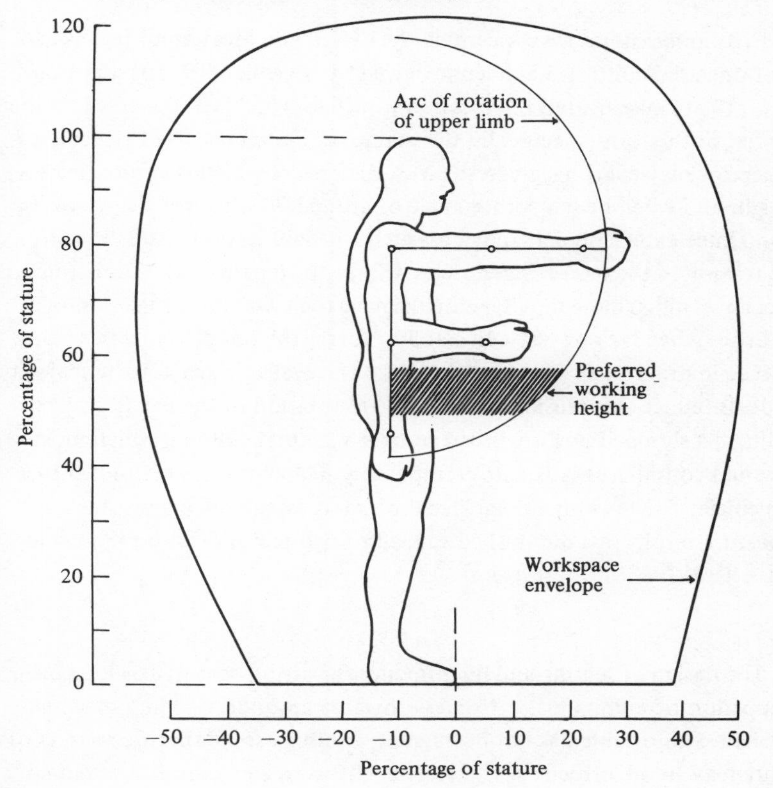

inclination of the trunk or bending of the knees, neither of which is acceptable. Controls or working surfaces which are too low aggravate the back problems of a range of people from lathe operators to housewives in the kitchen. A working surface which is too high (or a seat which is too low) will cause the subject to rotate his upper limb about the shoulder–hand axis (abduction and perhaps medial rotation at the shoulder joint). This imposes an additional load on the soft tissues of the shoulder region. Those hand locations which are most advantageous with respect to the shoulders (i.e. close to the chest) are likely to be unsatisfactory for manipulative tasks because of the visual requirements described above.

Sloping work surfaces, provision of support for the upper limb and raised reading stands may all help to solve the postural problems of sedentary workers. Experimental studies of simulated assembly tasks have shown that performance is improved and postural stress is lessened by a sloped working surface (Less, Eickleberg and Palgi, 1973; Less and Eickleberg, 1975).

The feet

Ergonomic literature concerning the design and location of pedals and other foot-operated controls has been reviewed by Kroemer (1971*b*). Pheasant and Harris (1981) investigated the location of pedals which are operated by the instep or flat of the foot (specifically the brake and clutch pedals of tractors). A seated operator may exert his greatest thrust on a pedal which is approximately at hip height and when he has a knee angle of around 30°, but the degree of hip flexion and knee extension demanded by such a pedal location results in considerable tension in the hamstring muscles. (Since the length of the hamstring muscles changes much more rapidly with hip angle than with knee angle, a similar problem arises when lack of seat adjustability forces the tall driver to flex both hip and knee in order to fit his lower limb into the available space.) The majority of individuals reduce this tension by a backward rotation of the pelvis. This in turn results in a slumped posture of the lumbar vertebral column, which most experts would consider unsatisfactory, especially in a vibrating environment like a motor vehicle. The lowering of the pedal which is essential for a satisfactory driving posture results in a diminished capacity for forceful exertion (Pheasant and Harris, 1981).

The seat

The design of seating and its influence on sitting posture has long been a preoccupation of ergonomists. Progress towards an understanding of what makes a chair comfortable has not been great, although some specific sources of discomfort may be identified. If the seat height is greater than the popliteal height of the sitter, compression of the thighs with consequent ischaemic pain results. If the seat length is greater than buttock–popliteal length, the back rest of the seat cannot be fully used; this results in slumped postures in which the lumbar vertebral column is flexed. In the erect position, the small of the back exhibits a modest concavity toward the posterior (the 'lumbar lordosis'), and it is generally accepted that the back-rests of chairs (especially working chairs, and vehicle seats) should be shaped so as to assist in the maintenance of the concavity. Fitzgerald (1973) achieved a striking amelioration of the back problems of air-crew by supplying them with customised individually contoured fibre-glass lumbar supports.

The reader wishing to consider seating research at greater length is referred to the ERS Symposium on the subject (Grandjean, 1969), to Branton (1974) and to Kroemer (1971*a*). Branton and Grayson (1967) is a particularly interesting study of sitting behaviour conducted by 'unobtrusive' methods.

Workspace geometry and body linkages

The design of a workplace 'from the man outwards' requires the coherent synthesis of separate items of information concerning working posture (together with relevant anthropometric considerations). Some authors, (Ely, Thomson and Orlansky, 1963; Wisner and Rebiffe, 1963; Rebiffe, Zayana and Tarriere, 1969) have used the linked body model of man to this end. It is necessary for this purpose to know (*a*) the link lengths of the human body and (*b*) the acceptable range of angles subtended by any given pair of segments. Areas of space suitable for the placement of control devices may then be determined by simple geometrical construction. Workspaces designed in this manner are commonly specified with respect to the seat reference point (SRP). Although the link lengths of adult males are known (as a proportion of stature) with tolerable accuracy, the constraints which should be placed upon the angles between segments are speculative and rest much more on the informed value judgements of individual authorities than on any substantial body of empirical evidence.

A suitable set of link lengths and angular constraints are given in Table 3.4. The link lengths are based on Dempster (1955) supplemented from the data in Table 3.3. Slight modifications have been made for the sake of internal consistency. The angular constraints are a compromise of the sources listed above: i.e. better than arbitrary, but in no sense definitive. Fig. 3.28 shows preferred zones for the placement of hand- and foot-operated controls according to these criteria. (The hand zone refers to an object at the centre of a closed grip; the foot zone to the ball of the unshod foot.) The zones were constructed by simple geometrical drawing. The reader is invited to repeat this construction and experiment with various combinations of angular constraints.

Bonney and Williams (1977) described a computer program called CAPABLE (Control And Panel Arrangement By Logical Evaluation) which includes routines for applying constraints of the type described above.

Conclusion

The above discussion shows that the design of a workspace to suit the user is not a simple problem. Many variables interact in a manner which is complex and not always predictable. Recommendations concerning 'good' working postures are often based on supposition, conjecture and inadequate experimental

Table 3.4. *Dimensions of linked body man and angular constraints*

(a) Dimensions of linked body man (% stature)	
Upper limb	
Arm length	17.3
Forearm length	15.5
Hand length (to grip centre)	4.0
Lower limb	
Thigh length	24.8
Leg (shank) length	23.4
Ankle height	4.7
Foot length	15.3
Heel to ball length	11.2
Heel to horizontal location of ankle joint	3.3
Subischial height (stature minus sitting height)	47.7
Trunk[a]	
Hip joint above SRP	5.2
Hip joint anterior to SRP	6.9
Shoulder joint above SRP	32.0
C7 above SRP	37.8
Eye above SRP	45.9
Vertex above SRP	52.3

(b) Angular constraints				
Ankle angle			=	90°
40°	<	Knee angle	<	100°
90°	<	Hip angle	<	105°[b]
25°	<	Shoulder angle	<	45°[c]
70°	<	Elbow angle	<	100°
Wrist angle			=	0°

[a] All dimensions refer to the fully erect sitting position. SRP, seat reference point.

[b] The upper limit is determined by the point at which the seat compresses the thigh.

[c] The lower limit is determined by the line of the trunk.

or epidemiological evidence. The reader is therefore urged to test every design by conducting fitting-trials using an anthropometric range of subjects in a life-sized realistic mock-up of the proposed workplace.

3.6 Manipulative tasks

The anatomy and mechanics of the human hand are extremely complicated. Analyses of hand function based on the individual muscles and articulations are likely to be unrewarding for ergonomic purposes. The interested reader will find a succinct account of the anatomy of the hand in Basmajian (1970). The present account will concentrate on function rather than structure.

The classification of manipulative tasks

The human hand is capable of an infinite variety of coordinated purposive activities. As a first step towards their (as yet unaccomplished) analysis it is of interest to divide these activities into categories. A primary division may be made between gripping ('prehensile') and non-gripping activities. In gripping activities the elements of the hand (i.e. the digits and various portions of the

Fig. 3.28. Preferred zones for the location of hand- and foot-operated controls, according to the criteria discussed in the text. SRP, seat reference point.

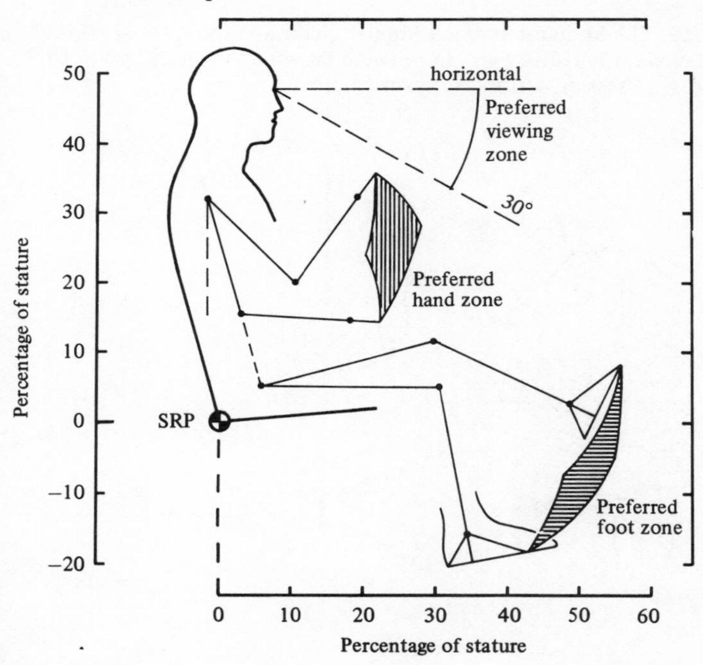

palm) form a closed kinetic chain and act in opposition to each other to exert compressive forces on the object gripped. In non-gripping actions the forces are exerted through the hand in an open chain of linkages. (For the present purposes the kinetic chain involved may be held to terminate at the wrist joint; kinetic chains involving other parts of the body and its environment are not relevant to the present discussion. See Fig. 3.29.)

Attempts have been made to subdivide gripping actions. Taylor and Schwartz (1955) described six categories: cylindrical grasp; spherical grasp; hook-grip (as in carrying a suitcase); tip grip (between the tips of the thumb and fingers); palmar grip (between the pads of the thumb and fingers); and lateral grip (between the pad of the thumb and side of the first finger, as in using a key). Napier (1956) divided all gripping actions into the 'power grip', in which the thumb and fingers clamp the objects against the palm, and the 'precision grip', in which the object is pinched between the flexor surfaces of the fingers and thumb. This terminology is slightly confusing as both types of grip may be deployed with variable degrees of strength and precision.

A classification based exclusively on hand posture and ignoring the mechanics of the manual task is inadequate for the purposes of ergonomic discussion. A provisional functional classification is presented in Fig. 3.30. Categories of hand function have been located with respect to two notional axes: the vertical

Fig. 3.29. The human hand as a kinetic chain. (*a*) Cylindrical grip: a closed chain. (*b*) Hook grip: an open chain which is on the point of closing. (*c*) Pressing action: an open chain.

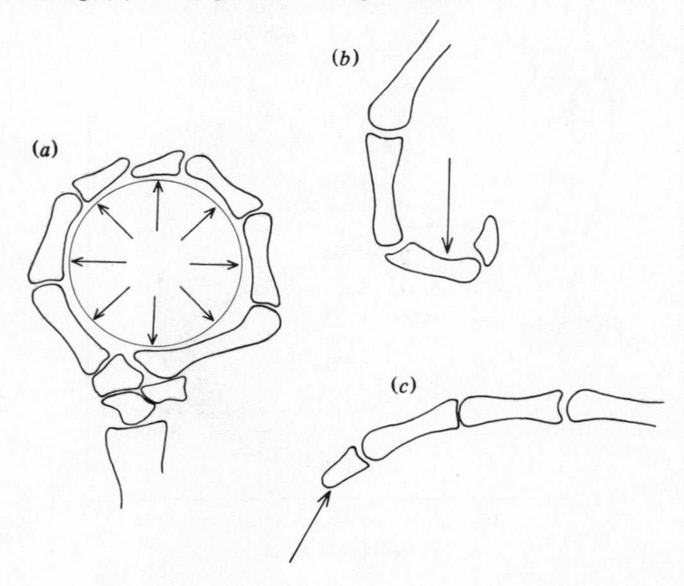

axis represents the degree to which the hand is used in closed or open chain configuration, and the horizontal axis the degree of contact of the hand with the object held.

Percussive activities are the most extreme examples of open chain function. Hook actions and scoop actions (as in picking up a handful of loose small objects) are intermediate between closed and open chain functions. The degree of contact between the hand and the object manipulated increases from left to right. Reading across the chart we can identify a sequence of gripping actions from that of holding a needle (tip grip) to that of holding a sledgehammer (cylindrical grip), in which increasing degrees of hand–object contact occur in a closed chain configuration. Reading vertically we see the closed chain spherical grip transformed into the open chain action of pushing or slapping whilst the degree of hand–object contact remains similar. A fist is essentially a cylindrical grip hand configuration used for open chain activity; although the hand postures are the same, the functions, and hence the locations on the chart are quite different. The chart emphasises the qualitative comparisons that may be made between various devices and hand functions. In practice the range of implements in a particular class may overlap with other classes (e.g. small screwdrivers may overlap with pens).

Fig. 3.30. A provisional classification of hand functions

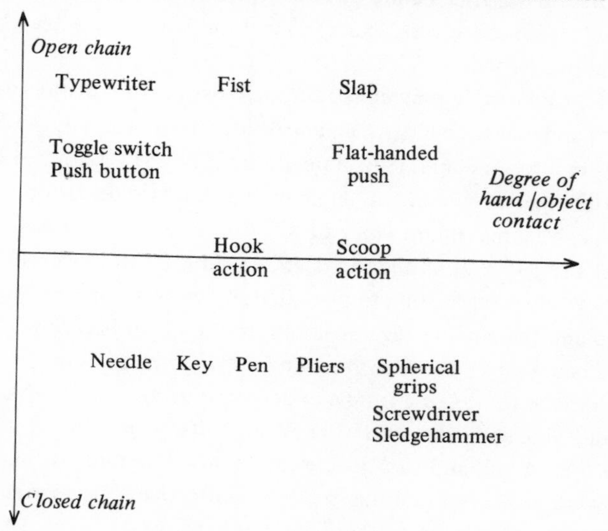

The ergonomic evaluation of handles

The functions of any handle, be it that of a hand tool, a domestic utensil, or a control device in a complex man–machine system, is to transmit forces between the musculo-skeletal system of the human operator and the outside world. The following comments refer principally (but not necessarily exclusively) to those hand–handle interfaces where the forces involved are substantial.

It is probable that, in common with many other human artefacts, hand tools evolve over a period of time by a process akin to natural selection. It is therefore unlikely that any ergonomic analysis could improve the design of, for example, the carpenter's saw. In contemporary situations where the designer of a tool is remote from its users the possibility of finding major design flaws is greater. Mis-matches can also arise when the user population of a tool changes; as, for example, when women take up heavier, traditionally male jobs (Ducharme, 1977).

A prerequisite for the design and evaluation of handles is anthropometric data for the target population of potential users. Garrett (1971) has published a useful compilation of such data.

A common, and easily avoided source of discomfort in handle design is the excessive pressure which arises when the area over which the mechanical load is distributed is inadequate. (The carrying handles of suitcases are often unsatisfactory in this respect.) The presence on the handle surface of unnecessary raised edges gives rise to undesirable pressure hot-spots. 'Finger-shaping' on handles has this effect if it fits the designer's hand but mis-fits the hands of others. Greenberg and Chaffin (1976) cite several examples of tools which have been badly designed in the above respect.

The design of a tool handle may constrain the wrist posture of the tool user. A rod held in the hand makes an obtuse angle (γ) with the long axis of the forearm when the wrist is in a relaxed and neutral position (Fig. 3.31). Barter *et al.* (1957) give a value of 102° mean value of this angle with a standard deviation of 7°. The same authors give a maximum range of 47° (mean) ± 7° (S.D.) adduction from the neutral and 27° ± 9° abduction. Hence a typical full range of values would be 75°–149°. It is interesting to note that, by convention, the angle between the grip and the cutting edge of saws is in the upper half of this range, with tenon saws being set at a more obtuse angle than cross-cut saws.

Fig. 3.32 shows the results of a simple experiment using a saw with an adjustable-angled blade. Nine subjects (who were all students of furniture design and hence skilled in the use of hand tools) were timed on a standard cutting task. It must be emphasised, however, that many other factors (such as work surface height) which may be equally important have not been investigated. The ways in

Fig. 3.31. The neutral position of the wrist. E is the centre of rotation of the elbow joint; W is the centre of rotation of the wrist joint; γ is the angle of grip.

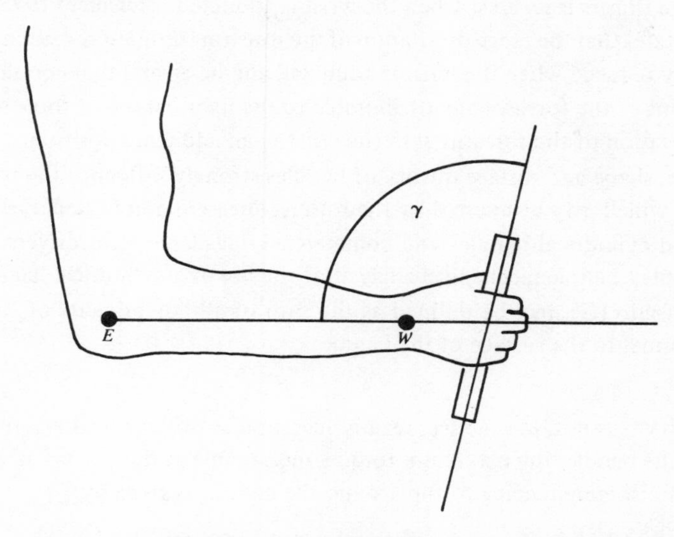

Fig. 3.32. The effects of handle/blade angle on speed of sawing. Relative speed of performance of a standard cutting task (mean ± 1 S.D.) is plotted against the angle (θ) subtended between the hand-grip and blade of the saw. Obtuse angles imply adduction of the wrist.

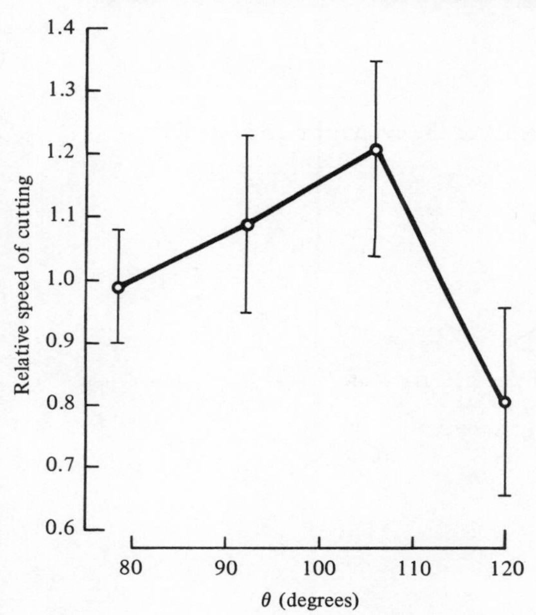

which the design of tools have adapted to biomechanical aspects of their use would merit further study. Hazelton *et al.* (1975) have shown that the flexor force of the fingers is greatest when the wrist is adducted. Tichauer (1975), however, states that the range of rotation of the forearm (pronation – supination) is markedly reduced when the wrist is adducted and he asserts that certain soft tissue lesions of the forearm are attributable to the habitual use of tools which demand rotation of the forearm with the wrist in an adducted position.

The size, shape and surface quality of handles strongly influence the forces or torques which may be exerted by their users. Pheasant and O'Neill (1975) investigated cylindrical handles and commercially available screwdrivers. Fig. 3.33 presents a simple mechanical analysis of the use of a cylindrical handle. The strength of grip (G) may be defined as the sum of all components of forces exerted normal to the surface of the handle:

$$G = \Sigma g$$

The quantity G is not, in practice, readily measurable. When exerting a turning action on the handle, the maximum torque, measurable at the instant at which the hand is just commencing to slip around the handle, is given by:

$$\tau = S \cdot D$$

where τ is the maximum torque,
S is the total frictional force (shear) at the hand–handle interface, and
D is the handle diameter.

Fig. 3.33. The mechanics of the cylindrical grip (see text).

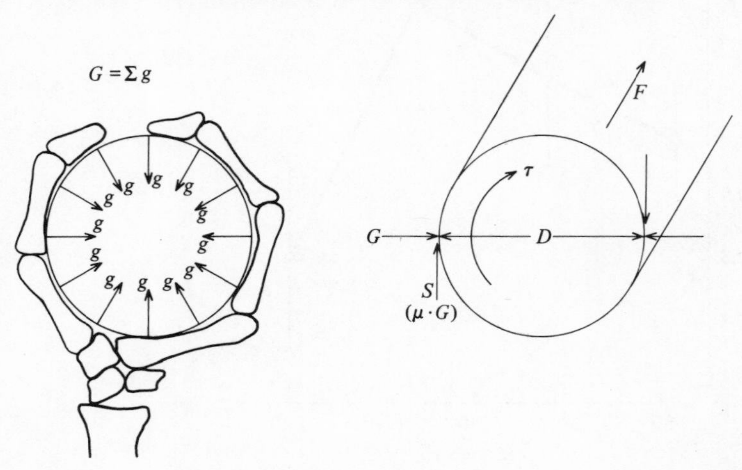

Furthermore,

$$S = \mu \cdot G$$

where G is the grip strength and defined above, and μ is the coefficient of limiting friction at the hand–handle interface. In thrusting actions, however (exerted in the direction of the long axis of the handle), the diameter is not directly involved and the maximum force (F) is given by the relationship

$$F = \mu' \cdot G'$$

(Due to the structural complexity of the hand we cannot be certain that $G = G'$ or that $\mu = \mu'$.)

The maximum supinator torque (i.e. clockwise with the right hand) which a sample of fit young subjects could exert on a range of smooth cylindrical handles is shown in Fig. 3.34(*a*), plotted as a function of handle diameter. Fig. 3.34(*b*) shows the same torques divided by the diameters to provide values for the shear forces. The shear force (and by inference the strength of grip) has a statistically significant optimum in the 30–50-mm range of handle sizes. Fig. 3.35 shows the torques which were exerted on the same set of smooth cylinders compared with the torques which could be exerted on a set of cylinders which had been knurled to give extra 'purchase', a range of commercially available screwdrivers, and a T-bar handle placed in its optimum orientation (i.e. gripped in the hand with the forearm fully prone, as described on p. 143). Although some screwdrivers are significantly better than smooth cylinders of a similar diameter, none are better than knurled cylinders. None of the cylindrical or screwdriver handles exploits the full capacity of the body for supinator torque exertion as measured by the T-bar. It is interesting to record that no commercially available screwdriver approaches the optimal diameter for torque exertion ($\triangleq$ 50 mm). The authors have heard anecdotes of craftsmen who custom-build their own, larger screwdriver handles, which are said to be very satisfactory. The effectiveness of a screwdriver is, of course, a function of more factors than the available torque, and the manner of use varies with the orientation of the tool with respect to the axis of the forearm.

The forces which could be exerted in a forward thrusting action using a range of cylindrical handles are plotted in Fig. 3.34. Their optimal range is similar to that shown for the shear forces during rotation. Similar results were found for a downward thrusting action (Fig. 3.36). The subjects in the latter experiment were invited to rank the handles in order of preference for the diameters which they supposed would give them the best grip if employed in the hand-rails besides staircases or on buses. The mean choice intensity for these rankings is also shown in Fig. 3.36. Neither the optimum handle diameter (performance-wise) nor the subjectively preferred handle diameter was significantly correlated in

Fig. 3.34. The strengths of gripping, turning and thrusting actions, plotted against handle diameter. In this and subsequent figures, the measured torques or forces have been normalised with respect to each individual subject's average performance. (For further experiment details, see Pheasant and O'Neill, 1975.) The mean and S.D. of the normalised scores of the sample of subjects are plotted. (*a*) Clockwise torques exerted on a set of smooth cylindrical handles. (*b*) Tangential (shearing) forces at the hand–handle interface. Twisting shears (filled circles) were calculated from the above data; forward thrusting forces (open circles) were measured directly.

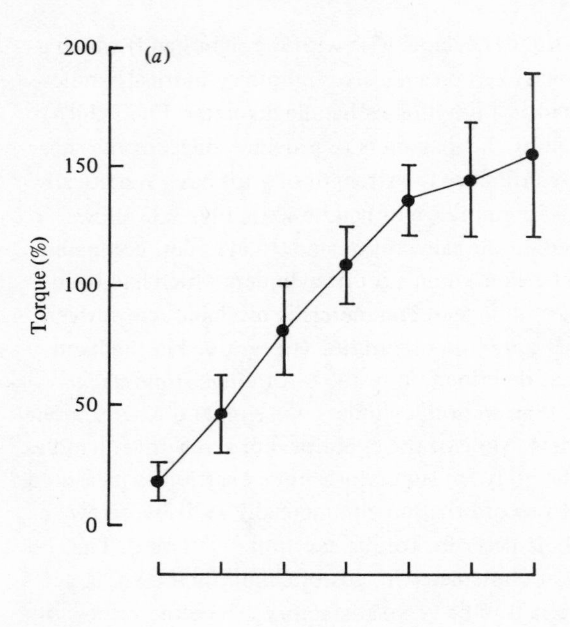

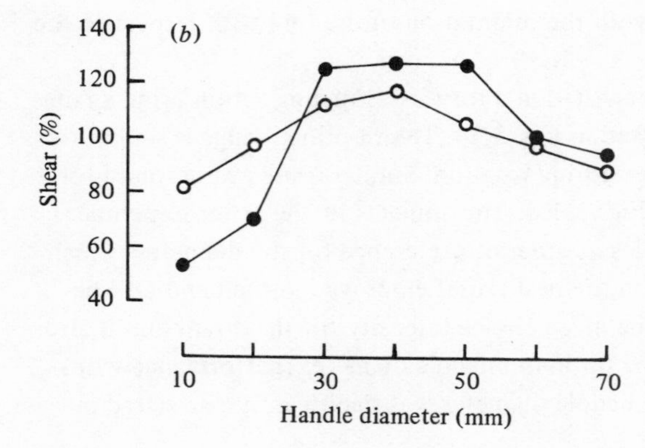

this study with either the hand length or the maximum closed grip diameter of the subjects. (Maximum closed grip diameter was measured using a graduated wooden cone down which the subjects moved their hands until the tips of the thumb and middle finger could only just touch each other.) This lack of correlation is to be anticipated from the fact that the total range of maximum closed grip diameters of the study subjects was small compared with the range of handle sizes presented. A study which included a greater range of hand sizes (e.g. juveniles) would possibly give different results. All subjects exerted their optimum downward thrusting force on a handle which was smaller than their maximum closed grip diameter; this suggests that the strongest grip is established when the thumb and fingers can wrap around the object and slightly overlap each other (Pheasant, 1978).

The above results are self-consistent and compatible with the electromyographic findings of Ayoub and Lo Presti (1971). It is of interest also to compare them with some recommendations based on subjective judgements that are to be found in the literature, which are summarised in Fig. 3.37. Brooks, Ruffel-Smith and Ward (1973) asked a sample which included a high proportion of elderly and disabled subjects to select a suitable diameter for the vertical

Fig. 3.35. A comparison of the strengths of turning actions using a variety of handle types (see Pheasant and O'Neill, 1975). Squares, knurled cylinders; filled circles, smooth cylinders; open circles, screwdrivers. Means ± 1 S.E.M. are shown.

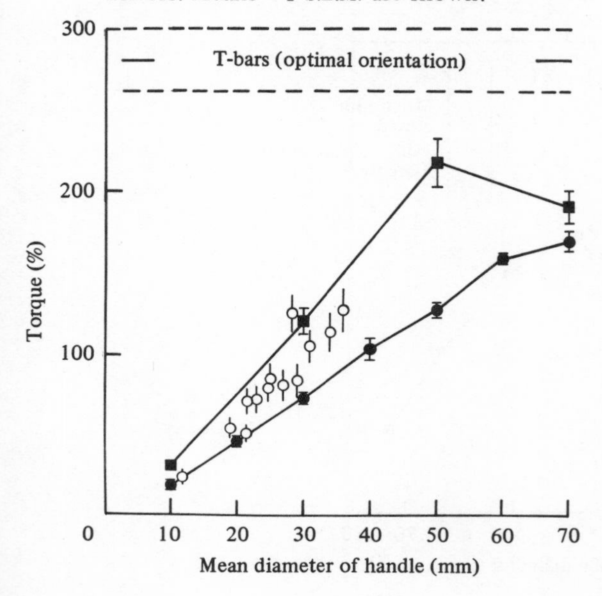

hand-rail of buses; Hall and Bennett (1956) asked able-bodied subjects to select a stair-rail; neither Dreyfuss (1967) nor Goldsmith (1976) give any indication of the origin of their recommendations.

Torque measurements have also been made using spherical and disc-shaped handles (Pheasant, 1978). The results are summarised in Fig. 3.38. At any given handle diameter, greater torque can be exerted on a cylindrical handle than on a spherical one. The optimal diameters for the development of torque were 50–70 mm for cylinders, 60–75 mm for spheres and 90–130 mm for discs. These differences presumably reflect the way in which the hand 'opens' as it grasps first a

Fig. 3.36. Cylindrical handles used in a downward thrusting action. (*a*) Strength of thrust (mean ± 1 S.D.). (*b*) Mean choice intensity of the preferred handle sizes of the subjects. The mean and the total range of the maximum closed grip diameters of the subjects are also shown.

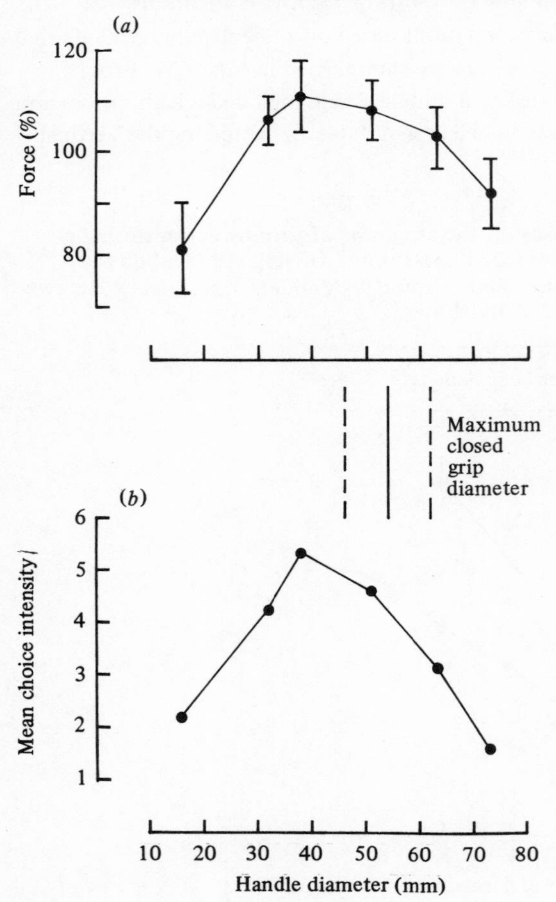

Fig. 3.37. Sizes of cylindrical handles recommended by various authorities.

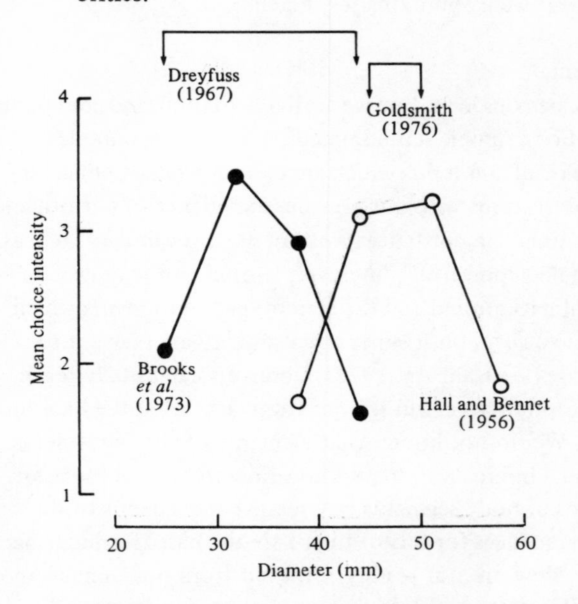

Fig. 3.38. A comparison of the torques which can be exerted on cylindrical (open circles), spherical (filled circles) and disc-shaped (crosses) handles of various sizes. Mean of 12 subjects (six males, six females). Results have all been expressed as a percentage of the performance on a cylindrical handle of 50 mm diameter and the same surface quality as the handle concerned.

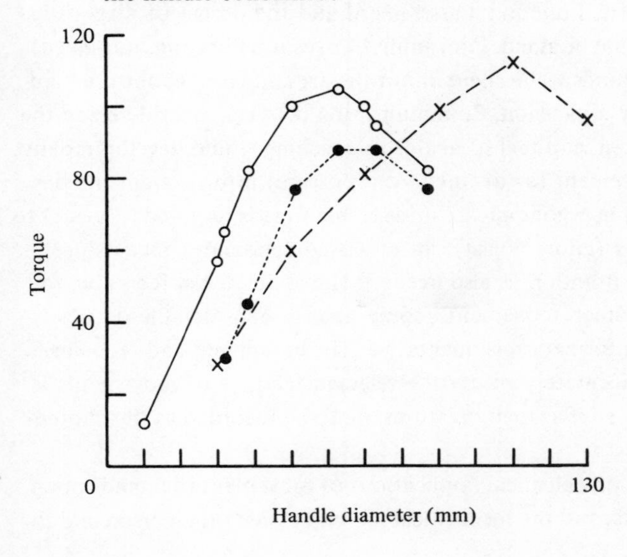

cylinder, then a sphere then a disc. The torques exerted on cylinders and discs of their respective optimal sizes were approximately equal.

3.7 Human movement

When we observe persons in motion we are likely to think and communicate in adjectival terms. For example 'coordinated', 'clumsy', 'economical'. 'efficient', 'precise' and 'skilful' are terms which are used in a relative manner when we compare our observations with our previous experience of performance by ourselves or others. In some circumstances we may use a vocabulary such as 'graceful', 'modest', 'languid', 'powerful', 'incisive', 'erotic' and 'exhausted', which reflects our cultural background and the stereotyped movements which serve for a language in our society. Both types of vocabulary are relevant to ergonomics in a broad sense (see Hallberg, 1976). Observers can usually reach agreement about the appropriate term but it is far from clear what the bases of our communications are. We do not know what elements of the movements provide the important visual inputs, nor do we know how to process the information. If the spatial paths of body segments, the relative movements of the parts and their temporal sequences (or sets of them) are the bases for judgement (and what else is there?), they are not readily retrieved from our minds. An experienced clinician, with anatomical language at his command, can only recount a few elements of what he has seen in real time if he concentrates upon using anatomical terms and ignores other elements.

It would be a great achievement to relate the normal vocabularies of perceived movement to the measurable elements, and one which would have immediate practical benefit. Functional assessment and the design of diagnostic aids in orthopaedics would be aided. Film animation (especially computer-aided), in which the animator tends to be silent about the art and vocal about the hardware, would have a new dimension. Communication between an athlete and the coach would be enhanced, and technical aids to coaching would develop rapidly.

The analysis of movement (as distinct from posture) into its elements has received little attention in ergonomics. Considerable time is required (2 years) to train an expert notator (choreologist); an educated (choreate) readership, committed to the same notation, is also needed. The choreologist feels able to notate all manner of human movements, sometimes in considerable detail, although much of the information is qualitative. The equipment and technical backing necessary for accurately *measuring* selected features of movements is expensive, prohibitively so if as many features are to be recorded as the choreologist can encompass.

Understanding the biomechanical implications of movement (demands upon muscle groups and joints, and the mechanical match between the person and the task) awaits more development than the techniques of observation and recording. An example of biomechanical analysis in orthopaedics, and its inherent assumptions, may be found in Paul, Hughes and Kenedi (1972). Our knowledge of control of even the simplest movements, let alone industrial tasks, is still the subject of widely varying hypotheses (Stelmach, 1976); it will be some time before we can postulate or recognise movements which make optimal use of neuromuscular control systems. The techniques for data acquisition have, for the moment, outstripped our ideas for analysis. Whatever type of information is obtained, the subject of movement analysis is too young for us to say how it should be used in ergonomics.

The origins of choreology are to be found in eighteenth-century ballet, when the movements were first scored in order to preserve and re-create the choreographer's work. If the need had arisen recently *de novo,* it seems likely that special filming techniques, with records of archival quality, would be developed, since re-creation, not analysis, is the objective. The notations that are introduced below were mostly developed from ballet and are gradually evolving as methods of *analysis.* In view of their *potential* importance we shall briefly review some of the major approaches to movement analysis so that the reader will at least be aware of what is currently available.

Notations based on visual observation and intuition

Many movement notations exist; most major ballet companies employ choreologists to write and interpret notated works, which are analogous to musical scores. The best known notations are Laban, Benesh and Eshkol–Wachman, all of which have been further developed for application to movement studies in general. Movements are frequently brief and elusive and the choreologist may sometimes resort to film in order to examine movement in greater detail. The details recorded are either qualitative or semi-quantitative, but not usually in a form that anatomists, physiologists and biomechanists would use to draw conclusions in their own terms.

Laban notation and Observable Motion Data Recording (OMDR)

The technique of OMDR (Preston-Dunlop, 1969), which developed from Laban notation, provides three complementary methods of notating observed features, with various degrees of detail that are selected according to the requirements. Motif writing, for example, is a broad description, using symbols, of the purposeful elements or 'motivations' in the sequence: e.g. 'walk to chair, turn to face the other way, sit down'; detail is deliberately omitted. A kinetogram, which appears alongside the motif writing, describes the movements of selected body parts, i.e. type of movement and direction, and may be extremely detailed. Each finger movement, for example, may be individually notated in a task involving manual dexterity, while the notation of walking could be confined to simpler statements concerning the successive placements of the feet (see Fig. 3.39). A third chart, the linear effort graph, expresses the perceived efforts made to control and produce the actions, i.e. how the subject uses available time and space and applies pressure, as perceived and sensed by the observer, expressed relative to the concept of an 'effortless norm'. A chapter would be required to explain fully any notation, and the reader is referred to the bibliography of Preston-Dunlop (1969). Recent developments in the use of computers to create the movements of a stick figure on a screen from Laban-notated instruction (Professor Calvert and colleagues at Simon Fraser University, B.C.), could (as with other notations) be instrumental in recognising how adequate a notation is and what must be added in order to create a satisfactory illusion on 'playback', and to provide a full description.

Fig. 3.39. Walking as described by a kinetogram. Observable Motion Data Recording (OMDR) normally provides additional information by setting motif writing and an effort graph alongside the kinetogram (Preston-Dunlop, 1969).

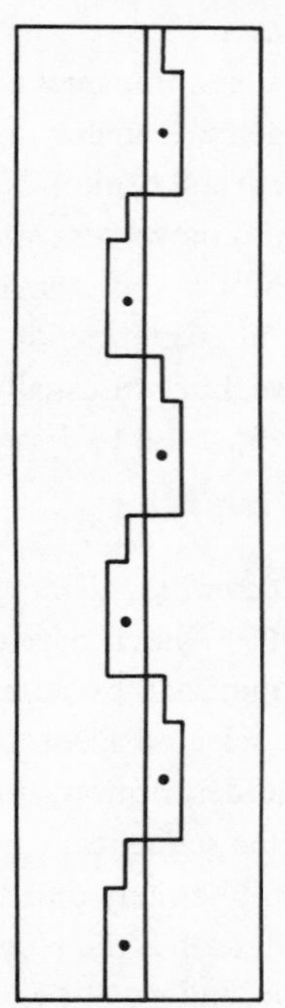

Benesh movement notation

The Benesh notation captures selected features of the person as they would be seen in rear projection. The information is presented in square frames of notation which are divided horizontally by five staves. The symbols indicate positions of parts in the rear view, and whether the parts are positioned (and limbs are bent) level with, in front of, or to the rear of the body. Further cursive symbols indicate paths taken during the movement. A succession of frames is used to represent the parts of a movement sequence. For some purposes the staves represent levels of the top of the head, the shoulders, waist, knees and ground. If the upper parts of the body alone are of interest (as in sitting at a bench), the staves may be used to represent levels between the hips and the head instead. A brief introduction is given by Benesh and McGuiness (1974), and a description in greater depth by Causley (1967). An example of Benesh notation applied to walking is shown in Fig. 3.40. As with Laban systems, the Benesh notation has been applied to fields outside ballet: e.g. Milani-Comparetti and Gidoni (1968) in studies of cerebral palsy and Kember (1976) for recording of seated postures. It is reported that work-study applications and the use of the notation in conjunction with computers have been tried.

Undoubtedly more use will be made of movement notations as the number of trained choreologists increases, but the long training period, apart from the

Fig. 3.40. Outline analysis of stave in Benesh movement notation of swing, double stance and full weight-bearing phases of a 'normal gait'. Square at left end of stave: weight-bearing on left foot, position and joint angles of right hip, knee and ankle in preparation for swing phase are notated. Position of arms, hands and fingers are notated. Gait speed per minute is stated above stave. Movement of legs, arms and torso are notated in the remainder of stave, the stave being read from left to right. Right swing phase is notated analysing at every point in time the angles of hip, knee and ankle as the swing phase moves forward to the double asterisk. Double weight-bearing is analysed showing centre of gravity transferring, trunk rotating and the changing angles of left hip, knee and ankle. Triple asterisk is the position arrived at when full weight-bearing takes place, analysing righting trunk rotation, hip tilt, length of stride and arm swing. The brackets indicate how the notated information is read for the left swing phase. Note: the asterisks are for reference only and are *not* part of the notation. (Figure and notes kindly provided by Ms Julia McGuiness, A.I.Chor. Benesh Movement Notation © 1955.)

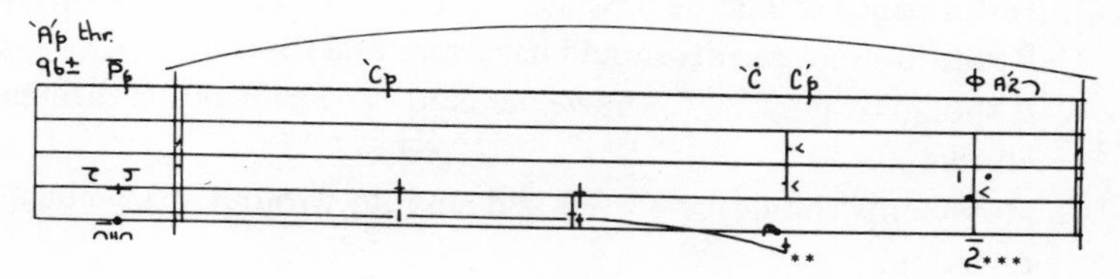

time required to become oriented to the objectives of a particular study, limits this increase. The capacity to record many aspects of an activity in a 'neutral' manner clearly satisfied the needs of clinicians in studies of cerebral palsy (Benesh and McGuiness, 1974). A neurologist said of the Benesh notation: 'It is flexible enough to be used for recording any type of movement; it allows the accurate recording of the finest details; and its use of symbols is minimal, ensuring efficiency, simplicity and speed'. A notation suitable for clinical purposes, where a premium is placed upon 'efficiency, simplicity and speed', may prove to have desirable qualities for ergonomic application. But with the analysis of movement being in its infancy, we do not know what value to place on such information when we have it. As far as biomechanics is concerned, quantitative, not qualitative, information is required *before* analysis. Movement notations best suited to particular applications may then evolve from those existing, and with evidence from quantitive analysis of the importance of particular qualitative features.

Method study and micromotion analysis

If a cine film is available of the movements made during work, preferably with a second view to give three-dimensional information, and with time presented in each frame, the movements can be timed and classified. A standard symbolism (see Fig. 3.41) of Therblig symbols is used by the work-study engineer which can be applied to the body as a whole, or to body parts. The charted result (Sino chart) presents a vertical time scale with the Therblig symbols and annotation against it. A similar scale and the same symbols may be used to describe the use made of machines in the workplace also. A combination of the statements about the man and the machine is called a Multiple Activity Chart. Larkin (1969) suggests that the chart be examined in detail to decide on the necessity for movements, their sequence and combination, and whether simplifications are possible. When the body movements are examined in detail (micromotion analysis), Larkin, echoing Barnes (1968), states that they should be judged according to some 'principles of motion economy', i.e.

1. Movements should be natural and the minimum for the job.
2. Movements of the limbs should be simultaneous rather than one at a time.
3. Performance should be possible using the left or right side of the body.
4. Repetitive movements should have a natural rhythm.
5. It should be possible for movements to become habitual through repetition.
6. Movements should be curved and smooth without sharp changes of direction.

Larkin suggests that a practical knowledge of ergonomics is necessary in order to use the principles efficiently! The subject does not seem sufficiently developed to do other than rely on commonsense (embodied in the principles) and experience.

Quantitative descriptions of movement

The position of an isolated segment of the body is defined by six coordinates. The reasoning is as follows. Suppose three points are considered on one segment. The first point requires three coordinates (X_1, Y_1, Z_1) for its definition, but without further information any orientation of the segment centred on that point would be possible. Fixing a second point with three coordinates (X_2, Y_2, Z_2) leads to redundant information, since as the two points are a fixed distance apart, Z_2 need not be stated. The segment is still not fully defined, since its rotation about an axis between points 1 and 2 is not defined. The third point (X_3, Y_3, Z_3) fully defines the segment's position, but since point 3 is a fixed distance from points 1 and 2, only X_3 is required. The argument can be expressed in terms of polar or cylindrical coordinates, but in each

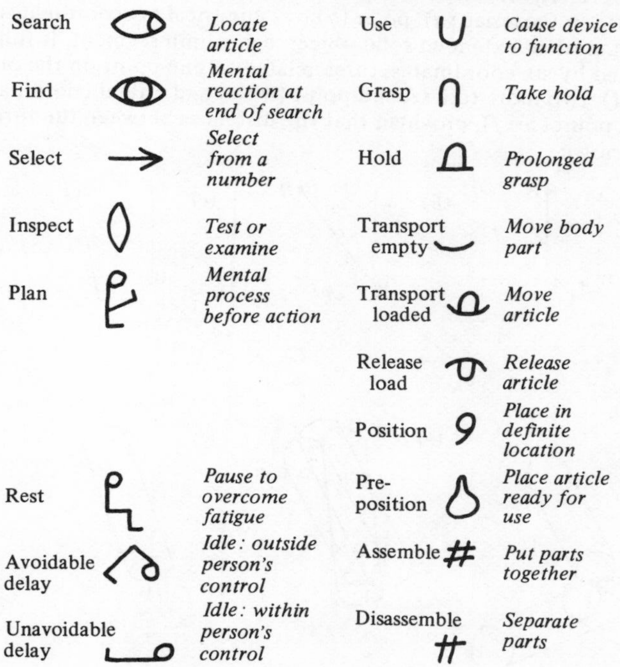

Fig. 3.41. Therblig symbols and a key to their meaning. They were introduced by W. Gilbreth for analysis of motion in work study.

case a total of six coordinates is required, and no more (see Fig. 3.42).

The description of the posture of the whole body might appear an intractable problem, i.e. an unmanageable amount of data, especially when the temporal variations of each variable are introduced to describe movement. However, the segments are not independent, and constraints such as the approximately fixed lengths of the segments, regions of contact of the body with the floor and workplace, and anatomical restrictions on movements at joints reduce the number of independent variables in a given circumstance (Fig. 3.43).

Eshkol–Wachman movement notation

A semi-quantitative description of movement (Eshkol and Nul, 1968), again originating from ballet, considers the proximal joints of each segment of the body to be at the centre of a sphere. The sphere is cut by a vertical reference plane, e.g. perpendicular to the line of the footlights, and polar coordinates ('latitude' and 'longitude') are used to describe the direction of the 'long axis' of the segment. A stave system of notation, in which a line is devoted to each of the relevant segments of the body, is then used to code postures. Symbols are

Fig. 3.42. The location of a point D in three-dimensional space may be defined in Cartesian (a), polar (b) or cylindrical (c) coordinates. The position of an extended solid object, e.g. a limb segment, is fully defined by six coordinates, three related to one point on the object (see d), two more to a second point (see e) and a final coordinate to a third point (see f), provided that the distances between the three points are known.

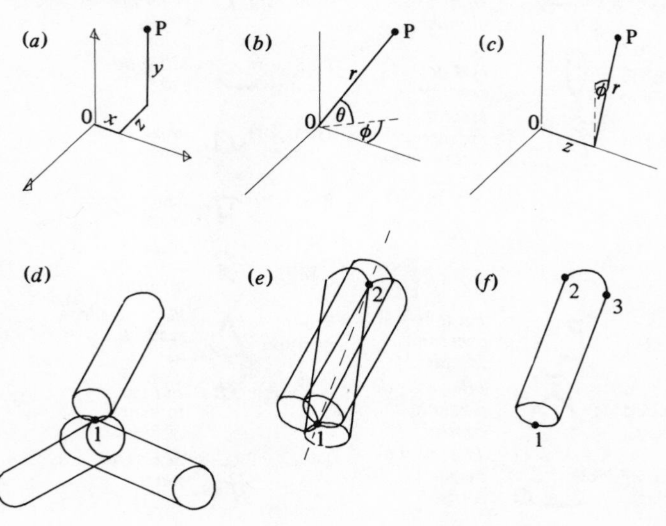

used to indicate the types of movement (conical, planar, etc.) which occur between the notated postures (see Fig. 3.44).

The system referred to, which divides a circle into eight 45° sectors only, is too crude for direct application in biomechanics, but computer studies have been made which employed finer gradations. A description of movement can obviously be made at any desired level of detail, but the complexity of the general description, when degrees rather than hemi-quadrants are used, renders the computer obligatory and detracts from the simplicity originally sought (Eshkol *et al.*, 1970).

Another method, the posturegram, for recording posture in relation to equip-

Fig. 3.43. The number of cordinates required to define the posture of the whole body is less than that required to define each segment separately, because of the several constraints relating to segment lengths, properties of the articulations and contact with the surroundings. In the illustration here, the constraints which reduce the number of variables required are: (1) knee joint on the surface of a sphere, centred on the hip joint (or vice versa), or on the ankle and subtalar joint (or vice versa); (2) the foot in contact with the ground has an effective rolling surface, defining ankle position with respect to the ground for any given foot orientation; (3) shank, thigh and pelvis are of fixed lengths (ignoring minor movements of centres of rotation due to combined sliding and rolling or articular surfaces and ligamentous constraints); (4) the knee joint is a hinge type to a first approximation so has only one degree of freedom of rotation.

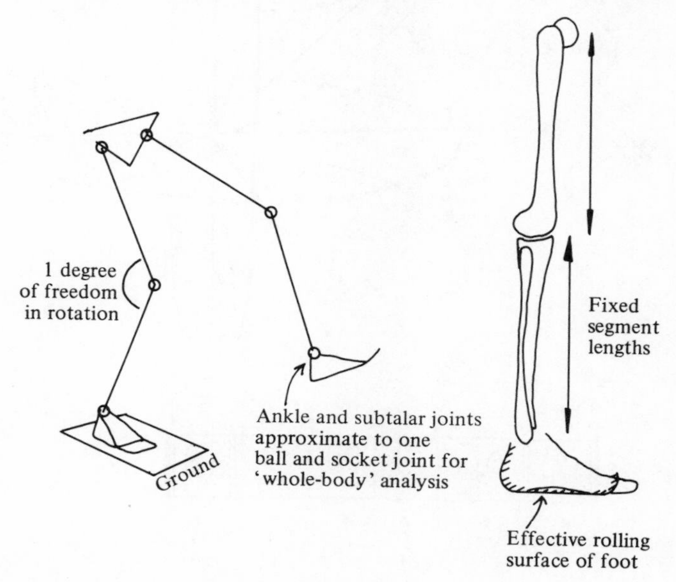

ment design (Priel, 1974), has some features very similar to those found in the Eshkol–Wachman notation.

The presentation of measured movements

Measurements of movements are made by the analysis of cine film, or with on-line techniques such as Selspot, Polgon or video systems. In each case a

Fig. 3.44. Eshkol–Wachman movement notation. The inset figure illustrates the body segments and their axes which are notated. The system for describing posture of a (forearm) segment by latitude and longtitude in units of 45° is shown. The line of notation (one is required for each segment) describes the movements of the wrist–elbow axis, relative to the elbow centre, when the path PQRST is described. (i) indicates that 45° units are in use; (ii) and (vii) describe the initial and final postures; (iii) indicates a vertical movement in the (0) plane of 1 unit; (iv) indicates a conical movement of 1 unit; (v) represents a movement which takes the limb into position $\binom{2}{2}$; (vi) indicates the final horizontal movement of 2 units i.e. 90°.

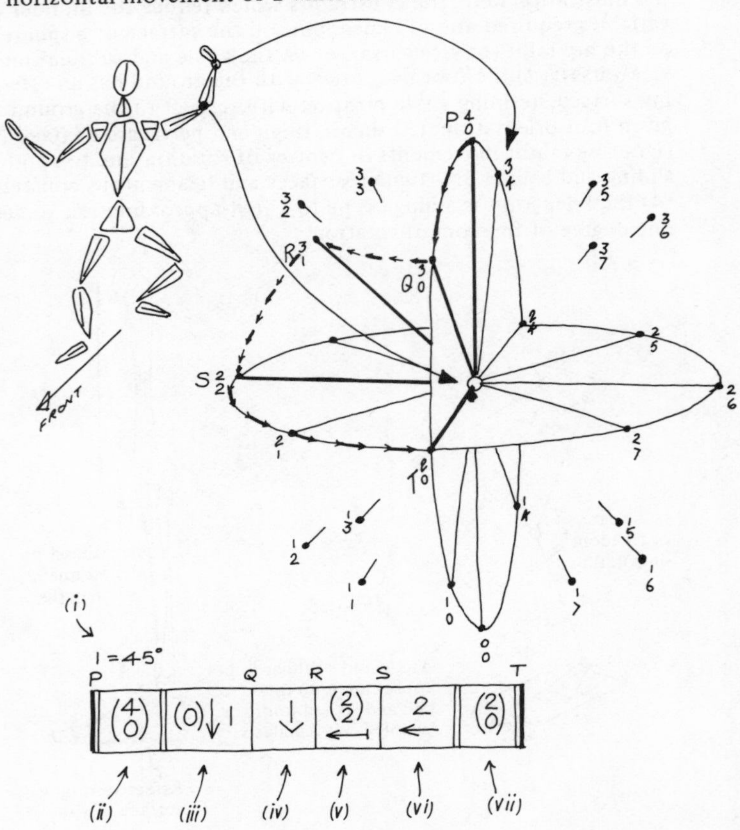

selection of a limited number of variables will have been made. There is at present no indication that 'total' quantitative descriptions of body movement are required because we do not know what to do with the data; this also applies to the movement notation.

Quantitative studies are mostly confined to clinical and sports research applications when there is a special interest in one region of the body, e.g. the use of the lower limbs in locomotion. The amount of information is always considerable and problems arise relating to the presentation of the information in compact and intelligible form. Charts of the variables as functions of time are the most common presentation, but the coordination of movement in several segments of the body is not easily appreciated by considering several plots, stacked one above the other.

The angle diagram, in which variables describing the angular positions of functionally related segments are plotted against each other (rather than against time), is a method of presenting the information as shapes which can be examined quantitatively if required. The shapes are sensitive to differences of performance. An example of thigh, knee and ankle movements in walking, using angle diagrams, is given in Fig. 3.45, which may be compared with the notation in Figs. 3.39 and 3.41. Fig. 3.46 shows some movements of the lower limbs engaged in common activities, as they appear on angle diagrams. It should be noted that only a small

Fig. 3.45. Angle diagrams (thigh–knee and knee–ankle) describing sagittal rotations of the lower limbs when walking at a slow-to-moderate speed. Mean data obtained from cine analysis of five men having stride lengths in the range 70 ± 2% of their standing heights.

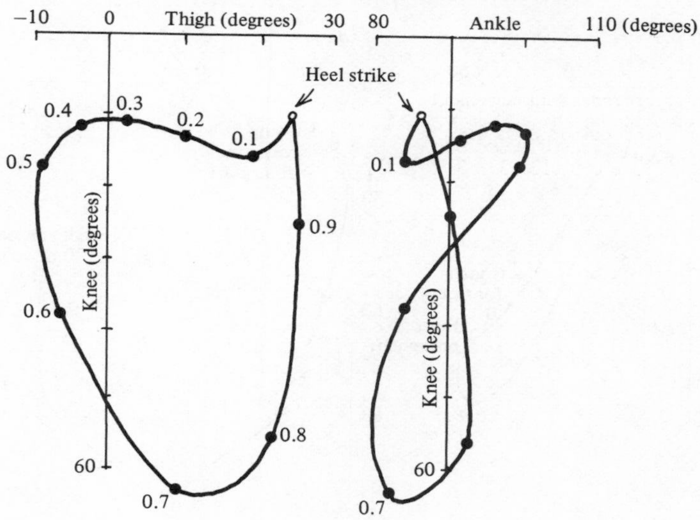

part of the total recordable information is presented. As we saw earlier, the existence of a posture described in one plane has many implications: e.g. lengths of muscles, possible torques at the joints, and distances between the ends of a limb. Angle diagrams may be used as vehicles for conveying some of this information (Grieve and Cavanagh 1973; Grieve *et al.*, 1978*b*).

Fig. 3.46. Thigh–knee diagrams describing sagittal motion during: (*a*) mounting one step equal to 10% of standing height (Grieve *et al.*, 1978*a*); (*b*) climbing a 75° ladder, based on 22 subjects (Dewar, 1977); (*c*) cycling at 95 r.p.m. (Nordeen and Cavanagh, 1976); (*d*) kicking a football in a fast kick (based on tracings from Zernicke and Roberts, 1976). HOA, hip over (i.e. vertically above) ankle.

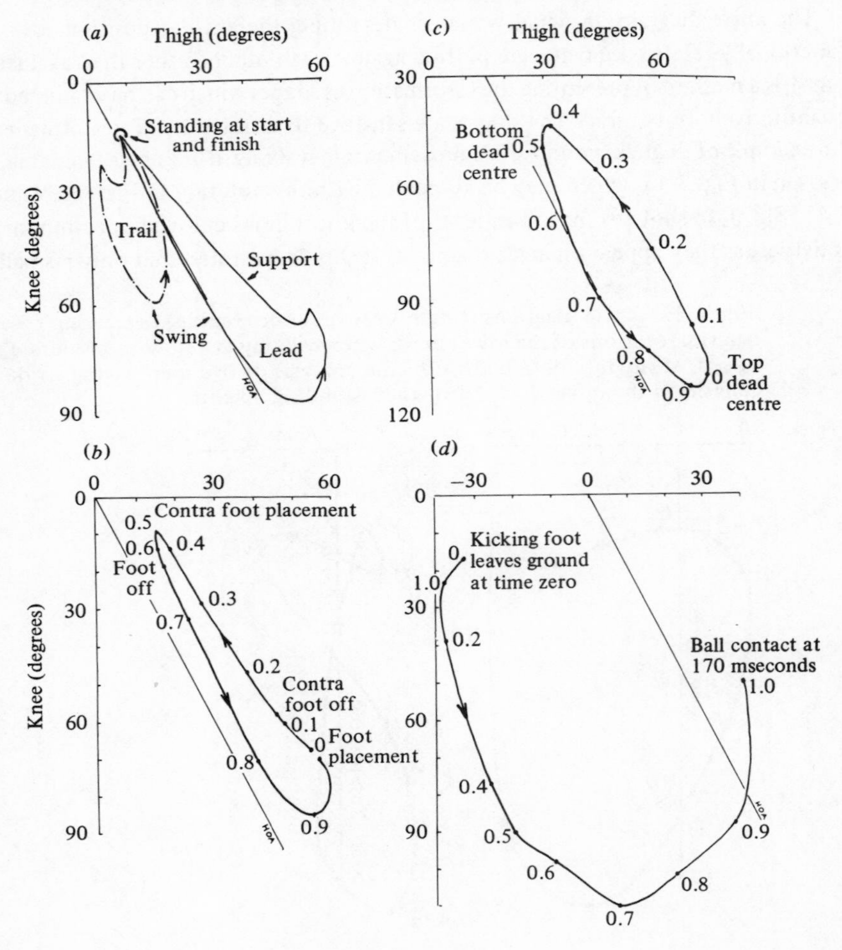

3.8 Heavy work

Heavy work involves the whole body; the manner in which contacts with the ground and the workpiece, the working posture and the intrinsic strength of the individual combine to influence the suitability of the execution to the task will be discussed. The ability of the trunk to act as a linkage is of paramount importance and the trunk is generally regarded as the region most at risk, especially during lifting. We will therefore discuss truncal mechanisms and their limitations before considering the whole body.

Mechanics of the trunk

The trunk is inevitably involved in all forceful activity, whether the subject is seated or standing. Several truncal muscles either influence or assist in the circulation of blood, retention of gut and bladder contents, and speech, in addition to fulfilling roles in posture and movement. Accommodation and compromise between these roles is continually and unconsciously sought during daily living.

Well-known associations exist between lifting or heavy manual work and failures of truncal mechanisms. The associations are not unique; predisposing changes due to ageing and degeneration may exist in both sedentary workers and those doing heavy manual jobs, and the stresses to which the trunk is subjected may be considerably influenced by 'good' or 'bad' working practice.

Painful strains and sprains of the low back, i.e. failures of the musculature and ligaments of the back under tension, are common. The herniation of the nucleus of an intervertebral disc ('slipped disc') under excessive compression, especially when coupled with tortional loading, is less common but more disabling. Heavy exertion may also precipitate herniation of abdomino-pelvic contents ('rupture') through the walls of the cavity, often but not necessarily in the inguinal region (groin), which constitutes a failure of the body wall under pressure. Paradoxically, failure of the wall is promoted by the intra-abdominal pressures which reduce the chances of other types of failure. All three types of loading arise during heavy manual work, especially during lifting. Bartelink (1957), Morris, Lucas and Bresler (1961), Eie (1966), Davis (1967) and Grieve (1977) provide a framework for further reading.

Muscles of the back

The muscles of the back are required during lifting, pulling towards the body and pushing backwards with the shoulders, in order to resist flexion of the trunk as the load is applied and/or to extend the trunk as the load is moved. The erector spinae muscles used for this purpose are anatomically complex, and serve to exert tension between bony regions on the dorsal aspects of sacrum, pelvis, vertebrae, ribs and skull. The longest and most superficial strands span many segments while the deepest parts are short and serve to modify and control posture between adjacent segments. Tensile failures of fascicles of these muscles constitute some of the strains and sprains associated with heavy manual work. Referring to the transverse section through the lower back in Fig. 3.47, the fully active extensor musculature, acting symmetrically in an adult, has a centroid of tension in the mid-line approximately 50 mm behind the centroids of the vertebral bodies. When a load is applied at shoulder level, perpendicular to the trunk, the muscle loading in the region of the lower back may be as much as 8 times greater than the external load and the vertebral column is compressed by a similar force.

When a heavy load is lifted vigorously, the back actually flexes in the early stages as the forces at the feet pass through their peak (Grieve, 1974). A condition of powerful eccentric contraction then exists in the back akin, for example, to that which is the precursor of muscle soreness in the front of the thigh when deep squatting or downhill running. While the extensor muscles are being stretched, the active fascicles of them will be exposed to their most severe stresses, i.e. tensions greater than those exerted under isometric or concentric conditions. The strength of the back has usually been measured in an erect posture with the pelvis restrained and horizontal forces measured at shoulder level. Maximum loads that men and women are prepared to lift are of similar magnitude to the measured strength (Poulsen and Jørgensen, 1971). Strength data may be found In Jorgensen (1970), Asmussen and Heebøll-Nielson (1958) and Troup and Chapman (1969). In common with the behaviour of many other muscle groups,

Fig. 3.47. Cross-section through the trunk in the mid-lumbar region. Erector spinae (e.r.s.), latissimus dorsi (l.d.), psoas (ps.), quadratus lumborum (q.l.), external oblique (e.o.), internal oblique (i.o.), transversus abdominus (t.a.) and rectus abdominus (r.a.) muscles are shown. The extent of the abdominal cavity in which intra-abdominal pressures (IAP) are created is indicated.

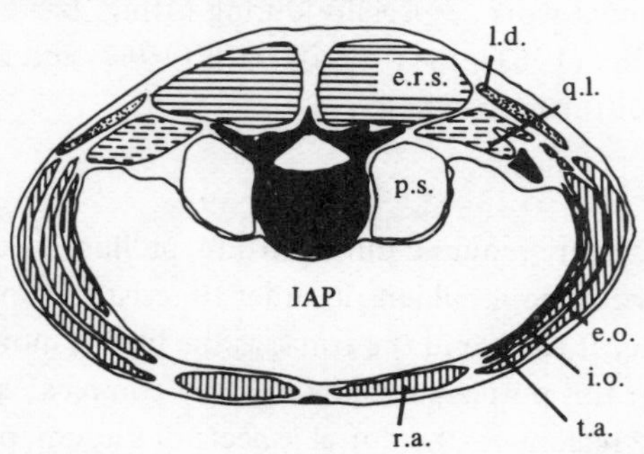

an increase in length of the extensor muscles increases the isometric strength, expressed as a torque about a mid-lumbar vertebral body (Grieve and Pheasant, 1976). The erect posture of the trunk is therefore weaker for external effort than for a flexed posture, as illustrated in Fig. 3.48. If the posture of the lumbar back is defined by the angle between pointers mounted at the sacral and 12th thoracic levels, the maximum torque about the region of the 4th lumbar vertebra increases approximately 1% for each degree of flexion (Pheasant, 1977).

The vertebral column

The vertebral column appears to be structured in rough proportion to the weights of body parts above the level concerned (Fig. 3.49), so that the unencumbered erect posture has a considerable safety factor. Volume 6(1) of

Fig. 3.48. Maximum isometric extensor torques due to pulls exerted at the hands and to the weights of upper limbs, head and upper trunk. Torques were calculated about the mid-lumbar vertebral column and expressed as percentages of the products of subjects' heights and weights. Lumbar angle is the angle between two pointers mounted normal to the back at 12th thoracic and sacral levels. The lines are the regression line of torque on angle ± 1 r.m.s. residual.

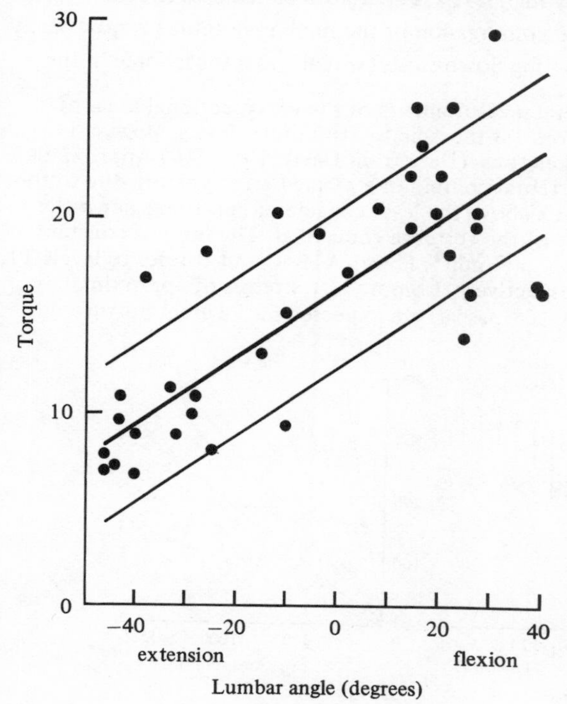

Orthopaedic Clinics of North America (1975), would serve as an introduction to the specialised area of spinal testing. Limiting strengths of vertebral bodies and intervertebral discs show large individual variations; for present purposes it will be assumed that failures may occur in adult vertebral columns under compressive stress in excess of 3 newtons/mm^2 (N/mm^2), and possibly as little as 1.5 N/mm^2 The safety factor in an erect posture may be halved by running, and slipping on a kerbstone can produce a jolt which leaves no safety margin. The safety of the back in lifting depends on both the load and the techniques adopted.

Let us consider a worker attempting to lift an object whose weight equals the International Labour Organisation (ILO) recommended adult male maximum of 490 N (50 kgf, 110 lb weight). We will assume, *for the moment*, that extensor torques in the trunk arise solely from tension in the back muscles acting at a distance from the accompanying compression of the vertebral column. If a stoop-lift is used, with the legs straight, an initial posture similar to that in Fig. 3.50(*a*) will be adopted. The turning moment about the lower lumbar region, due to the head, pendant arms and the trunk above the section considered, will cause vertebral compressive stresses well inside the danger zone. If the subject supports the load off the ground (let alone a lift with acceleration), the load on the back will rise to dangerously high levels. Fig. 3.50(*b*) illustrates the same man using a crouch-lift. The axial compression of the lumbar vertebrae is now partly due to the weight of parts acting downwards (which was a shear force in the

Fig. 3.49.(*a*) Skeletal measurements of the cross-sectional areas of vertebral bodies from 1st thoracic to 5th lumbar levels. Mean ± 1 S.D. for 14 adult skeletons. (Data from Davis, 1957.) (*b*) Approximate loadings on the vertebral column in a relaxed erect posture due to the weight of body parts above the level considered, plotted against the cross-sectional area of the vertebra concerned. The lines of constant stress are numbered in N/mm^2. Points A, B, C and D refer to levels T1, T4, T12 and L5 respectively. Compressive stresses of approximately 0.3 N/mm^2 (0.3×10^6 pascal) are expected in the erect posture.

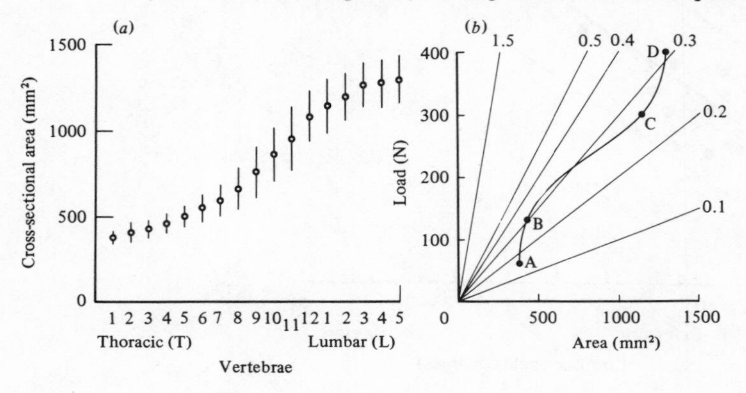

stoop-lift), and partly due to muscle action required to produce an extensor torque. Before he supports the load, the vertebral stress is a modest 0.95 N/mm^2 and even when the load is supported, the stresses are very much less than those in a stoop-lift. The comparison of stresses in these various conditions is made in Fig. 3.51.

High truncal stresses are by no means confined to lifting. During experiments on ten male students, a photograph was taken when each exerted a maximal static effort in one of eight directions upon a bar handle (1.0 or 0.25 m height), with toe placements (feet together) either under the bar or 0.5 m to the rear. The stick figures in Fig. 3.52 represent the mean postures adopted by the subjects in each manoeuvre, computed from the film analysis of the body markers. The measured manual force and the computed distribution of body

Fig. 3.50.(*a*) Man depicted in the initial posture of a stoop-lift. Horizontal distances of the centroids of the head, upper limbs, upper trunk and load with respect to the lumbosacral joint are shown. The weights (in N) of the parts and the turning moments (in N m) about the lumbosacral joint are indicated. The total turning moments before and after supporting the load are 151 and 366 N m respectively. (*b*) Man depicted in the initial posture of a crouch-lift. The quantities indicated correspond to those shown in (*a*). The total turning moments about the lumbosacral joint before and after supporting the load are 64 and 211 N m respectively.

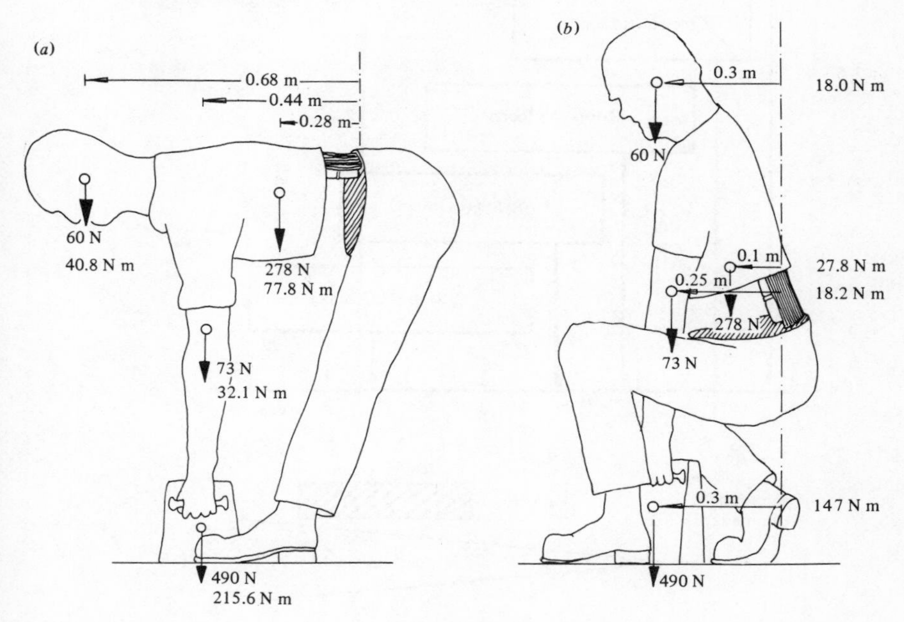

weight were used to calculate the net torque about the lumbosacral joint and the net force through the joint in the line between it and the 7th cervical vertebra. The total compressive stress on the lumbosacral disc was computed assuming the cross-section to be 1300 mm^2 and assuming, as above, that the torque arose solely from muscle tension acting 50 mm from the centroid of the disc. Histograms of the stresses obtained for high and low handles are shown. The calculated stresses frequently exceeded or were comparable with reported limiting strengths of the vertebral column. The result is not very sensitive to the numerical uncertainties inherent in the assumption. It is also clear, since Fig. 3.52 refers to maximal voluntary exertion in every case, that truncal stress should not be considered a limiting factor in every exertion.

The discussion of vertebral stress in both the hypothetical crouch – and stoop-lifts, and in the experiments with maximal exertion, assumed that extensor torque is attributable to the action of back muscles alone. However, the

Fig. 3.51. Approximate compressive stresses on the vertebral column in the lumbosacral region, based upon the models presented in Figs. 3.49(*b*) and 3.50 compared with limiting compressive strengths of cadaveric material.

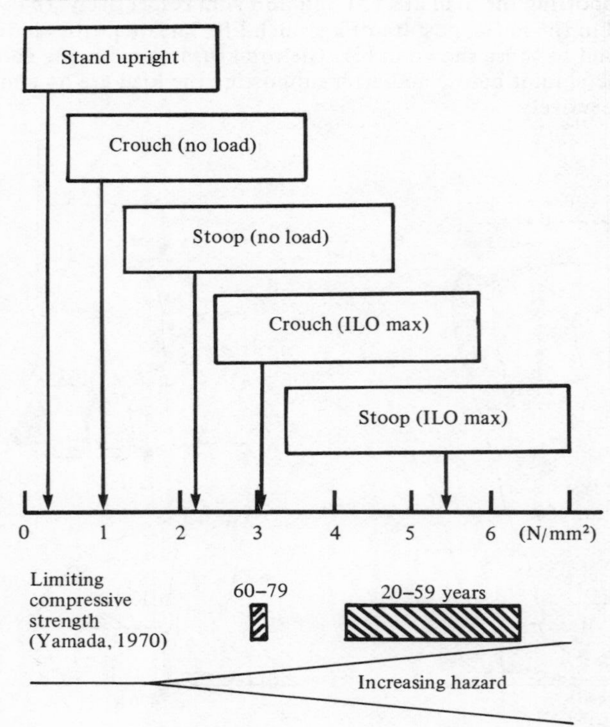

interruption of breathing and talking and the involuntary grunts and farts which may accompany a heavy exertion are evidence of pressure within the trunk. Intratruncal pressure alleviates the compressive stress upon the vertebral column and it is therefore relevant to consider its importance.

Intra-abdominal pressure and the anterior abdominal wall

The hydrostatic pressure within the abdomino-pelvic cavity rises during manual exertion due to the combined muscular actions of the diaphragm, pelvic floor and anterior abdominal wall upon the fluid contents. Pressures as high as 300 mm Hg (40 kilo pascals) have been reported in competitive weight-lifting; industrial work by lesser mortals involves more modest pressures which are regarded as undesirable if they exceed 100 mm Hg. Davis and Stubbs (1977) report that tasks, particularly in the construction and mining industries, in which pressures above 100 mm Hg are frequently encountered are those associated with a high incidence of back injury. The pressures, which may be readily monitored

Fig. 3.52. Histograms of compressive stresses (N/mm^2) on the lumbosacral disc during maximum exertions in eight directions. Based on 10 males, maximal exertion in symmetrical postures (handle heights 0.25 and 1.0 m, toe placements 0 and 0.5 m behind handle). The stick figures are computed means of postures observed during the exertions. The wedges, widening to the right, suggest increasing hazard as in Fig. 3.51. Histograms above and below the base-lines refer to handle heights of 1.0 and 0.25 m respectively.

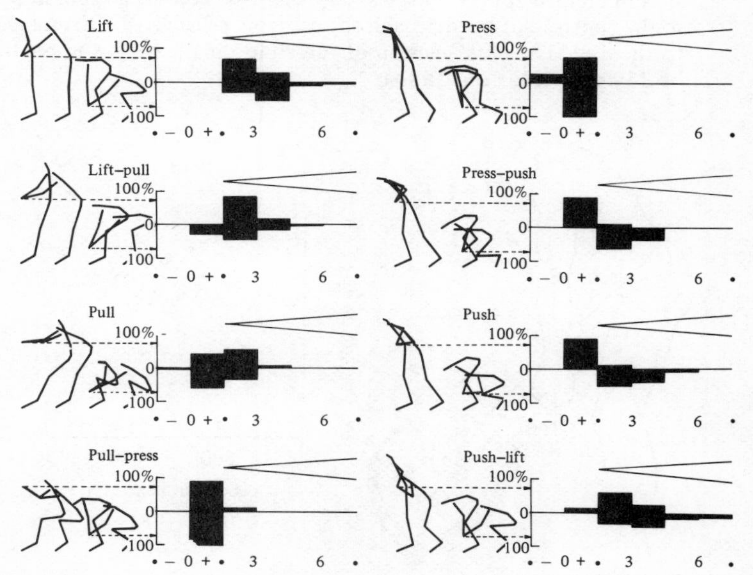

by a swallowed pressure-sensitive radio pill under field conditions, are considered by Davis to indicate the intensity of truncal stress. The most recent development is the provision of data on vertical and horizontal forces that can be exerted with one or both hands in various postures (erect trunk while sitting, standing or squatting) without causing pressures in excess of 90 mm Hg (Davis and Stubbs, 1977 *a, b*; MHRU 1980). The literature on intra-abdominal pressure (IAP) is reviewed by Grieve (1977); some key papers are by Bartelink (1957), Eie and Wehn (1962), Davis and Troup (1964) and Stubbs (1973).

If an IAP of 90 mm Hg acts in the man illustrated in Fig. 3.49(*b*) it is equivalent to an extensor force of 377 N acting approximately 0.1 m anterior to the vertebral bodies. The resulting extensor turning moment is 38 N m, i.e. one that is capable of appreciably relieving the compressive stress in the vertebral column and the tensile stress in the erector spinae muscles (Fig. 3.53). The anterior abdominal wall is curved in both horizontal and vertical planes. If it were detached and laid flat (see Fig. 3.54) it would be seen to contain muscle fibres and aponeuroses (flat sheets of tendon) which can transmit tension in

Fig. 3.53. Compressive stress on the lumbosacral disc of a man in the posture indicated, for various loads held in the hands. The intra-abdominal pressure (IAP) relieves part of the stress on the disc. Peak IAPs under static loading, when above zero, were calculated from IAP (Pa) = 21.4 × (load in hands (N)) − 3088, which was derived from a summary of experimental observations (see Grieve, 1977). The centroid of tension in erector spinae muscles (T_e) was assumed to lie 50 mm posterior to the centroid of the disc, and the centroid of the IAP 100 mm anterior to the disc. The cross-section of the abdominal cavity was assumed to be 314 cm^2 (Morton, 1944).

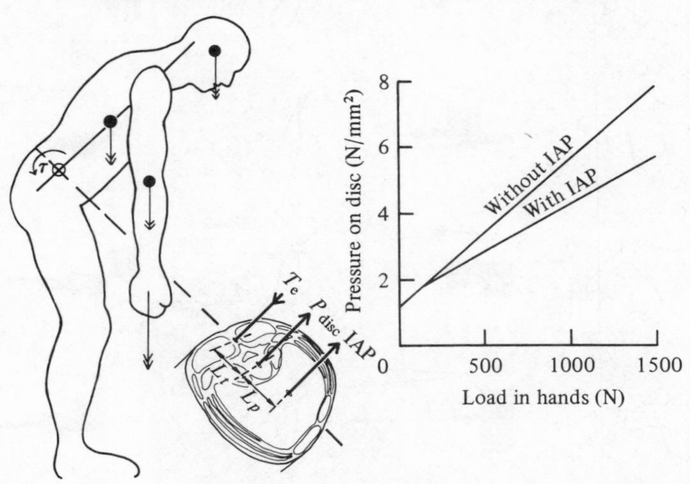

any direction. The vertical components of tension in the wall in front of the vertebral column tend to flex the trunk as well as pressurise the cavity. The horizontal components can only pressurise the cavity and therefore have a pure extensor effect. Depending on the circumstances, the anterior wall appears to be

Fig. 3.54. Half of the anterior abdominal wall, as it would appear if rolled flat, showing the lines of tension that can be created by the action of the various muscles in the wall. Every part of the wall is curved both vertically and horizontally. Both vertical and horizontal tensions (T_v, T_h : see equation) contribute to an equilibrium with the IAP. Vertical tensions, however, may contribute a flexor moment about the vertebral column. The pattern of tension development and pressure created within may cause a net flexor, or extensor, or neutral effect about the vertebral column.

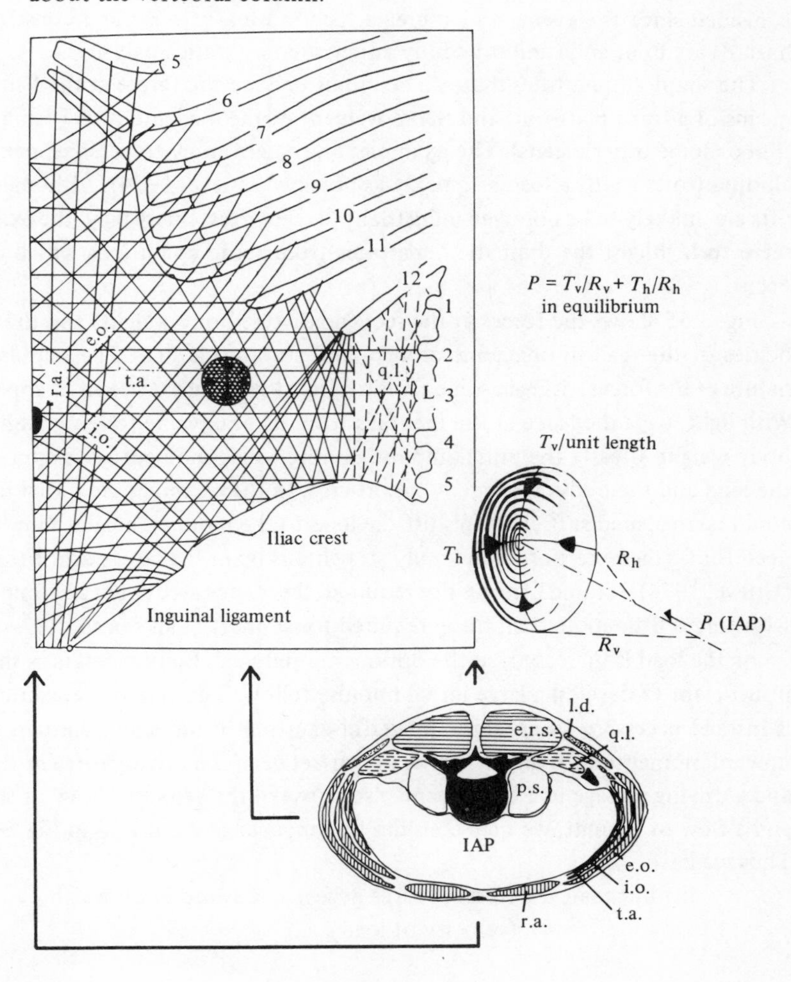

capable of pressurising the cavity and thus relieving the vertebral compression while the trunk is engaged in flexion or extension or even 'neutral' forceful activity (as in straining while defecating). Raised IAP is observed in pushing, pulling and lifting manoeuvres, and the oblique abdominal and rectus abdominis muscles are active in a variety of manoeuvres, some of which require truncal extension and other truncal flexion.

Dynamics of lifting

The dynamic analysis of lifting has received little attention, possibly because of the technical problems of performing a comprehensive analysis of even one lift in the laboratory, and the daunting variety of lifts which must be studied to reach general conclusions (Grieve, 1977). Research into the dynamics is needed since the events which create high internal stresses and constitute hazards are brief and cannot be fully anticipated by static analysis.

The simplest quantities that can be monitored are the forces at the feet (by means of a force platform) and the movement of the load and back (using cine film or some other means). The examples given here apply to persons making all-out efforts to lift a load as quickly as possible (Grieve, 1974). Although such lifts are unlikely to be observed industrially, except in an emergency, the examples serve to highlight the dramatic departures from static conditions which can occur.

Fig. 3.55 shows the forces at the feet (relative to body weight) and the velocities of the load during some crouch- and stoop-style lifts. The impulsive nature of the forces, rising in some cases to almost twice body weight, is apparent. With light loads the force at the feet falls from its peak value to levels well below body weight. Clearly the initial impulse, during which momentum is applied to the load and the body, is of great importance. In the extreme, the initial impulse could be imagined sufficient to shift the load to the required height with little need for further exertion as the body straightens up in the later part of the lift (Grieve, 1974). If rapid lifting is not required, the manoeuvre could be completed with forces little more than those required for support at all stages.

As the load is increased, so the options are reduced. Body strength is then insufficient to develop a large initial impulse followed by relative relaxation. It is instead necessary to continue powerful exertion in order to maintain the upward momentum. If an analogy is drawn between the driving force at the feet and a driving voltage in a circuit, and also between the velocity 'flow' of the load and a flow of current, we may consider the impedance the lift, as in Fig. 3.56. Thus we have

lift impedance = (driving force at feet relative to body weight) / (velocity of load)

The vigorous lifting of a light load is accompanied by an initial high, positive lift impedance, which falls to zero and becomes strongly negative as the lift proceeds. This is characteristic of a ballistic style of lifting. In contrast, the lift impedance remains positive when a leavy load is lifted, characteristic of lifts in which a ballistic style is not or cannot be adopted. With a very heavy load, e.g. 40–50-kg mass, the lifting force must be exerted continuously and, with little momentum available to carry through awkward postural configurations, the lift becomes limited by the weakest posture encountered.

The rate at which the back straightens (extends) during a vigorous lift, and the forces at the feet (which must reflect but not equate with the stresses on the back musculature), are strongly related. Fig. 3.57 shows that the back even flexes if the forces at the feet are sufficiently great, and straightening is delayed until the forces have passed their peak. Since lifting cannot be achieved without straightening, it seems likely that lifts would become impossible for the subjects featured in Fig. 3.57 if loads greater than 70% of body weight were

Fig. 3.55. Vertical forces at the hands (F_H) and feet (F_F), velocities of the centre of gravity of the body (V_G) and load in the hands (V_L), during crouch- and stoop-lifts with maximal effort. The hand and foot forces are expressed relative to the weights of the load and body respectively. Note the impulsive nature of the forces, and the contrast in patterns of foot forces when loads of 4 kg and 29 kg were lifted.

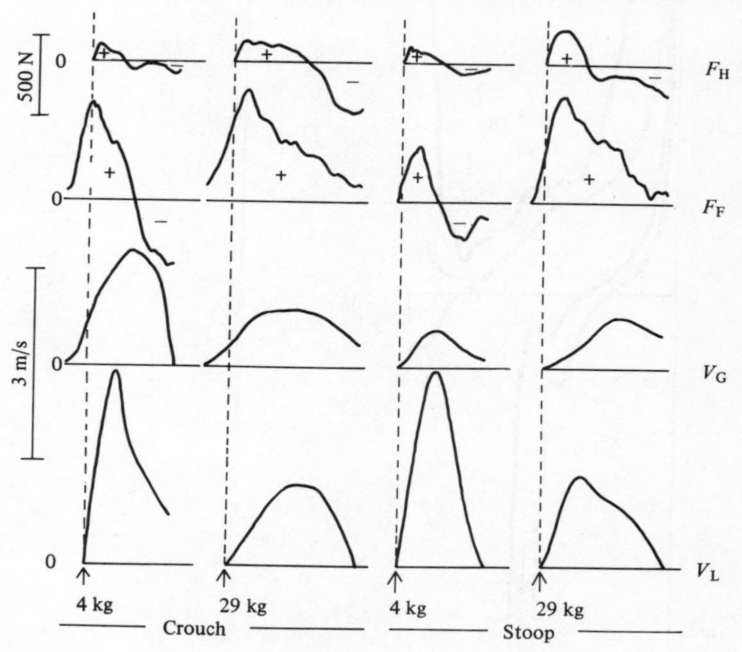

attempted. It is suggested that the back may have the capacity to extend under more severe loads, but does not do so because straightening would interfere with the maintenance of intra-abdominal pressure, which is protective to the back muscles and the vertebral column. The contents of the abdomino-pelvic cavity are, after all, mainly fluid and incompressible, apart from any contained gas. Relatively small changes of truncal posture, i.e. extension, during the few critical tenths of a second when peak forces are developed might cause depressurisation and the transfer of loads to the vertebral column and back muscles alone.

Industrial lifting

The exponents of the kinetics method of lifting emphasise six main features in order to avoid injury, i.e. good grip, chin in, back straight, arms close to the body, the use of body weight, and foot positions in which one foot is alongside the object and pointing in the direction of movement while the other

Fig. 3.56. Envelopes for two subjects in both crouch- and stoop-lifts showing the changes of lift impedance (see text) while lifting 4 kg and 29 kg loads, with maximal effort. The lighter load can be tackled ballistically whereas the heavier one cannot.

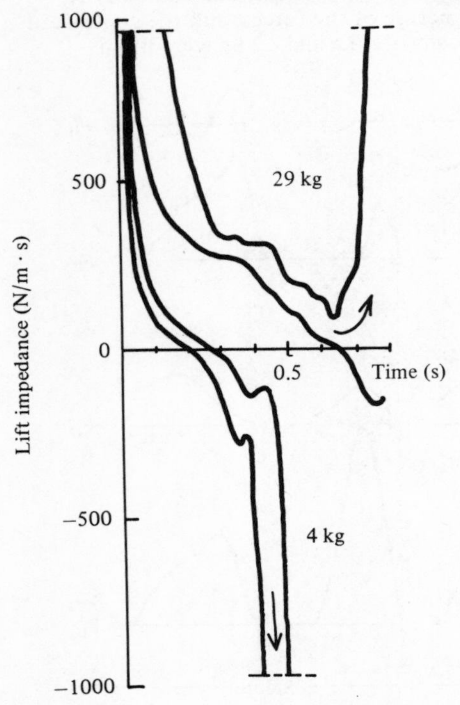

is behind the object to be lifted (Himbury, 1967). A recent compilation of good advice has appeared as an Instructors Manual from the Back Pain Association (1978).

The mechanical argument for lessening back stress by keeping the back upright and holding the load as close to the body as possible is correct, simple and appealing. The apparent advantages of 'bent-knees' over 'stoop' lifting are so dramatic that one might expect everyone to have a personal preference for the former. It appears to be normal for lifts which commence with bent knees to convert to stoop-lifts as the lift progresses, and the instinctive manner of reaching for an object from the floor is to stoop, thereby placing the back at risk. It should be questioned whether 'correct' techniques can be considered normal. The benefits of some industrial lifting programmes may derive from a Hawthorne effect, or by making the worker so conscious of the task that there is less chance of the habits of a lifetime asserting themselves. Unobtrusive photographic studies do not appear to have been made of adults and children engaged in lifting. If 'correct' and instinctive lifts differ, the emergency (and possibly the paced) situation is likely to evoke the latter, with their attendant hazards, in spite of exhortations to lift 'correctly'. If the techniques are similar the implication is that industrial demands are too high.

Three desiderata of a lift are the minimisation of stress on body parts, a stable base of support and the possibility of completion once started. These may prove

Fig. 3.57. Relationships between the vertical force at the feet and the rate of extension of the back (angular velocity between 1st thoracic and sacral vertebrae), during crouch-lifts (open squares) and stoop-lifts (filled circles) with maximal effort. Note that the back flexes rather than extends when forces at the feet exceed 1.7 times body weight.

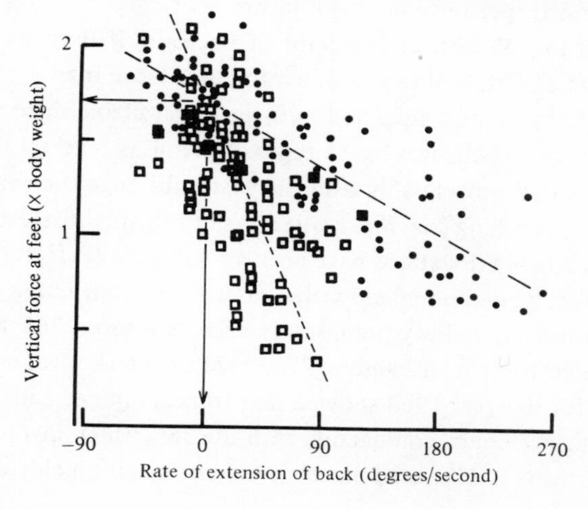

conflicting requirements. Minimising truncal stress may maximise stress on the knees. Knee ligaments may be injured in deep crouches (Klein, 1971); static upward pulls with an upright trunk and with the hands between the thighs are most powerful at 56°–68° of knee flexion but weaken rapidly with further flexion (Carpenter, 1938). Symmetry of body action, which is desirable under heavy loading, especially for the lumbar vertebral column, is more readily achieved in stooped postures than with bent knees. Restrictive clothing such as skirts, tight trousers, and protective aprons, the extent to which the knee obtrudes into the 'load-space', and a possibly limited mobility of the lower limbs in the older worker, are all factors which may persuade the individual not to crouch. A *symmetrical* bent-knee lift is unstable; the body is supported on the metatarsals with the ankles in full dorsiflexion. If a disturbing force or misjudgement of the load occurs, an attempt will be made to increase the foot-base. Rocking back upon both heels causes a backward topple, while moving one foot to the rear will cause internal asymmetry of loading. A third possibility is to change rapidly to a stoop, which increases the risks to the back but leaves the lower limbs better placed for further corrective movements. One foot should be placed initially behind the other to provide an adequate base. A knee-down posture is very stable (see Fig. 3.58) for lifts of restricted height. In general, all 'correct' bent-knee lifts (except those of competitive weight-lifters) are asymmetric. The completion of a lift is essential. A very slow lift (which is unavoidable with a heavy load) is limited by the weakest posture encountered during the manoeuvre. With modest loads, an alternative approach is possible by imparting momentum to the load which will carry it past 'critical' postures.

Epidemiology of injuries due to handling

Accidents in Britain to persons employed in premises subject to the Factories Act, are reported to HM District Inspector of Factories if they are either fatal or, through disablement, lead to loss of earnings for more than 3 days. The reporting procedure is standardised and statistics are available since 1924. Because of social change and changes in the types of premises covered, the total accident rate has fluctuated considerably over the years, although the percentage attributed to manual handling (see Fig. 3.59) remains remarkably steady. In recent years separate statistical tabulations have been available for (i) factory processes; (ii) docks, wharves, quays and inland warehouses; (iii) construction processes; or (iv) offices, shops and railway premises. In each case about 26% of the accidents are attributable to manual handling. A statistical breakdown according to body regions for the year 1964 showed that truncal injuries constituted 35% of the handling accidents, comparable with the 38% which involved the upper limb and hand. Brown (1971) noted the relatively unchanging incidence of reportable back injuries since the Second World War, commenting that the incidence had not lessened in spite of intensive efforts to introduce 'correct' methods of handling.

Fig. 3.60 summarises some of the recommendations that have been made concerning maximum weights to be lifted (Konz, Dey and Bennett, 1973). The differences in recommendations are so great that one suspects that political and social pressures and expediency sometimes outweigh biomechanical or medical argument. Codes of practice are expected shortly in the United Kingdom. It will be interesting to see how codes of practice reflect existing practices, sex differences in lifting capacity and the principles embodied in the Health and Safety at Work Act 1974, the Sex Discrimination Act 1975, and the Factories Act 1961.

Snook, Campanelli and Hart (1978) studied 191 low-back injuries, one from each organisation insured with Liberty Mutual Insurance in the USA. They concluded that firms which practised common selection techniques (based on medical histories, examinations and X-rays), or adopted training on safe lifting procedures, experienced as many injuries as the firms who practised neither. The importance of designing the job to fit the worker was emphasised but the evi-

Fig. 3.58. Bases of support (hatched) available while crouching and stooping. S, stable; U, unstable; VS, very stable.

Crouching

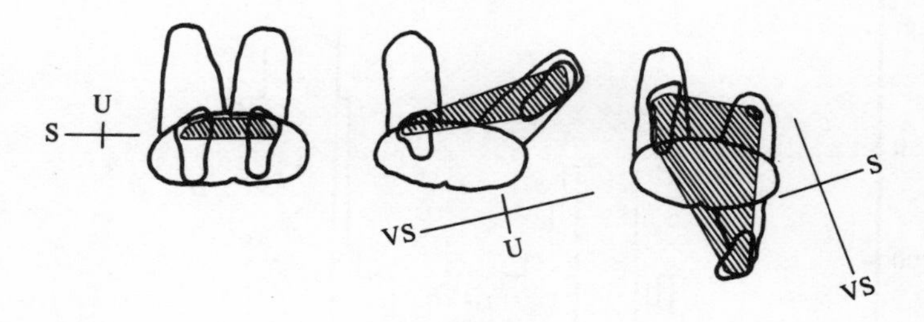

Stooping

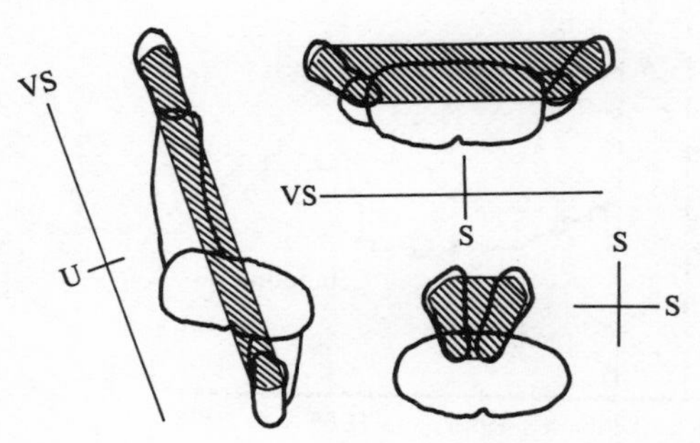

dence suggests that one-third of the back injuries associated with manual handling will happen anyway, regardless of the job. In current British terms, Snook's findings suggest that a number in the order of 30 000 reportable trunk injuries per annum are linked to task demands.

The static analysis of forceful exertion (PSD analysis)

This section deals with steady exertions which are considered to be in static equilibrium. The technique (Grieve, 1979*a, b*) is mainly graphical, and uses the Postural Stability Diagram (PSD). PSD analysis is so named because it may be used to discuss the stability of a working posture (in the sense of satisfactory, safe and stable) by considering the interaction of all factors affecting

Fig. 3.59. Annual incidence of all reportable industrial injuries since 1924, those associated with manual handling, and the latter as a percentage of the total.

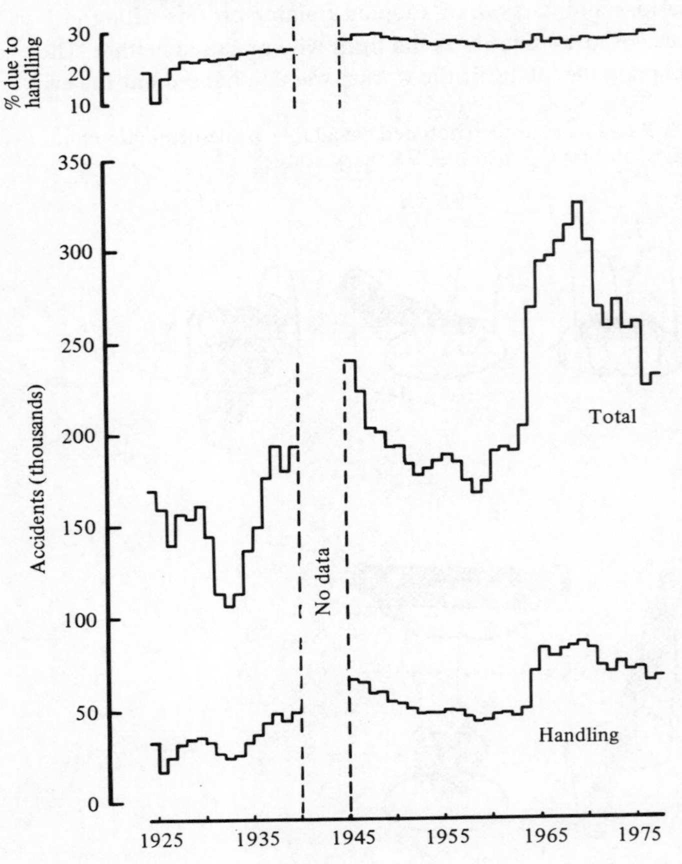

forceful exertion. Some novel forms of anthropometric data are required for its use.

The PSD represents forces between the person and the environment in a chosen plane of analysis; the vertical fore-and-aft plane is considered here. Forces are attributed to either 'hands' or 'feet'. All contacts of the body with the environment at which the task is being effected are grouped into the 'hands' category, not necessarily implying the hands anatomically. Similarly, all other interfaces are grouped together as having a supporting role and are called 'feet' (Fig. 3.61).

The square charts in Fig. 3.62 constitute a PSD when they are superimposed. The axes which have origin at the centre, labelled LIFT, PUSH, PRESS and PULL, refer to forces at the hands and are scaled from zero (centre) to body weight at the edge. PRESS rarely exceeds body weight because the feet would then normally leave the ground. Lift forces greater than body weight are sometimes possible but are rarely required in everyday life.

A static force at the 'hands' in any direction must be equal and opposite to the force at the 'feet' (plus body weight if the vertical is considered). Scales around the edge of the diagram represent forces of DOWN (vertically) and

Fig. 3.60. Recommended maximum lifts for various industries (referred to by Konz *et al.*, 1973). Labels A, B, C and D refer to the approximate weights of a full 2 gallon watering can, a large 'party' can of beer and children of ages 1½ and 3 years respectively. The range of laden weights of shopping baskets at the checkout of a food supermarket (Grieve, 1972) is indicated at E.

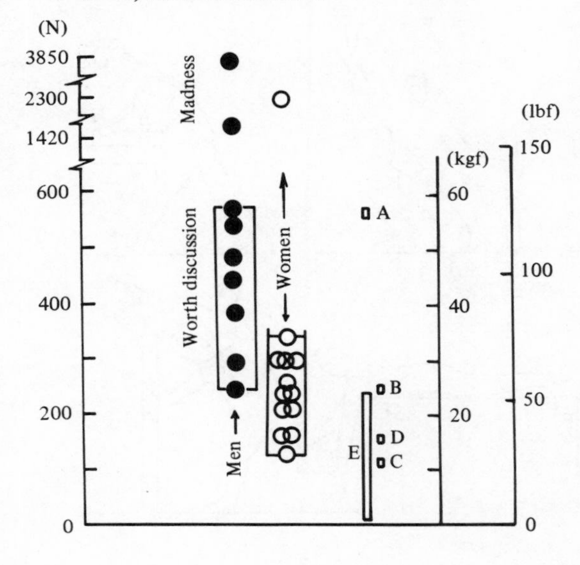

FORWARD or BACK (horizontally) at the 'feet'. The sketch in Fig. 3.63 depicts a man exerting component forces of PUSH and LIFT at the hands, whose resultant is an angle λ to the horizontal. On the accompanying PSD, *QP* represents the manual force in magnitude and direction, *RP* from the centre-base of the PSD represents the force vector at the feet and, completing the triangle of forces, *QR* represents body weight.

Anything that we may know about the limited ability of the person to exert force, the demand for force imposed by the task, or any other constraint upon forces between the person and the environment, may be stated graphically on the PSD providing that it can be quantified. The statements that can be made relate either to the person's ability to exert force (personal statements) or to the environmental constraints and task demands (environmental statements).

The equation of static exertion

Empirical observations of whole-body strength under various conditions will be presented in the next section. To the extent that laws of statics apply to any steady exertion, and indicate a relationship between the exertion and the placements of the foot, hand and weight centroids, the influences of posture on exertion has a theoretical foundation, which will be stated below. However, theory only accounts for observed strength under special limiting conditions;

Fig. 3.61. The regions of contact of the person with the environment may be allocated to 'hands' (H) or 'feet' (F) for purposes of PSD analysis, depending on whether they are regions where the task is effected or where the body is supported, respectively.

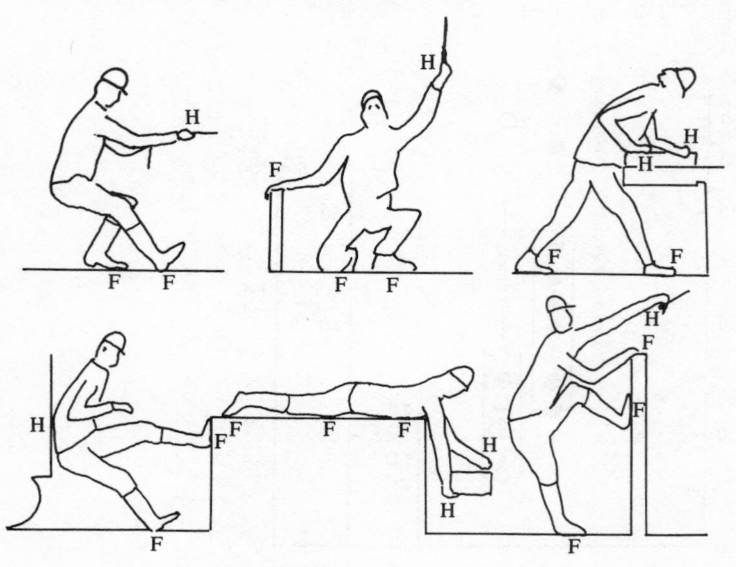

Fig. 3.62. The postural stability diagram is obtained by superimposing the diagrams representing horizontal and vertical forces at the 'hands' (top) and 'feet' (bottom). Forces are scaled as fractions of body weight, W. LIFT, PUSH, UP and FORW represent positive forces, and TWIST and TURN represent positive torques exerted *by* the person *upon* the environment.

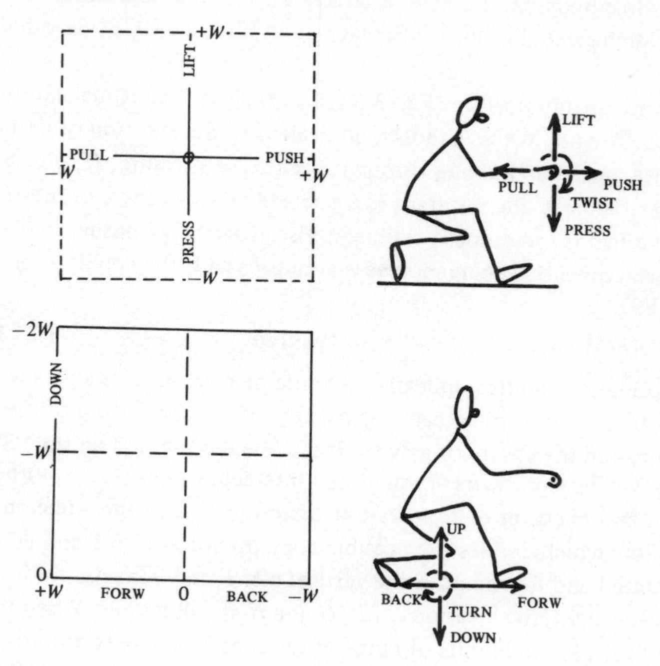

Fig. 3.63. PSD showing how the static exertion of force vector QP at the hands is accompanied by a vector RP at the feet. Triangulation of forces is completed by the body weight vector, QR.

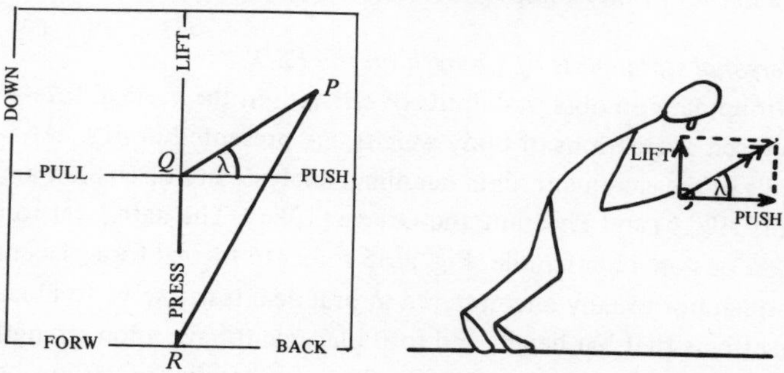

physiological knowledge about the articular chain must be combined with the principles of statics to provide a general prediction of strength. This synthetic approach is not sufficiently proven that we may dispense with empirical observations of strengths in many postures. Much progress has, however, been made with models of whole-body exertion and is largely attributable to Chaffin and his colleagues in Michigan (Chaffin, 1969; Martin and Chaffin, 1972; Garg and Chaffin, 1975).

Consider a person in any posture (Fig. 3.64) and analyse the manual exertion in a vertical plane. The use of the 'hands' is equivalent to the exertion of a force (components LIFT and PUSH) acting through a centre of pressure, H, plus a torque, TWIST. At the 'feet' the resultant force acts through a centre of pressure, F. A torque at the feet is sometimes possible but is sufficiently unusual to be omitted from the argument. Taking moments about F yields the equation of static exertion (ESE):

$$\text{LIFT} = (h/b) \cdot \text{PUSH} - a \cdot W/b + \text{TWIST}/b \qquad (3.13)$$

The ESE is important and its implications should be examined. It states what *combinations* of forces and torques can co-exist for a given disposition of the centres of pressure and the centre of gravity. It may be represented on the PSD as a straight line, having slope h/b (or tan δ) and intercept $(\text{TWIST} - a \cdot W)/b$. Postural changes and exertion of torques can therefore change the slope and intercept of the line which defines the possible combinations of LIFT and PUSH. With hands separated and feet apart in the vertical plane, and with room for the person to shift his weight, the locations of F, G and H are all mobile. When the extremes of the foot-base and limits of range of the centre of gravity anteroposteriorly are measured, the observed limits of strength, the limiting ESEs, partly predict the observed limits of strength when presented as personal statements on a PSD. Although measurements of centroids of gravity and pressure can establish the slope and intercept of the ESE that is in use at any instant, the limiting position of the working point, i.e. of the manual force vector, *along* the line is a matter of physiology, not statics.

Personal statements of strength on the PSD

Group data on observed limits of strength in the vertical fore-and-aft plane, expressed as fractions of body weight, are presented in Figs. 3.65–3.68. The methods of measurement, data handling, analysis and discussion are given in Grieve (1979*a*, *b*) and Pheasant and Grieve (1981). The data refer to two-handed exertions on a bar handle. Fig. 3.65 refers to formal foot placements which although not usually encountered in practical tasks, serve to illustrate the systematic effects that bar height and foot placement have upon strength. Free-style foot placements are featured in Fig. 3.66, for handles of various heights.

The effects of low ceilings and restrictive walls to the rear are featured in Figs. 3.67 and 3.68. Mean strengths are shown in the diagrams. A task-designer would obviously consider persons who were weaker than average for most purposes. Statistical statements are in preparation.

Fig. 3.64.(*a*) *F*, *G* and *H* represent the centres of pressure of the feet, the centre of gravity, and the centre of pressure at the hands, respectively. The ratio of distances a/b and of h/b (or tan δ) determine the combinations of forces that can be exerted horizontally and vertically in the absence of torque. (*b*) Taking moments about *F* leads to the equation of a line on the PSD giving all possible combinations of horizontal and vertical forces which satisfy conditions of equilibrium. The line is displaced vertically for a given value of TWIST.

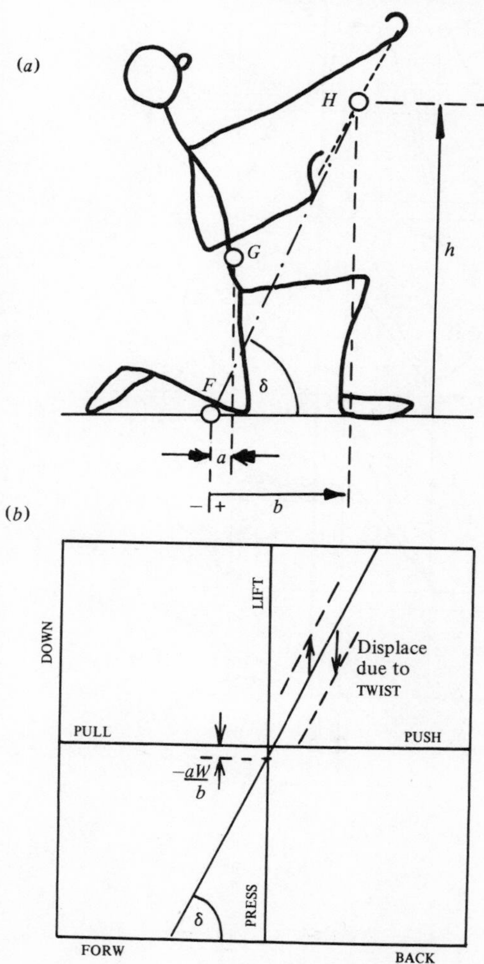

Fig. 3.65. Mean strengths as fractions of body weight (based on measurements of 10 adult males) for static, bi-manual exertions at three handle heights, with four combinations of formal foot placements, (L, leading; T, trailing) relative to the handle. The lines on the PSD represent average limiting equations of static exertion. Those on the left of each PSD arise from a combination of the *anterior* limit of the feet for support and the *posterior* limit of the body's centre of gravity; those on the right represent the opposite combination. Note that the limiting combinations are only employed over a narrow range of directions of exertion.

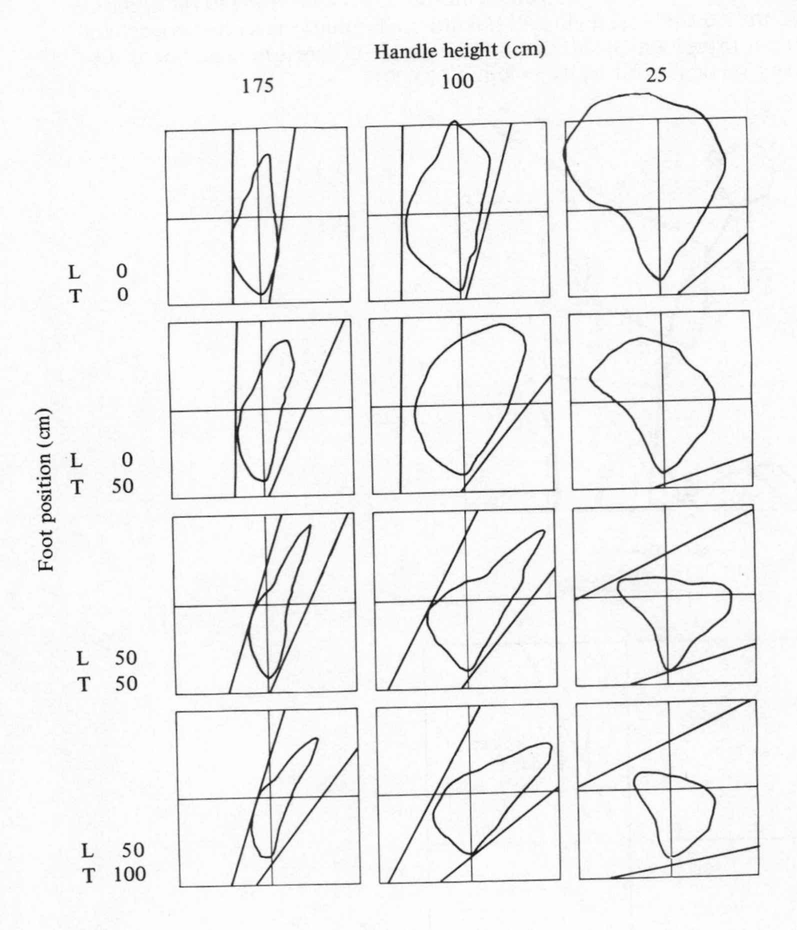

Preferential directions for exertion

Figs. 3.65–3.68 clearly demonstrate that the direction of exertion is an important determinant of manual strength. Consequently, providing that the task-demands permit, it may be advantageous to exert a force in a 'strong' direction simply to achieve a 'useful' component in some other direction which is larger than could be achieved by *accurate* exertion in that direction. Ayoub

Fig. 3.66. Mean strengths as fractions of body weight during static, bimanual exertion at five handle heights (B), with freestyle placement of the feet. Based on measurements of 10 adult males.

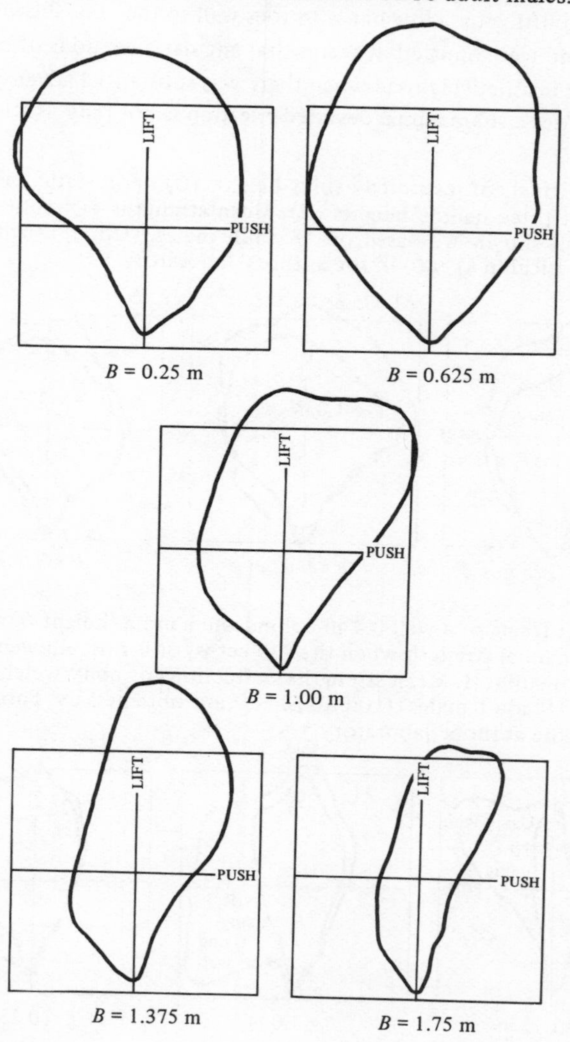

and McDaniel (1974) regarded a misaligned manual force as inefficient; they defined an efficiency of pushing as the cosine (as a percentage) of the angle between the resultant and the horizontal. This concept is task-oriented. From the operator's point of view, the Advantage of using a Component of Exertion (ACE) may be substantial. Fig. 3.69 shows the angular deviations of exertion from the vertical (or horizontal) that naive subjects employed when asked to maximally pull, push, lift or press, plotted against the deviations which analysis of PSD data (Fig. 3.65) suggests would yield the maximum values of ACE (MACE). Observation of a few extremely weak exertions, e.g. pushing on a high bar with toes underneath, and lifting on a low bar with toes well to the rear, showed very variable behaviour and were omitted; it seems that angular deviations of up to half a quadrant were instinctively used when there was substantial advantage to be gained. In such cases a submaximal deviated exertion is not only as strong as a

Fig. 3.67. Effects of restricted ceiling heights (C) upon static, bi-manual strength, at three handle heights (B). Mean strengths as fractions of body weight shown are based on 18 adult males. (Unpublished data obtained by Rubin (1980) in the authors' laboratory.)

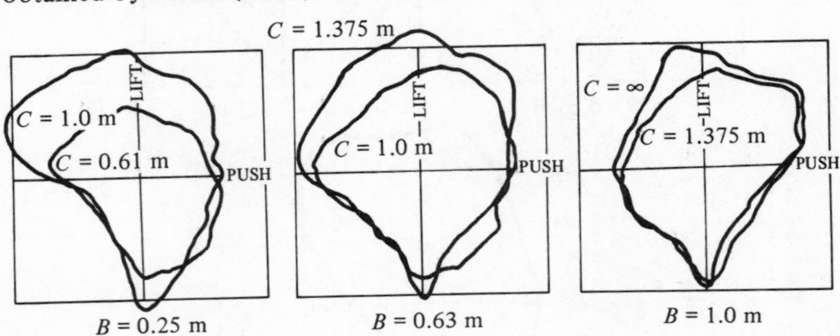

Fig. 3.68. Effects of a wall 0.4 m behind the handle (height B) upon static, bi-manual strength when the subject is, or is not, allowed to use the wall for support. Mean strengths as fractions of body weight are based on 18 adult males. (Unpublished data obtained by Thompson (1980) in the authors' laboratory.)

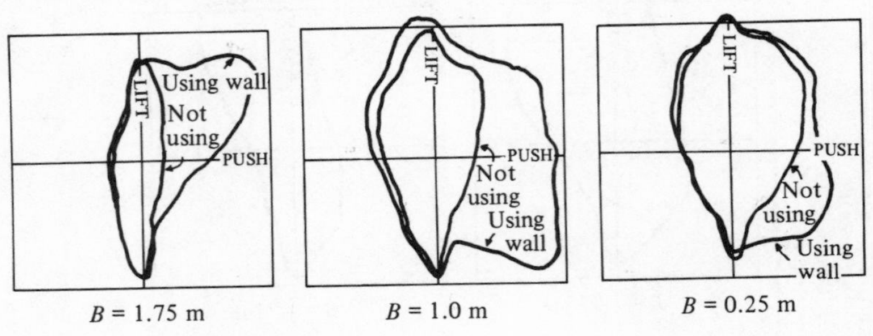

maximal directed effort, but it would decrease more slowly due to fatigue. Tasks which are tolerant of forces exerted in other than the useful direction, i.e. other than that in which minimal force will accomplish the task, should be designed in preference to those which demand a precise direction of exertion.

Friction with the ground

Most exertions depend upon sufficient friction between the feet and the ground. The coefficient of limiting static friction, μ, defines the maximum possible ratio of the tangential compared with the normal force at an interface if slip is to be avoided. On level ground, normal and tangential refer to DOWN and FORW (or BACK) forces on the PSD. On sloping ground, angle ϕ to the horizontal, both normal and tangential forces contain components of DOWN and FORW (or BACK). In the general case (see Fig. 3.70) and introducing a new coefficient, μ', which we will call the apparent coefficient of friction, the maximum

Fig. 3.69. Angular deviations from the vertical or horizontal of forces exerted by naive subjects when asked to push, lift, press and pull maximally, plotted against the deviations which will yield maximum components in the vertical or horizontal (as predicted from data shown in Fig. 3.65). The subjects (10 males) used the same foot and hand placements as in Fig. 3.65.

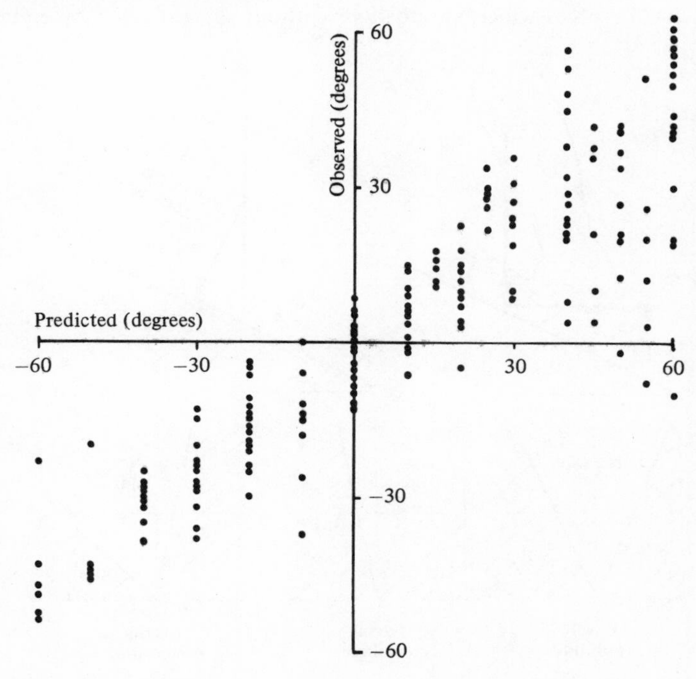

possible ratio of *horizontal* compared with *vertical* force is given by μ', where

$$\mu' = \frac{\mu - \tan\phi}{1 + \mu \tan\phi} \tag{3.14}$$

when the horizontal force at the feet faces down a slope, or,

$$\mu' = \frac{\mu + \tan\phi}{1 - \mu \tan\phi} \tag{3.15}$$

when the horizontal force at the feet faces up a slope.

Kroemer (1974) and Grandjean (1973) may be consulted for values of μ to be expected for various conditions and types of floor and shoe materials. The surprising feature is the small values of μ which should be assumed by the workplace designer if footwear, flooring and cleanliness are outside his control. Fig. 3.71 shows values μ, for slopes up to 20° and indicates 5th percentile values of the coefficients of limiting static friction for the most slippery shoe–floor combination that Kroemer (1974) reported.

Fig. 3.70. Foot on a slope, showing how the horizontal and vertical components of force contribute to the normal and tangential forces. The lines on the PSDs show the limiting combinations of DOWN and BACK forces which are possible without slip (as given by equation 3.14).

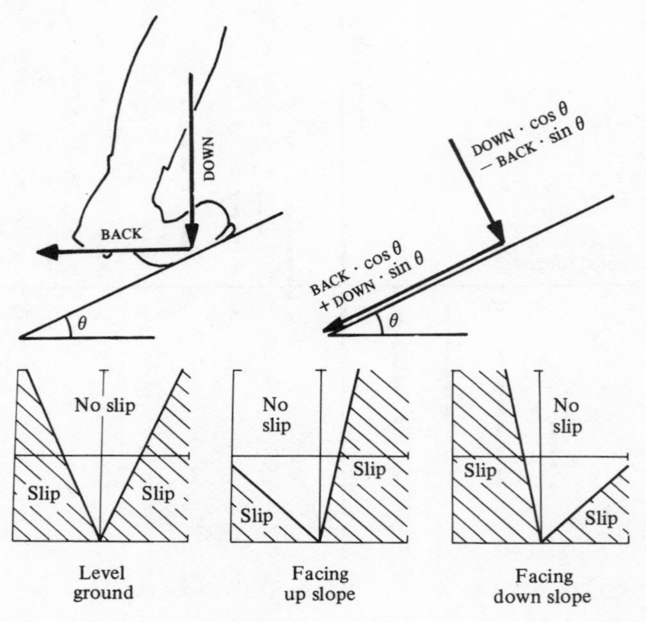

Combinations of personal and environmental statements

One hypothetical example will serve to illustrate the use of the PSD for analysis of a task; other examples are given in Grieve (1979*b*). The problem is to recommend a maximum permissible weight of trolley which a user could just prevent from running back on a slope without assistance. All relevant factors can be stated on the PSD and are summarised below.

(*a*) *Selection of a suitable personal statement about strength*

(i) *Handle height.* Assume: 0.95 m, which is a common specification.

(ii) *Restricted toe access due to trolley design.* Assume: Trolley A, which on level ground allows toes to be 0.28 m in front of the handle. When

Fig. 3.71. Chart showing values of the apparent coefficient of friction, μ', for inclines between $\pm$ 20° and various values of the coefficient of limiting static friction, μ. Fifth percentile values of μ for various shoe-floor combinations (Kroemer, 1974) are indicated.

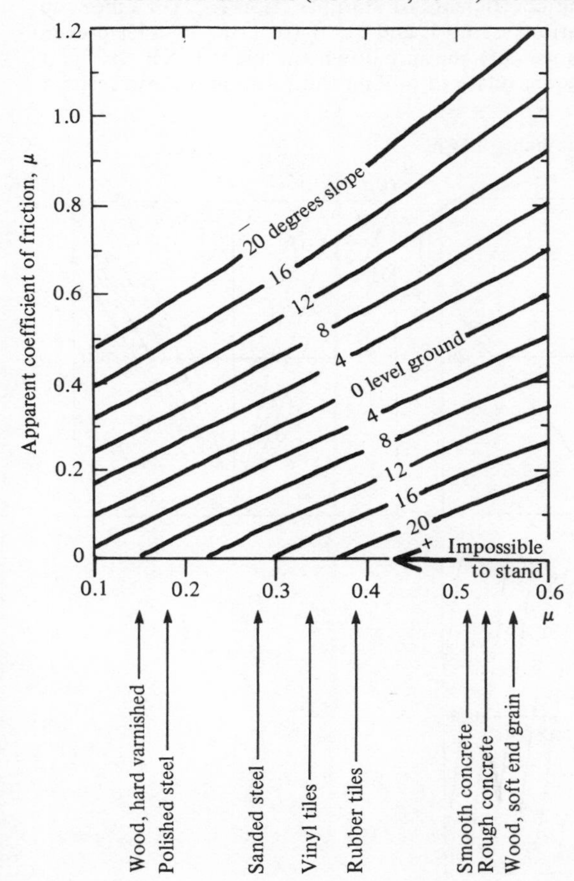

pulling on a gradient of one in 7 (a likely maximum) the toes will be restricted to 0.1 m horizontally in front of the handle (0.9 m/7 = 0.14 m). Trolley B, fitted with a tail-board, prevents toe placements in front of the handle on level ground and restricts them to 0.14 m behind the handle when pulling on a slope. Pushing the trolley requires feet support well behind the handle and the trolley and feet do not interact.

(iii) *Percentile of population considered.* Assume: Design for 50th percentile in this example, although 5th percentile would be more likely choice in practice.

Fig. 3.72. PSDs relevant to use of trolleys on inclines. (*a*) Personal statements of strength (handle height 0.95 m, anterior limits of toe placement relevant to pulling a trolley with and without tail-board). (*b*) Environmental statement concerning foot-slip on a gradient of 1 in 7 for various limiting coefficients of static friction (μ). (*c*) Lines representing combinations of LIFT and PUSH (or PULL) which prevent trolleys of various weights running down the incline. NB: left and right halves of diagram relate to pulling and pushing trolleys respectively.

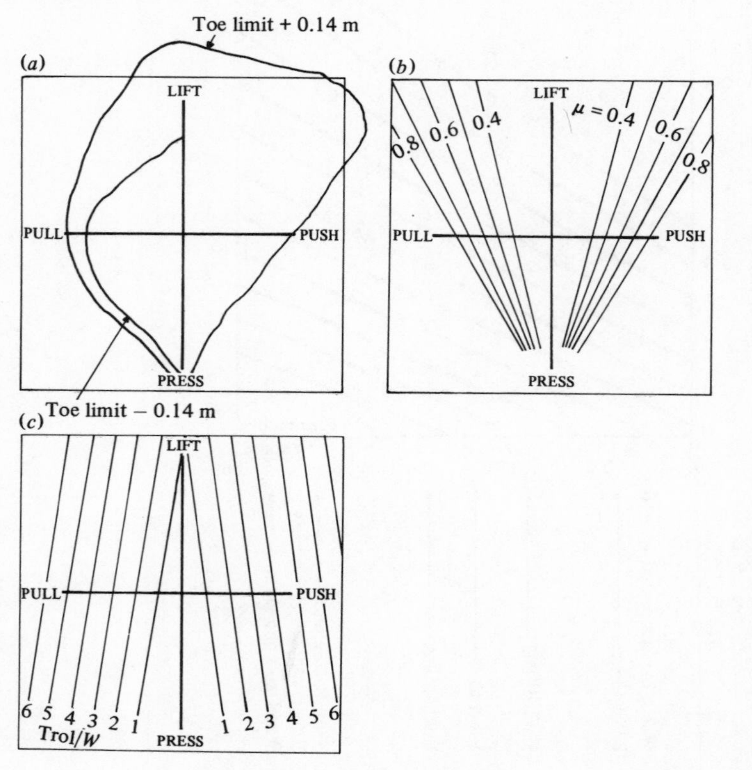

ACTION: Select personal statement of strength/weight for hand height of 0.95 m, freestyle foot placement, with toe limit of 0.14 m in front of hands for pulling and pushing of trolley A. For pulling (only) of trolley B a toe limit of 0.14 m behind the hands is chosen. Suitable PSD (mean strength/weight of 10 adult males) is shown in Fig. 3.72(*a*).

(b) Possibility of foot-slip during exertion

(i) *Limiting coefficient of static friction.* Assume: Range of 0.4 to 0.8.

(ii) *Gradient.* Assume: Worst case of 1 in 7.

ACTION: Note apparent coefficient of friction (see p. 185) and draw limiting lines on the PSD (Fig. 3.72*b*).

(c) Force required to prevent run-back

(i) *Weight of trolley*

(ii) *Gradient.* Assume: Worst case of 1 in 7.

ACTION: By considering manual and trolley weight forces resolved parallel with the ground, the lines on the PSD representing forces which prevent run-back can be constructed: i.e. slopes of ± 7 and intercept equal to trolley weight as shown in Fig. 3.72(*c*).

When the PSD statements are superimposed, limiting combinations of foot-slip, strength and trolley loading can be recognised (Fig. 3.73). The findings are summarised in Table 3.5.

The result indicates that the heaviest trolleys are best handled by pushing

Fig. 3.73. Superposition of PSDs in Fig. 3.72 allows limiting combinations of trolley loading, strength and foot-slip to be determined.

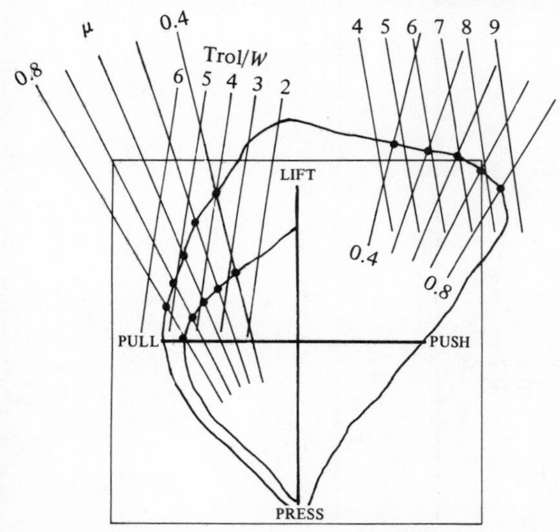

Table 3.5. *Maximum trolley weight as a fraction of body weight*

Limiting coeff. of static friction, μ	Apparent coeff. of friction, μ	Pull up slope		Push up slope
		With tail-board	Without tail-board	With or without tail-board
0.4	0.24	2.7	4.0	4.9
0.5	0.33	3.3	4.6	5.7
0.6	0.42	3.7	4.8	6.9
0.7	0.51	3.9	5.0	7.7
0.8	0.59	4.2	5.2	8.3

them up slopes and under these conditions it does not matter whether a tail-board is fitted or not. In practice, the loadings given in Table 3.5 are too high because they apply to a 50th percentile maximum strength. It would be safer to consider a person of 5th percentile strength and to make an allowance for the possibility of sustained support. No data on the fatiguability of whole-body strength are available, but arguing from the behaviour of isolated muscle groups (see section on fatigue, p.92), a reduction of loading by 30% would lead to an expectation of approximate holding times of 20 seconds.

References

A.J. (1963) *Manikins.* A.J. Information Sheet 1186. *Architects Journal,* 13 Feb.

Andersson, B.J.G. and Ortengren, R. (1974) Myoelectric back muscle activity during sitting. *Scandinavian Journal of Re-habilitation Medicine, Supplement,* **3**, 1–18.

Andrae, B., Ekmark, J. and Laestadius, H. (1971) *Anthropometry of Flying Personnel in the Swedish Air Force.* Library Translation 1502, RAE. Farnborough, Hants.: RAE.

Asmussen, E. and Heebøll-Nielson, K. (1958) Posture mobility and strength of the back in boys aged 7 to 16 years old. *Acta Orthopaedica Scandinavica,* **28,** 174–89.

Asmussen, E. and Heebøll-Nielson, K. (1962) Isometric muscle strength in relation to age in men and women. *Ergonomics,* **5**, 167–76.

Asmussen, E. and Sorensen, N. (1971) The 'wind-up' movement in athletics. *Travail Humain,* **34,** 147–55.

Asmussen, E., Hansen, O. and Lammert, O. (1965) The relation between isometric and dynamic muscle strength in man. *Communications of the Danish Association for Infantile Paralysis,* report 20.

Astrand, P. and Rodahl, K. (1970) *Textbook of Work Physiology.* New York: McGraw-Hill.

Ayoub, M.M. and Lo-Presti, P. (1971) The determination of an optimum size cylindrical handle by means of electromyography. *Ergonomics,* **14,** 509–18.

Ayoub, M.M. and McDaniel, J.W. (1974) Effects of operator stance on a pushing and pulling task. *AIEE Transactions,* **6,** 185–95.

Back Pain Association (1978) *Lifting Instructors' Manual.* Teddington, Middx: Back Pain Association.

Bainbridge, S. (1972) *British School Population Dimensional Survey, 1971.* Dept of Education and Science Building Bulletin 46. London: HMSO.

Barnes, R.M. (1968) *Motion and Time Study: Design and Measurement of Work.* New York: Wiley.

Bartelink, D.L. (1957) The role of abdominal pressure in relieving pressure of the lumbar intervertebral dics. *Journal of Bone and Joint Surgery,* **39B,** 718–25.

Barter, T., Emmanuel, I. and Truett, B. (1957) *A Statistical Evaluation of Joint Range Data.* WADC Technical Note 53-311. ASTIA Document AD 131028. Dayton, Ohio: Wright Patterson Air Force Base.

Bartlett, F.C. (1953) Psychological criteria of fatigue. In *Symposium on Fatigue,* ed. F.W. Floyd and A.T. Welford. London: H.K. Lewis.

Basmajian, J.V. (1967) *Muscles Alive: Their Functions Revealed by Electromyography.* Baltimore: Williams and Wilkins.

Basmajian, J.V. (1968) The present status of electromyographic kinesiology. In *Medicine and Sport 2. Biomechanics,* ed. J. Wartenweiler, E. Jokl and M. Hebbelink. Basel and New York: Karger.

Basmajian, J.V. (1970) *Primary Anatomy.* Baltimore: Williams and Wilkins.

Battye, C.K. and Joseph, J. (1966) An investigation by telemetering of some muscles in walking. *Medical and Biological Engineering,* **4,** 125–35.

Benesh, R. and McGuiness, J. (1974) Benesh movement notation and medicine. *Physiotherapy,* **60,** 176–8.

Bennett, E. (1963) Product and design evaluation through the multiple forced choice ranking of subjective feelings. In *Human Factors in Technology,* ed. E. Bennett, J. Degan and J. Spiegel, chapt. 33. New York: McGraw-Hill.

Bolton, C.B., Kenward, M., Simpson, R.E. and Turner, G.M. (1973) *An Anthropometric Survey of 2000 Royal Airforce Aircrew, 1970/1971.* RAE Technical Report 73083, Farnborough, Hants.: RAE.

Bonney, M.C. and Williams, R.W. (1977) CAPABLE. A computer program to layout controls and panels. *Ergonomics,* **20,** 297-316.

Bouisset, S., Lestienne, F. and Maton, B. (1976) Relative work of main agonists in elbow flexion. In *Biomechanics VA,* ed. P.V. Komi. Baltimore: University Park Press.

Branton, P. (1974) Seating in industry. In *The Applied Ergonomics Handbook,* ed. B. Schackel. Guildford, Surrey: IPC.

Branton, P. and Grayson, G. (1967) An evaluation of train seats by observation of sitting behaviour. *Ergonomics,* **10,** 35–51.

Broman, H. (1977) An investigation of the influence of sustained contraction on the succession of action potentials from a single motor unit. *Electromyography and Clinical Neurophysiology,* **17,** 341–58.

Brooks, B.M., Ruffel-Smith, H.P. and Ward, J.S. (1973) *An Investigation of Factors Affecting the Use of Buses by Both Elderly and Ambulant Disabled Persons.* British Leyland UK/transport and Road Research Laboratory Report, Contract No. CON/3140/32.

Brown, J.R. (1971) *Manual Lifting and Related Fields: An Annotated Bibliography.* Ontario: Labour Safety Council.

Burwell, R.G. (1978) Biostereometrics, shape replication and orthopaedics. In *Orthopaedic Engineering,* ed. D. Harris and K. Copeland. London: Biological Engineering Society.

Caldwell, L.S. (1963) Relative muscle loading and endurance. *Journal of Engineering Psychology,* **2,** 155–61.

Caldwell, L.S. (1964) The load–endurance relationship for a static manual response. *Human Factors,* **6,** 71–9.

Carlson, F.D. and Wilkie, D.R. (1974) *Muscle Physiology.* Englewood Cliffs, NJ: Prentice-Hall.

Carpenter, A. (1938) A study of the angles in the measurement of the leg lift. *Research Quarterly,* **9,** 70–2.

Case, J. and Chilver, A.K. (1971) *Strength of Materials and Structures.* London: Edward Arnold.

Cathcart, E.P., Hughes, D.E.R. and Chalmers, J.E. (1935) *The Physique of Man in Industry.* IHRB Report 71. London: HMSO.

Causley, M. (1967) *An Introduction to Benesh Movement Notation.* London: Parrish.

Cavagna, G.A., Dusman, E. and Margaria, R. (1968) Positive work done by a previously stretched muscle. *Journal of Applied Physiology,* **24,** 21–32.

Chaffin, D. B (1969) A computerised biomechanical model: development of and use in studying gross body action. *Journal of Biomechanics,* **2,** 429–34.

Chaffin, D.B. (1973) Localised muscle fatigue: definition and measurement. *Journal of Occupational Medicine,* **15,** 346–54.

Chapanis, A. (1959) *Research Techniques in Human Engineering.* Baltimore: Johns Hopkins Press.

Clauser, C.E., Tucker, P.E., McConville, J.T., Churchill, E., Laubach, L.L. and Reardon, J.A. (1972) *Anthropometry of Airforce Women.* Report KMRL-TR-70-5. Ohio: Wright Patterson Air Force Base.

Clements, E.M.B. and Pickett, K.G. (1957) Stature and weight of men from England and Wales in 1941. *British Journal of Preventive and Social Medicine.* **11,** 51–60.

Close, R.I. (1972) Dynamic properties of mammalian skeletal muscles. *Physiological Reviews,* **52,** 129–97.

Corlett, E.N. (1981) Postural consideration in workspace design. In *Proceedings of the NATO Symposium on Anthropometry and Biomechanics,* ed. R.S. Easterby, K.M.E. Kromer and D. Chaffin. New York: Plenum Press.

Corlett, E.N. and Bishop, R.P. (1976) A technique for assessing postural discomfort. *Ergonomics,* **19,** 175–82.

Corlett, E.N. and Manenica, I. (1980) The effects and measurement of working postures. *Applied Ergonomics,* **11,** 7–16.

Craik, K.J.W. (1947) Theory of the human operator in control systems. I. The human operator as an engineering system. *British Journal of Psychology,* **38,** 56–61.

Dalessio, D.J. (ed.) (1980) *Wolff's Headache and other Head Pain.* Oxford: Oxford University Press.

Damon, A.D. and Stoudt, H. W. (1963) The functional anthropometry of old men. *Human Factors,* **5,** 485–91.

Damon, A.D., Stoudt, H.W. and McFarland, R.A. (1966) *The Human Body in Equipment Design.* Cambridge, Mass.: HUP.

Daniels, G.S., Meyers, H.C. and Worrall, S.H. (1953) *Anthropometry of WAF Basic Trainees.* WADC Technical Report. 53–12. Ohio: Wright Patterson Air Force Base.

Davis, P.R. (1957) *Studies on the functional anatomy of the human vertebral column with special reference to the thoracic and lumbar regions.* Phd Thesis, University of London.

Davis, P.R. (1967) The mechanics and movements of the back in working situations. *Physiotherapy,* **53,** 44–7.

Davis, P.R. and Stubbs, D. A. (1977*a*) Safe levels of manual forces for young males: I. *Applied Ergonomics,* 8, 141–50.

Davis, P.R. and Stubbs, D.A. (1977*b*) Safe levels of manual forces for young males: II. *Applied Ergonomics,* **8,** 219–28.

Davis, P.R. and Troup, J.D.G. (1964) Pressures in the trunk cavity when pushing, pulling and lifting. *Ergonomics,* **7,** 465–74.

Dempster, W.T. (1955) *Space Requirements of the Seated Operator: Geometrical kinematic and Mechanical Aspects of the Body with Special Reference to the Limbs.* WADC Technical Report 55–159. Ohio: Wright Patterson Air Force Base.

Dempster, W.T., Gabel, W.C. and Felts, W.J.L. (1959) The anthropometry of the manual workspace for the seated subject. *American Journal of Physical Anthropology,* **17,** 289–317.

DES (1970) *Furniture and Equipment Dimensions, Further and Higher Education:* 18–25 *Age Group.* Des Building Bulletin 44. London: HMSO.

De Vries, H.A. (1968) Method of evaluation of muscle fatigue and endurance from electromyographic curves. *American Journal of Physical Medicine,* **47,** 125–35.

Dewar, M.E. (1977) Body movements in climbing a ladder. *Ergonomics,* **20,** 67–86.

Doebelin, E.O. (1966) *Measurement Systems: Application and Design.* New York: McGraw-Hill.

Doss, W.S. and Karpovitch, P.V. (1965) A comparison of concentric, eccentric and isometric strength of elbow flexors. *Journal of Applied Physiology,* **20,** 351–3.

Dreyfuss, H. (1967) *The Measure of Man. Human Factors in Design,* 2nd ed. New York : Whitney Library of Design.

Ducharme, R.E. (1977) Women workers rate 'male' tools inadequate. *Human Factors Society Bulletin,* **20**(4), 1–2.

Dyson, G.H.G. (1977) *The Mechanics of Athletics.* London : Hodder and Stoughton.

Edman, K.A.P., Elzinga, G. and Noble, M.I.M. (1978) Enhancement of mechanical perform-

ance by stretch during tetanic contractions of vertebrate skeletal muscle fibres. *Journal of Physiology (London),* **281,** 139–55.
Eie, N. (1966) Load capacity of the low back. *Journal of the Oslo City Hospitals,* **16,** 74–8.
Eie, N. and Wehn, P. (1962) Measurements of intra-abdominal pressure in weight bearing of the lumbo-sacral spine. *Journal of the Oslo City Hospitals,* **12,** 205–17.
Ellis, D. (1951) Speed of manipulative performance as a function of work surface height. *Journal of Applied Psychology,* **35,** 289–96.
Ely, J.H., Thomson, R.M. and Orlansky, J. (1963) Layout of workplaces. In *Human Engineering Guide to Equipment Design,* ed. C.T. Morgan, J.S. Cook, A.T. Chapanis and M.W. Lund, chapt. 7. New York : McGraw-Hill.
Eshkol, N. and Nul, R. (1968) *Eshkol–Wachman Movement Notation. Classical Ballet.* Tel Aviv: Israel Music Institute.
Eshkol, N., Melvin, P., Michl, J., Von Foerster, H. and Wachmann, A. (1970) *Notation of Movement.* BCL Report 10.0. Urbana, Illinois : Biological Computer Laboratory, University of Illinois.
Feinstein, B., Lindegaarde, B., Nyman, E. and Wohlfart, G. (1955) Morphologic studies of motor units in normal human muscles. *Acta Anatomica* **23,** 127–42.
Fick, R. (1911) *Anatomie und Mechanik der Gelenke.* Jena : Fischer.
Fitzgerald, J.G. (1973) *The IAM Type Aircrew Lumbar Support.* Aircrew Equipment Group Report 304. Farnborough, Hants. : Institute of Aviation Medicine.
Floyd, W.F. and Silver, P.H.S. (1955) The function of erectores spinae muscles in certain movements and postures in man. *Journal of Physiology (London),* **129,** 184–203.
Floyd, W.F. and Welford, A.T. (eds.) (1953) *Symposium on Fatigue : The Ergonomics Research Society.* London: H.K. Lewis.
Garg, A. and Chaffin, D.B. (1975) A biomechanical computerised simulation of human strength. *AIIE Transactions,* **2,** 1–15.
Garrett, J.W. (1971) The adult human hand : some anthropometric and biomechanical considerations. *Human Factors,* **13,** 117–31.
Garrett, J.W. and Kennedy, K.W.A. (1971) *A Collation of Anthropometry,* AMRL-TR-68-1. Ohio : Wright Patterson Air Force Base.
Goldsmith, S. (1976) *Designing for the Disabled,* 3rd ed. London : RIBA
Gooderson, C.Y. and Beebee, M. (1976) *Anthropometry of 500 Infantrymen 1973–1975.* Report APRE 17/76. Farnborough, Hants. : APRE.
Gordon, A.M., Huxley, A.F. and Julian, F. (1966) The variation in isometric tension with sarcomere length in vertebrate muscle fibres. *Journal of Physiology (London),* **184,** 170–92.
Grandjean, E. (ed.) (1969) *Sitting Posture.* London : Taylor and Francis.
Grandjean, E. (1973) *Ergonomics of the Home.* London : Taylor and Francis.
Greenberg, L. and Chaffin, D.B. (1976) *Workers and their Tools : A Guide to the Ergonomic Design of Hand Tools and Small Presses.* Michigan : Pendal Press.
Grieve, D.W. (1974) Dynamic characteristics of man during crouch and stoop lifting. In *Biomechanics IV,* ed. R.C. Nelson and C. Morehouse, pp. 19–24. Pennsylvania : University Park Press.
Grieve, D.W. (1975) Electromyography. In *Techniques for the Analysis of Human Movement,* ed. H.T.A. Whiting, pp. 109–49. London : Lepus Books.
Grieve, D.W. (1977) The dynamics of lifting. *Exercise and Sport Sciences Reviews,* **5,** 157–79.
Grieve, D.W. (1979*a*) The postural stability diagram (PSD): personal constraints on the static exertion of force. *Ergonomics,* **22,** 1155–64.
Grieve, D.W. (1979*b*) Environmental constraints on the static exertion of force: PSD analysis in task design. *Ergonomics,* **22,** 1165–75.
Grieve, D.W. and Arnott, A.W. (1970) The production of torque during axial rotation of the

trunk. *Journal of Anatomy,* **107,** 147–64.
Grieve, D.W. and Cavanagh, P.R. (1973) The quantitative analysis of phasic electromyograms. In *New Developments in Electromyography and Clinical Neurophysiology,* vol. 2, J.E. Desmedt, pp. 489–96. Basel : Karger.
Grieve, D.W. and Cavanagh, P.R. (1974) The validity of quantitative statements about surface electromyography recorded during locomotion. *Scandinavian Journal of Rehabilitation Medicine, Supplement* **3,** 19–35.
Grieve, D.W. and Pheasant, S.T. (1976) Myoelectric activity, posture and isometric torque in man. *Electromyography and Clinical Neurophysiology,* **16,** 3–21.
Grieve, D.W. and Rennie, R. (1967) Measurement of electromyographic activity over long periods. In *Instrumentation in Medicine,* pp. 48–51. London : Hanover Press.
Grieve, D.W., Leggett, D. and Wetherstone, B. (1978*a*) The analysis of normal stepping movements as a possible basis for locomotor assessment of the lower limbs. *Journal of Anatomy,* **127** 515–32.
Grieve, D.W., Miller, D.I., Mitchelson, D.B., Paul, J.P. and Smith, A.R. (1975) *Techniques for the Analysis of Human Movement.* London : Lepus Books.
Grieve, D.W., Pheasant, S.T. and Cavanagh, P.R. (1978*b*) Prediction of gastrocnemius length from knee and ankle joint posture. In *Biomechanics VIA,* ed. E. Asmussen and K. Jorgensen, pp. 405–12. Baltimore : University Park Press.
Grieve, J.I. (1972). Heart rate and daily activities of housewives with young children. *Ergonomics,* **15,** 139–46.
Hall, N.B. and Bennett, E.M. (1956) Empirical assessment of handrail diameters. *Journal of Applied Psychology,* **40,** 381–2.
Hallberg, G. (1976) A system for the description and classification of movement behaviour. *Ergonomics,* **19,** 727–39.
Harding, R.H. and Sen, R.N. (1969) A new simple method of quantifying the electromyogram to evaluate total muscle activity. *Journal of Physiology (London),* **204,** 66*P*.
Haslegrave, C.M. (1979) An anthropometric survey of British drivers. *Ergonomics,* **22,** 145–54.
Hazelton, F.T., Smidt, G.L., Flatt, A.E. and Stephens, R.I. (1975) The influence of wrist position on the force produced by the finger flexors. *Journal of Biomechanics,* **8,** 301–6.
Hellebrandt, F.A. (1938) Standing as a geotropic reflex. *American Journal of Physiology,* **121,** 471–4.
Hermansen, L. (1974) Individual differences. In *Fitness, Health and Work Capacity,* ed, L.A. Larson, chapt. 21. London and New York : Nacmillan.
Herron, R.E. (1972) Biostereometric measurement of body form. *Yearbook of Physical Anthropometry,* **16,** 80–120.
Hertzberg, H.T.E., Churchill, E., Dupertuis, C.W., White, R.M. and Damon, A. (1963) *Anthropometric Survey of Turkey, Greece and Italy.* Oxford University Press.
Hertzberg, H.T.E., Daniels, G.S. and Churchill, E. (1954) *Anthropometry of Flying Personnel, 1950.* WADC Technical Report 52–321. Ohio : Wright Patterson Air Force Base.
Hettinger, T. (1961) *Physiology of Strength.* Springfield, Ill. : Charles C. Thomas.
Hill, A.V. (1938) The heat of shortening and the dynamic constants of muscle. *Proceedings of the Royal Society of London, Series B,* **126,** 136–95.
Hill, A.V. and Howarth, J.V. (1969) The reversal of chemical reactions in contracting muscle during an applied stretch. *Proceedings of the Royal Society of London, Series B,* **151,** 169–93.
Himbury, S. (1967) *Kinetic Methods of Manual Handling in Industry.* ILO Occupational Safety and Health Series 10. Geneva : ILO.
HMSO (1965) *Abatement of Electrical Interferences.* Hospital Technical Memorandum 14. London : HMSO.

Horder, A. (ed.) 1958 *The Ilford Manual of Photography*, Ilford, Essex: Ilford Ltd.

Hulton, B., Thorstensson, A., Sjodin, B. & Karlsson, J. (1975) Relationship between isometric endurance and fibre type in human leg muscles, *Acta Physiologica Scandinavica*, **93**,

Hünting, W., Grandjean, E. and Maeda, K. (1980) Constrained postures in accounting machine operators. *Applied Ergonomics*, **11**, 145–50.

Ince, N.E., Redrup, R. and Piper, J. (1973) *Anthropometry of 500 Armoured Corps Servicemen 1972*. Report APRE 36/73. Farnborough, Hants : Ministry of Defence, Army Personnel Research Establishment, c/o RAE.

Jonsson, B. (1968) Wire electrodes in electromyographic kinesiology. In *Medicine and Sport, 2. Biomechanics*, ed. J. Wartenweiler, E. Jokl and M. Hebbelinck, pp. 123–7. Basel and New York : Karger.

Jonsson, B. (1974) Function of the erector spinae muscle on different working levels. *Acta Morphologica Neerlando-Scandanavica*, **12**, 211–14.

Jørgensen, K. (1970) Back muscle strength and body weight as limiting factors for work in the standing, slightly-stooped position. *Scandinavian Journal of Rehabilitation Medicine*, **2**, 149–53.

Kadefors, R., Peterson, I. and Herberts, P. (1976) Muscular reaction to welding work: an electromyographic investigation. *Ergonomics*, **5**, 543–58.

Kapandji, I.A. (1974) *The Physiology of the Joints*, vol. 3. *The Trunk and Vertebral Column*. London: Churchill Livingstone.

Kember, P.A. (1976) The Benesh movement notation used to study sitting behaviour. *Applied Ergonomics*, **7**, 133–6.

Kemsley, W.F.F. (1950) Weight and height of a population in 1943. *Annals of Eugenics* **15**, 161–83.

Kennedy, K.W. (1964) *Reach Capability of the USAF Population. Phase I. Outer Boundaries of Grasping-reach Envelopes for the Shirt-sleeved, Seated operator*. Report AMRL-TDR-64-59. Ohio : Wright Patterson Air Force Base.

Khosla, T. and Lowe, C.R. (1968) Height and weight of British men. *Lancet*, **i**, 742–5.

Klein, K.K. (1971) *The Knee in Sports*. Austin and New York : Pemberton Press.

Komi, P. (1973) Measurement of the force–velocity relation in human muscle under concentric and eccentric contractions. In *Biomechanics III*, ed. C.S. Cerquiglini, A. Venerando and J. Wartenweiler, pp. 224–9. Basel : Karger.

Komi, P.V. and Buskirk, F.R. (1972) Effect of concentric muscle conditioning on tension and electrical activity of human muscle. *Ergonomics*, **15**, 417–34.

Komi, P.V., Vittasalo, J.T., Havu, M., Thorsensson, A. and Karlsson, J. (1976) Physiological and structural performance capacity. In *Biomechanics V*, ed. P.V. Komi, pp. 118–23. Baltimore : University Park Press.

Konz, S. (1967) Design of work stations. *Journal of Industrial Engineering*, **18**, 413–23.

Konz, S.A., Dey, S. and Bennett, C. (1973) Forces and torques in lifting. *Human Factors*, **15**, 237–45.

Kroemer, K.H.E. (1964) Heute zutreffende Korpermasse. *Arbeitswissenschaft*, **3**, 42–5.

Kroemer, K.H.E. (1971*a*) Seating in plant and office. *Journal of the American Industrial Hygiene Association*, **32**, 633–52.

Kroemer, K.H.E. (1971*b*) Foot operation of controls. *Ergonomics*, **14**, 333–61.

Kroemer, K.H.E. (1974) Horizontal push and pull forces. *Applied Ergonomics*, **5**, 94–102.

Larkin, J.A. (1969) *Work Study. Theory and Practice*. New York : McGraw-Hill.

Less, M., Eickleberg, W. and Palgi, S. (1973) Effects of work surface angles on productive efficiency of females in a simple manual task. *Perceptual and Motor Skills*, **36**, 431–6.

Less, M. and Eickleberg, W. (1975) Force changes in neck vertebrae and muscles. In *Biomechanics VA*, ed. P.V. Komi, pp. 530–6. Baltimore : University Park Press.

Le Veau, B. (1977) *Williams and Lissner : Biomechanics of Human Motion*. Philadelphia : W.B. Saunders.

Lindholm, L.E. (1974) An optoelectronic instrument for remote on-line movement monitoring. In *Biomechanics IV*, ed. R.C. Nelson and C.A. Morehouse, pp. 510–12. Baltimore : University Park Press.
Lindstrom, L., Kadefors, R. and Peterson, I. (1977) An electromyographic index for localised muscular fatigue. *Journal of Applied Physiology*, **43**, 750–4.
Lindstrom, L., Magnusson, R. and Peterson, I. (1970) Muscular fatigue and action potential conduction velocity changes studied with frequency analysis of EMG signal. *Electromyography*, **10**, 341–56.
Lippold, O.C.J., Redfearn, J.W.T. and Vuco, J. (1960) The electromyography of fatigue. *Ergonomics*, **3**, 121–33.
Lundervold, A. (1951) Electromyographic investigations during sedentary work, especially typewriting. *British Journal of Physical Medicine*, **14**, 32–6.
Martin, J.B. and Chaffin, D.B. (1972) Biomechanical and computerised simulation of human strength in sagittal plane activities. *AIIE Transactions*, **4**, 19–28.
MHRU (Materials Handling Research Unit, University of Surrey) (1980) *Force Limits in Manual Work*. Guildford, Surrey : IPC Press.
Milani-Comparetti, A.M. and Gidoni, E.A. (1968) A graphic method of recording normal and abnormal movement patterns. *Developmental Medicine and Child Neurology*, **10**, 633–6.
Miller, D.I. and Nelson, R.C. (1973) *Biomechanics of Sport. A Research Approach*, Philadelphia : Lea and Febiger.
Milner, M., Basmajian, J.V. and Quanbury, A.O. (1971) Multifactorial analysis of walking by electromyography and computer. *American Journal of Physical Medicine*, **50**, 235–58.
Mitchelson, J.B. (1975) Recording of movement without photography. In *Techniques for the Analysis of Human Movement*, pp. 59–65. London : Lepus Books.
Monod, H. (1972) How muscles are used in the body. In *The Structure and Function of Muscle*, 2nd edn, vol. 1, ed. G.H. Bourne, chapt. 2. London and New York : Academic Press.
Monod, H. and Scherrer, J. (1965) The work capacity of a synergic muscle group. *Ergonomics*, **8**, 329–38.
Montegriffo, V.M.E. (1968) Height and weight of a United Kingdom adult population with a review of the literature. *Annals of Human Genetics*, **31**, 389–99.
Morris, J.N., Heady, J.A. and Raffle, P.A.B. (1956) Physique of London busmen: epidemiology of uniforms. *Lancet*, **ii**, 569–70.
Morris, J.M., Lucas, D.B. and Bresler, B. (1961) The role of the trunk in the stability of the spine. *Journal of Bone and Joint Surgery*, **43A**, 327–51.
Mortimer, J.T., Magnusson, R. and Petersen, I. (1970) Conduction velocity in ischaemic muscle: effect on EMG frequency spectrum. *American Journal of Physiology*, **219**, 1324–9.
Morton, D.J. (1944) *Manual of Human Cross Sectional Anatomy*. Baltimore: Williams and Wilkins.
Murray, M.P. and Sepic, B.S. (1968) Maximum isometric torque of hip abductor and adductor muscles. *Physical Therapy*, **49**, 845–51.
Napier, J.R. (1956) The prehensile movements of the human hand. *Journal of Bone and Joint Surgery*, **38B**, 902–13.
Nordeen, K.S. and Cavanagh, P.R. (1976) Simulation of lower limb kinematics during cycling. In *Biomechanics VB*, ed. P.V. Komi, pp. 26–33. Baltimore : University Park Press.
Ortengren, R., Andersson, G., Broman, H., Magnusson, R. and Petersen, I. (1975) Vocational electromyography; studies of localised muscle fatigue at the assembly line. *Ergonomics*, **18**, 157–74.
Park, K.S. and Chaffin, D.B. (1974) A biomechanical evaluation of two methods for manual

load lifting. *AIIE Transactions,* **6,** 105–13.

Paul, J.P., Hughes, J. and Kenedi, R.M. (1972) Biomechanical compatibility of prosthetic devices. In *Biomechanics. Its Foundations and Objectives,* ed. Y.S. Fung, N. Perrone and M. Anliker pp. 531–48. Englewood Cliffs, NJ : Prentice-Hall.

Person, R.S. (1958) An electromyographic investigation on co-ordination of the activity of antagonist muscles in man during the development of a motor habit. *Zhurnal Vysshei Nervnoi Deyatel 'nosti imeni I.P. Pavlova,* **1,** 17–27.

Petrofsky, J.S. (1979) Frequency and amplitude analysis of the EMG during exercises on the bicycle ergometer. *European Journal of Applied Physiology,* **41,** 1–15.

Petrofsky, J.S., Dahms, T.E. and Lind, A.R. (1975) Power spectrum analysis of the EMG during static exercise. *Physiologist,* **18,** 350.

Pheasant, S.T. (1977) A biomechanical analysis of human strength. PhD Thesis, University of London.

Pheasant, S. (1978) Getting a grip: investigations of the hand–handle interface. *Ergonomics,* **21,** 393.

Pheasant, S. (1980) A technique for the rapid estimation of anthropometric data. *Journal of Anatomy,* **130,** 649.

Pheasant, S.T. and Grieve, D.W. (1981) The principal features of maximal exertion in the sagittal plane. *Ergonomics,* **24,** 327–38.

Pheasant, S.T. and Harris, C. (1981) Human strength in the operation of tractor pedals. In *Ergonomics in Agriculture and Forestry,* ed. D. O'Neill, pp. 51–64. Loughborough, Leics.: The Ergonomics Society. (Also in *Ergonomics,* in press.)

Pheasant, S.T. and O'Neill, D. (1975) Performance in gripping and turning: a study of hand/handle interactions effectiveness. *Applied Ergonomics,* **6,** 205–8.

Pottier, H., Lille, F., Phyon, M. and Monod, H. (1969) Etude de la contraction statique intermittente. I. La capacité de travail. *Travail Humain,* **32,** 27–7.

Poulsen, E. and Jørgensen, K. (1971) Back muscle strength, lifting and stooped working postures. *Applied Ergonomics,* **2,** 133–7.

Preston-Dunlop, V. (1969) A notation system for recording observable motion. *International Journal of Man–Machine Studies,* **1,** 361–86.

Priel, V.Z. (1974) A numerical definition of posture, *Human Factors,* **16,** 576–84.

Rack, P.M.H. and Westbury, D.R. (1969) The effects of length and stimulus rate on tension in the isometric cat soleus muscle. *Journal of Physiology (London),* **204,** 443–60.

Rebiffe, P.R., Zayana, O. and Tarriere, C. (1969) Détermination des zones optimales pour l'emplacement des commandes manuelles dans l'éspace de travail. *Ergonomics,* **12,** 913–24.

Roberts, D.F. (1960) Functional anthropometry of elderly women. *Ergonomics,* **3,** 321–7.

Roebuck, J.A. Jr, Kroemer, K.H.E. and Thomson, W.G. (1975) *Engineering Anthropometry Methods.* New York : Wiley.

Rohmert, W. (1960) Ermittlung von Erholungpausen für statische Arbeit des Menschen. *Internationale Zeitschrift für Angenandte Physiologie Einschliesslich Arbeitsphysiologie,* **18,** 123–64.

Rubin, A. (1980) Unpublished MSc report, University of London.

Ruch, T.C. and Patton, H.D. (1965) *Physiology and Biophysics.* Philadelphia : W.B. Saunders.

Sato, M. (1966) A problem in the frequency analysis of the electromyogram. *Electromyography,* **6,** 21–3.

Sinelnikoff, E. and Grigorowitsch, M. (1931) Die Beweglichkeit der Gelenke als sekundares geschlechtliches und konstitutionelle Merkmal. *Zeitschrift für Konstitutionslehre,* **15,** 679–95.

Smidt, G.L. (1973) Biomechanical analysis of knee flexion and extension. *Journal of Biomechanics,* **6,** 79–92.

Smith, J.W. (1956) Observations on the postural mechanism of the human knee joint.

Journal of Anatomy, **90,** 236–60.

Smith, J.W. (1957) The forces operating on the human ankle joint during standing. *Journal of Anatomy,* **91,** 545–64.

Snook, S.M., Campanelli, R.A. and Hart, J.W. (1978). A study of three preventative approaches to low back injury. *Journal of Occupational Medicine,* **20,** 478–80.

Snyder, R.G., Schneider, L.W., Owings, C.L., Reynolds, H.M., Golumb, D.H. and Schork, M.A. (1977) *Anthropometry of Infants, Children and Youths to Age 18 for Product Safety Design.* Report PB-270 227, Consumer Product Safety Committee. Bethesda, Maryland : United States Department of Commerce.

Stelmach, G.E. (1976) *Motor Control. Issues and Trends.* New York and London : Academic Press.

Stubbs, D.A. (1973) *Manual Handling in the Construction Industry.* Report of the Department of Human Biology. University of Surrey.

Tanner, J. (1962) *Growth at Adolescence.* Oxford : Blackwell Scientific.

Taylor, J.H. (1973) Vision. In *Bioastronautics Data Book,* ed. J.T. Parker and V.R. West, chapt. 13. Washington, DC : NASA.

Taylor, C.L. and Schwartz, R.J. (1955) The anatomy and mechanics of the human hand. *Artificial Limbs,* **2,** 22–35.

Thomas, D.P. and Whitney, R.J. (1959) Postural movements during normal standing in man. *Journal of Anatomy,* **93,** 534–39.

Thompson, D., Barden, J.D., Kirk, N.S., Mithchelsom, D.L. and Ward, J.S. (1973) *Anthropometry of British Women.* Report of the Institute of Consumer Ergonomics. Loughborough, Leics. : ICE.

Thompson, S.J. (1980) Unpublished MSc report, University of London.

Thorstensson, A. and Karlsson, J. (1976) Fatiguability and fibre composition of human skeletal muscle. *Acta Physiologica Scandanavica,* **98,** 318–22.

Thorstensson, A., Grimby, G. and Karlsson, J. (1976) Force–velocity relations and fibre composition in human knee-extensor muscles. *Journal of Applied Physiology* **40,** 12–16.

Tichauer, E.R. (1975) *Occupational biomechanics : The Anatomical Basis of Work-place Design.* Rehabilitation Monograph 51. New York University : Institute of Rehabilitation Medicine.

Tichauer, E.R. (1978) *The Biomechanical Basis of Ergonomics.* New York : Wiley.

Tiffin, J. and McCormick, E.J. (1966) *Industrial Psychology.* London : Unwin.

Timoshenko, S. and Young, D.H. (1956) *Engineering Mechanics,* 4th ed. New York : McGraw-Hill.

Travell, J. (1967) Mechanical headache. *Headache,* **7,** 23–9.

Troup, J.D.G. and Chapman, A.E. (1969) The strength of the flexor and extensor muscles of the trunk. *Journal of Biomechanics,* **2,** 49–62.

Ward, J. (1971) Ergonomic techniques in the determination of optimal work surface heights. *Applied Ergonomics,* **2,** 171–7.

Ward, J.S. and Kirk, N.S. (1968) The relation between some anthropometric dimensions and preferred working surface heights in the kitchen. *Ergonomics,* **11,** 410–11.

Waris, P. (1979) Occupational Cervico brachial syndromes : a review. *Scandinavian Journal of Work Environment and Health,* **5,** Supplement 3, 3–14.

Weston, H.C. (1953) Visual fatigue, with special reference to lighting. In *Symposium on Fatigue,* ed. W.F. Floyd and A.T. Welford. London : H.K. Lewis.

Whitney, R.J. (1962) The stability provided by the feet during manoeuvres whilst standing. *Journal of Anatomy,* **96,** 103–11.

Wickström, P. (1978) Effect of work on degenerative back disease. *Scandinavian Journal of Work Environment and Health,* **4,** Supplement 1, 1–12.

Wilkie, D.R. (1950) The relation between force and velocity in human muscle. *Journal of Physiology (London),* **110,** 249–80.

Winter, D.A. (1979) *Biomechanics of Human Movement*. New York : Wiley.

Winter, D.A., Greenlaw, R.K. and Hobson, D.A. (1972) Television computer analysis of kinematics of human gait. *Computer Biomedical Research*, **5**, 498–504.

Wisner, A. and Rebiffe, R. (1963) Methods of improving work-place layout. *International Journal of Production Research*, **2**, 145–67.

Yamada, H. (1970) *Strength of Biological Materials*. Baltimore: Williams and Wilkins.

Zernicke, R.F. and Roberts, E.M. (1976) Human locomotor extremities kinetic relationships during systematic variations in resultant limb velocity. In *Biomechanics VB*, ed. P.V. Komi Baltimore: University Park Press.

4 VIBRATION AND LINEAR ACCELERATION

Stephen Cole

This chapter aims to provide an outline of man's responses to vibration and how they can be measured and evaluated. A complete definitive description cannot be achieved within the allotted limits but further study of the references cited will provide an expanded description. For the general area of the human aspects of vibration, Guignard and King (1972) provides a detailed survey of current knowledge and literature. A detailed description of the mechanical and physical aspects of vibration is given in a series of handbooks edited by Harris and Crede (1961).

4.1 Knowledge

What is vibration?

The human being in his environment has to endure many physical stresses. Evolution has endowed him with the ability to withstand such stresses provided that they do not exceed certain not easily defined limits. When walking, running, and riding in vehicles he undergoes vibrations, buffetings and, especially in cases of accidents, very hard impacts. All these stresses have a mechanical effect on man, whose reaction depends on the dynamic properties of his body. As a consequence of these mechanical effects, physiological reactions will occur, and these will depend on both the magnitude and the duration of the imposed stress. Thus, the dynamic properties of the body system are relevant to any attempt to determine the physiological reactions to imposed mechanical stresses.

A system or body can be said to vibrate when its component parts oscillate about a reference position, usually the position of stationary equilibrium. Any system, by virtue of its inherent mass and elastic properties, can be made to vibrate by application of either internal or, more usually, external forces. The effect of the vibration will depend upon the relationship between the applied force and the system's mechanical properties.

Vibration may be described in terms of amplitude, frequency and phase. The amplitude, or intensity, is expressed in terms of the extent of motion, or displacement, about the equilibrium position. Frequency, which is the reciprocal of the duration of one complete cycle of motion, describes the repetition rate of the motion. Phase is important when comparing two vibration motions within the same system.

The simplest form of vibration is a regular periodic oscillation, a simple harmonic motion, which when plotted as a function of time is represented by a sinusoidal curve (see Fig. 4.1). There are some properties of waveforms which strictly are only applicable to sinusoidal waveforms but whose principles may also be used to describe more complex waveforms. It is often convenient to evaluate a waveform by measuring the root mean square (r.m.s.) value of its intensity. This is defined as the square root of the average of the squared values of any set of numbers (e.g. vibration magnitudes).

This value is related to the power or energy contained within the waveform. For a pure sinusoidal motion the r.m.s. value is $\sqrt{\frac{2}{2}}$ (0.707) times the maximum value of the magnitude (the peak value). The relationship between the

Fig. 4.1. Sinusoidal or simple harmonic waveform, showing the peak, root mean square (r.m.s.) and average values. T, period of one cycle; f, frequency.

$$x_{\text{average}} = \frac{1}{T}\int_0^T x \, dt$$

$$x_{\text{r.m.s.}} = \sqrt{\left(\frac{1}{T}\int_0^T x^2 (t) \, dt\right)}$$

$$\text{crest factor} = \frac{x_{\text{peak}}}{x_{\text{r.m.s.}}}$$

$$\text{for pure sinusoidal motion } x_{\text{r.m.s.}} = \frac{\pi}{2\sqrt{2}} x_{\text{average}} = \frac{1}{\sqrt{2}} x_{\text{peak}}$$

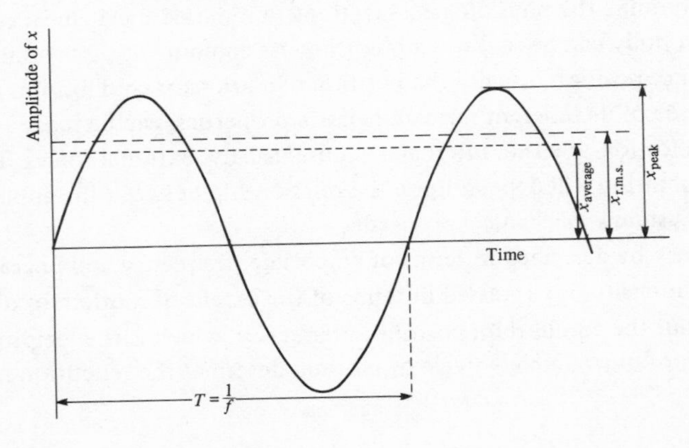

values shown in Fig. 4.1, for a magnitude value x, is as follows:

$$x_{\text{r.m.s.}} = \sqrt{\frac{1}{T}\int_0^T x^2(t)\,\mathrm{d}t} = \frac{\pi}{2\sqrt{2}}\,x_{\text{average}} = \frac{1}{\sqrt{2}}\,x_{\text{peak}}$$

The ratio between the r.m.s. and the peak values is called the crest factor. The magnitude (amplitude) may be defined in terms of either displacement (length) or its time derivatives velocity and acceleration. These are shown diagrammatically in Fig. 4.2.

The relationship between these three descriptors is, for a displacement amplitude x (often in terms of metres):

displacement amplitude x (m)

velocity amplitude $\dot{x} = \dfrac{\mathrm{d}x}{\mathrm{d}t} = 2\pi f x$ (m/s)

and

acceleration amplitude $\ddot{x} = \dfrac{\mathrm{d}(\dot{x})}{\mathrm{d}t} = \dfrac{\mathrm{d}^2x}{\mathrm{d}t^2} = (2\pi f)\,x^2$ (m/s^2)

for a frequency f, usually expressed as Hz (hertz, or cycles per second).

Fig. 4.2. The relationship between the displacement, velocity and acceleration waveforms.

$x = x_0 \sin 2\pi ft = x_0 \sin(\omega t + \phi)$ the displacement waveform

$\dot{x} = \dfrac{\mathrm{d}x}{\mathrm{d}t} = x_0 2\pi f \cos 2\pi ft = x_0\,\omega\,(\cos \omega t + \pi/2)$ the velocity waveform

$\ddot{x} = \dfrac{\mathrm{d}^2x}{\mathrm{d}t^2} = -x_0\,(2\pi f)^2 \sin 2\pi ft = -x_0\,\omega^2 \sin(\omega t + \pi)$ the acceleration waveform

where $2\pi f = \omega$ the angular frequency, and
T = the period of 1 cycle = $1/f$ = 2π radians = 360°.

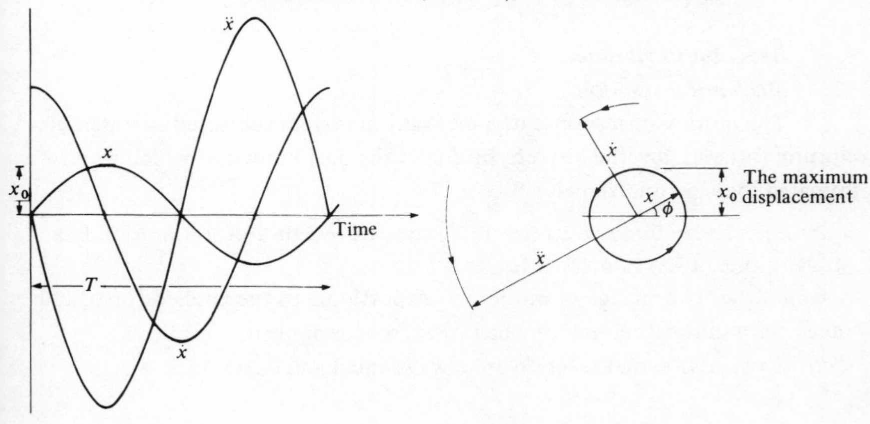

It can be seen that each successive time derivative is achieved by a multiplication by $2\pi f$ (or ω), and when plotted produces the cosine and sine curves shown in Fig. 4.2.

The vibrations met in real life are usually characterised by more complex waveforms. These vary from simple mixtures of sinusoidal motions, perhaps combining a fundamental frequency together with some of its harmonics, to a random type of motion. Sinusoidal waveforms are often known as deterministic waveforms. This term describes the fact that their amplitude can be predicted at any instant of time through knowledge of their characteristics. Random waveforms are said to be non-deterministic or irregular; their amplitudes cannot be so absolutely predicted. There are two types of random waveforms: stationary and non-stationary. In the case of the stationary waveforms the probability of their magnitude falling within a given range can be specified by a probability function. Non-stationary random waveform amplitudes cannot be specified in this or any other manner. Random vibrations frequently occur in nature and may be described as motions which move in irregular cycles which are never repeated exactly. Some examples of the different types of waveforms are shown in Fig. 4.3.

Other types of motion may not involve regular or irregular oscillations over a relatively long time, but rather a mechanical system or body may be subjected to a rapidly acting force or shock motion. These mechanical shocks or impacts are essentially transient vibrations, often of a fairly random nature. It is impossible to define a precise duration below which accelerations may be termed impacts, although there is a clear difference between long-duration and impact accelerations. A shock can be described as a transmission of energy into a system which takes place in a duration which is less than the natural period of oscillation of the system.

Reaction to vibration

Mechanical reaction

The motion or response of a mechanical system subjected to external excitation forces is governed by physical laws the best known of which were formulated by Newton. Namely:

First Law. Every body continues in its state of rest or uniform motion in a straight line, unless impressed forces act on it.

Second Law. The change of motion is proportional to the applied force, and takes place in the direction in which that force is applied.

Third Law. Action and reaction are always equal and opposite.

To understand such responses it is often easier to construct a simple mechanical system or model, and to determine its responses and thus formulate its equations of motion. One of the simplest mechanical models is that described as the single degree of freedom model, which is shown in Fig. 4.4. It contains a

Fig. 4.3. Different waveforms, showing their time histories and related frequency spectra. (*a*) Pure sinusoidal motion; (*b*) two sinuisoidal motions added together; (*c*) random motion with one sinusoidal motion, added together.

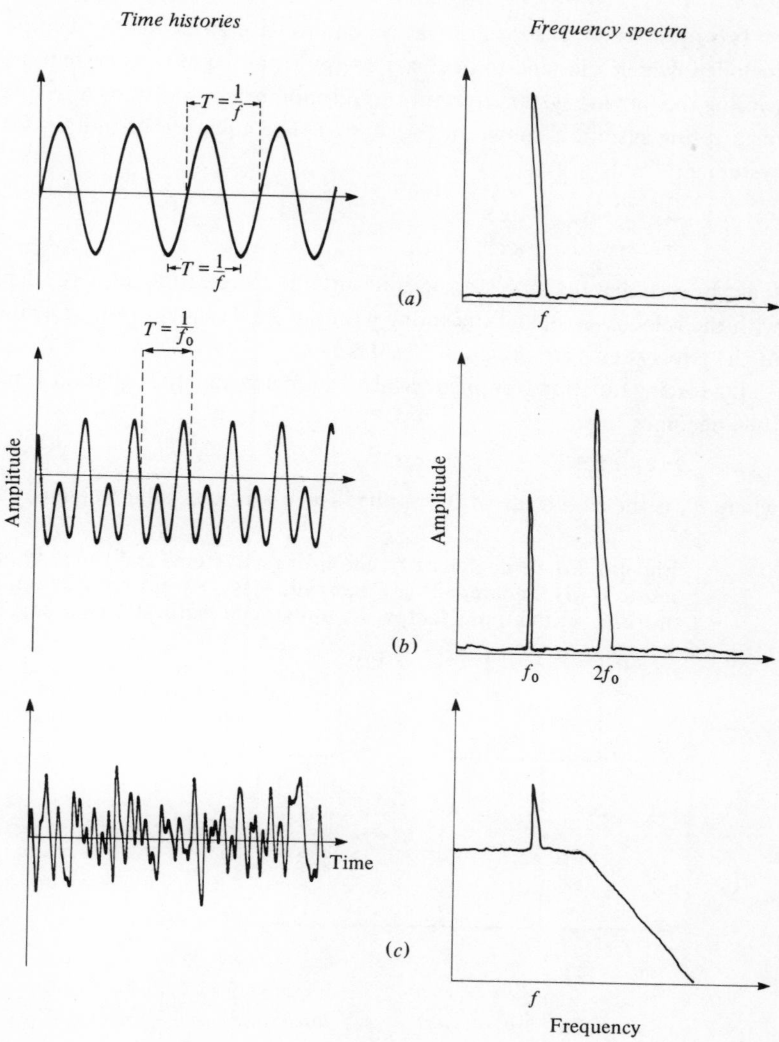

mass moving along one axis only. In addition to the mass there is an elastic coupling connecting it to the base (Fig. 4.4*a*). A free undamped mass spring system will demonstrate a sinusoidal motion when displaced from its static equilibrium position, and Newton's second and third laws can be directly applied to its motion. The mass m is supported on a spring element having a stiffness k. If the mass is displaced from its equilibrium position by the displacement x, the spring force is exactly equal and opposite to the mass force since the mass is balanced. Thus:

$$m\ddot{x} = -kx, \quad \text{or} \quad m\ddot{x} + kx = 0$$

It is more usual, in a mechanical system, to have an element of damping included, which will tend to dissipate energy from the moving system, thus causing the moving system to return to its equilibrium position. A simple damped mass spring system is shown in Fig. 4.4(*b*). The equation of motion for this system is:

$$m\ddot{x} = -(kx + c\dot{x})$$
$$\text{or} \quad m\ddot{x} + c\dot{x} + kx = 0$$

It can be seen that the mass is associated with the acceleration term, the damping with the velocity term and the spring with the displacement term. The motions of the two systems are shown in Fig. 4.5.

If a forcing function is applied to such a simple system the equation of motion then becomes

$$m\ddot{x} + c\dot{x} + kx = F_0 \sin \omega t$$

where F_0 is the amplitude of the applied force whose angular frequency is ω.

Fig. 4.4. Single degree of freedom linear systems and their equations of motion. (*a*) Undamped; (*b*) damped. Mass, m; spring, k; damper, c; motion, x; damping factor, h; undamped natural frequency, ω_n.

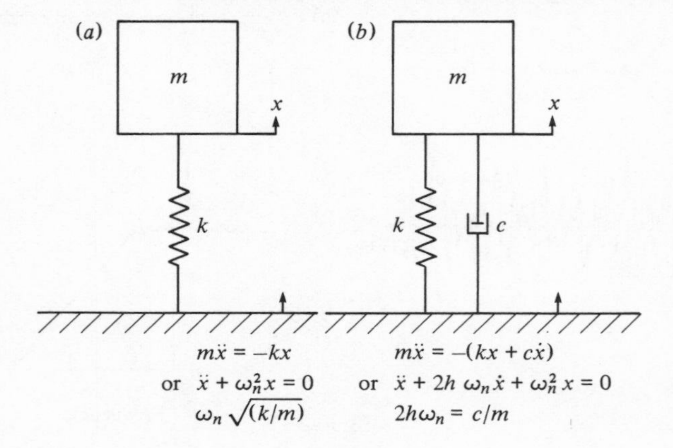

The system will vibrate for as long as the force is applied. The response of the system, the ratio between the applied and response amplitudes and their phase relationship, depends upon the frequency ratio ω/ω_n. The response is shown in Fig. 4.6.

From these curves several characteristics can be seen. There is always a peak amplitude of response at a particular frequency, the value depending on the level of damping. This frequency is called the 'system resonant frequency', and may be defined as: 'the condition in which any change in forcing frequency causes a decrease in the response of the system being vibrated'. In addition, the phase curve passes through a 90° lag at the resonant frequency. A characteristic of resonant systems which can be used in vibration isolators is that the transmissibility value falls below unity for a frequency ratio greater than 1.414 (i.e. $\sqrt{2}$). The effect of damping level on both the magnitude and phase can clearly be seen. Both the amplitude of the frequency response and its resonant frequency are reduced as the damping level is increased. For the phase lag, increasing the damping level reduces the rate of change, especially about the resonant frequency.

Although these simplified systems have been described in some detail, it should not be forgotten that real mechanical systems are more complex. Many real systems, including the human body, are composed of several vibrating masses each with its own differing spring and damping characteristics. This complicates the understanding of the mechanical system being investigated, a point

Fig. 4.5. Motion waveforms for the two systems shown in Fig. 4.4. T, period of undamped motion = $2\pi/\omega_n$, where ω_n (undamped natural frequency) = $\sqrt{(k/m)}$; T_0, period of damped motion = $2\pi[(\sqrt{1-h^2})\omega_n]$ ω_n, where h is the damping factor.

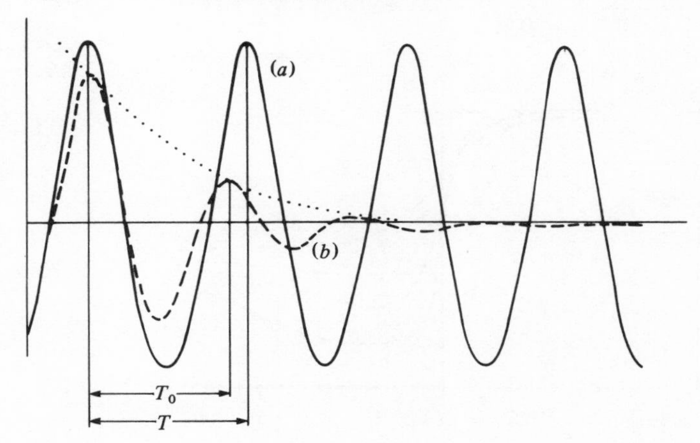

which is described in a later section of this chapter. It is also possible in real systems that the characteristics of the spring and damping elements may not be linear. For instance the stiffness of a spring may be dependent upon the square of its displacement value; this may possibly be true in the case of some human tissue. Non-linear characteristics can also be shown by damping elements.

Fig. 4.6. Theoretical response of a single degree of freedom system. The frequency ratio is the ratio between the forcing frequency and the natural frequency, ω/ω_n.

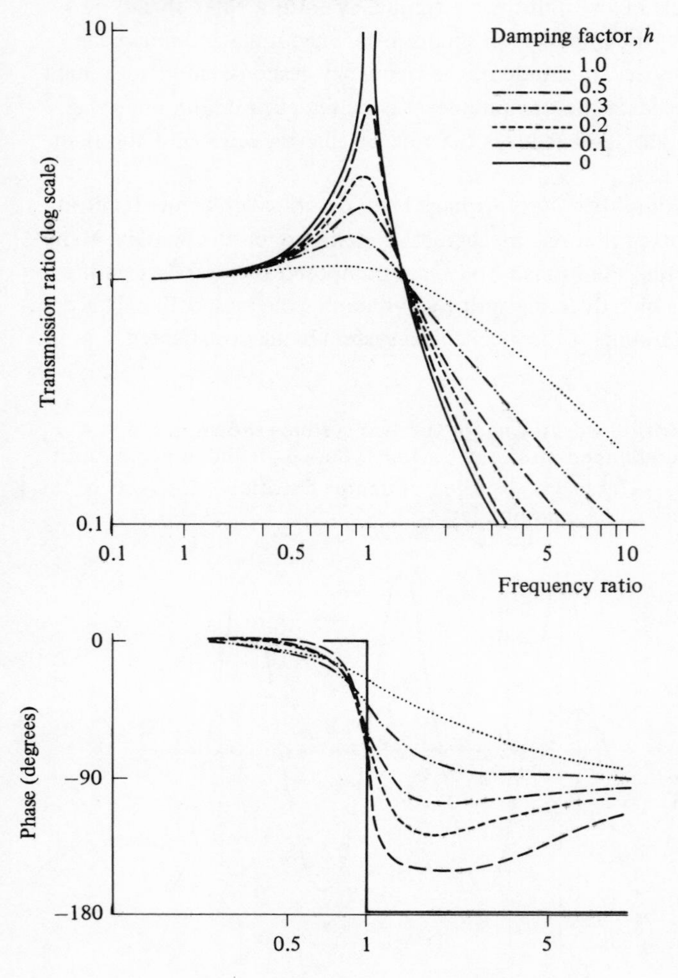

Human responses to mechanical stimuli

The human body, although biologically active, must also be considered as a mechanical system, albeit a very complex one. The human being exhibits mechanical as well as biological responses to applied stimuli.

Fig. 4.7. Typical function curves for three different mechanical systems. (*a*) Impedance functions (ratio of force to velocity at driving point).

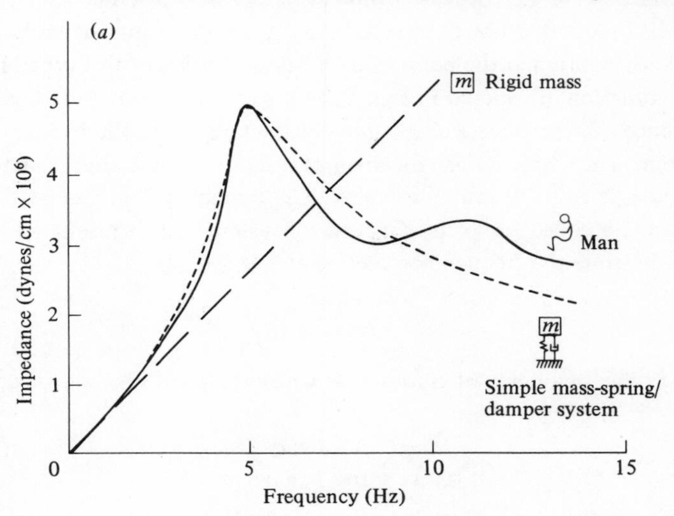

(*b*) Transfer functions (ratio of force to acceleration at driving point).

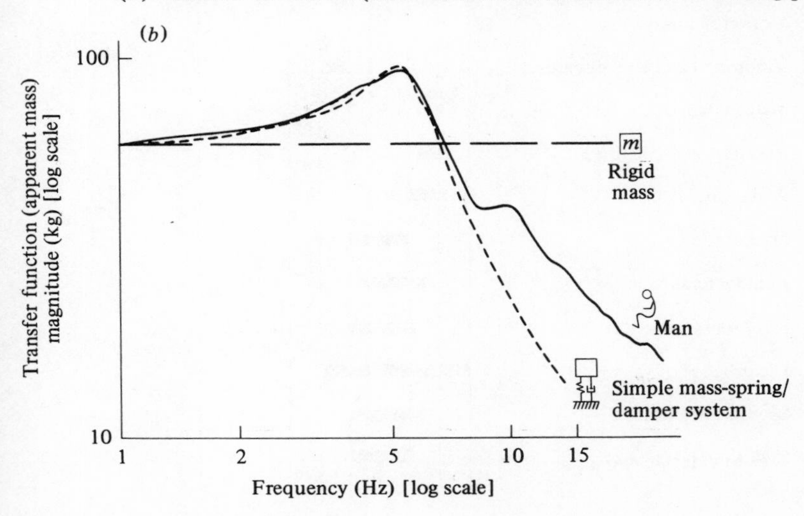

Mechanical effect. The physical, or mechanical, response is also known as the 'biodynamic' response. This response has been measured by the classical mechanical techniques, both for energy 'transfer' and for 'transmission'. The 'transfer' function is a measure of the mechanical impedance. (The impedance of a body is the ratio between the force and velocity at the point where energy is applied to the body.) Probably the most important early work in this field was due to Coermann and his co-workers (e.g. Coermann *et al*., 1960). Some impedance curves are shown in Fig. 4.7(*a*). Because of the difficulty of measuring velocity, a similar 'transfer' function can be estimated by using the force and the more easily measured acceleration at the point of application. Curves of this type of 'apparent mass' function are shown in Fig. 4.7(*b*). Both these curves demonstrate the existence of resonance and damping characteristics in the human mechanical system. The whole human mechanical response, whilst appearing to be basically simple, is actually rather complex. The human system can be thought of as being a collection of mechanical subsystems; some resonance phenomena for these are the frequency effects shown in Fig. 4.8.

Fig. 4.8. The frequency of vibration at which different physiological effects occur.

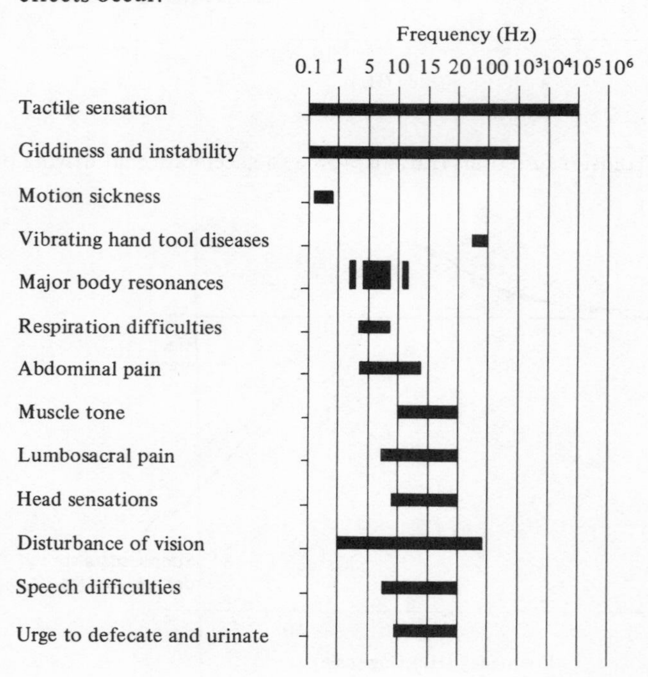

Biological effect. The biological responses to an externally applied force on the human body are generally greater the longer the force acts, and the magnitude of the force will also affect the magnitude of the response. The human body usually remains in a state of equilibrium with the forces acting on it; any shift of this state elicits a response which attempts to re-establish an equilibrium, perhaps at a different level. The force acting on the body may act either specifically at an organ or tissue site (as a resonance phenomenon or differential motion), or in a general stress-type action on the whole body, where the frequency dependence is less marked and the intensity of the force is more important.

Whole-body reactions tend to be relatively long term, mainly involving reflex actions. It is convenient to describe the main body systemic reactions separately.

1. *The respiratory system*. Sinusoidal vibration causes an increase in respiration, the greatest effect occurring at about 4–5 Hz, the major resonant frequency of the body. Changes in the minute volume of ventilation are usually affected by alterations in the tidal volume rather than by changes in ventilation rate. Part of this increase is due to an increased requirement for oxygen in subjects exposed to vibration, as a result of the increased muscular work needed to control the postural changes due to the vibration motion. Most research has shown that the increase in ventilation is more than enough to satisfy the increased oxygen demands, and that in fact hyperventilation takes place. It is probable that vibration causes hyperventilation through two main effects: a mechanical pumping of the diaphragm and breathing system, an effect strongly dependent upon resonant frequency phenomena; and an excessive stimulation of sensory proprioceptors causing an inappropriate reflex increase in respiration.

2. *The cardiovascular system*. Heart rate is a commonly recorded parameter for subjects exposed to vibration. There is usually a rise in heart rate at the onset of vibration, though the degree of rise is highly variable between subjects when compared with pre-exposure and resting levels. It is difficult to measure the electrocardiogram (ECG) during vibration, due to motion artefacts of the electrodes and their leads, so a detailed analysis of any change in waveform shape has not been carried out. It is, however, generally agreed that the shape of the ECG wave is not significantly altered by vibration. High levels of acceleration maintained for long durations have the most marked effect on this system, primarily through fluid pooling in the body extremities along the axis of the applied acceleration. Movement of blood in this fashion causes the sensations of 'black-out' and 'red-out'.

3. *The neuromuscular system*. This system is basically composed of two parts, the neurological and the muscular subsystems.

The sensation of vibration is difficult to quantify. Unlike hearing, there is no single specific organ which detects motion. Cutaneous appreciation of vibration (vibro-tactile sense) is variable over the body's surface. Whole-body vibration is universally perceived by a variety of sensory systems that include pressure and stretch receptors deep in the body's tissues and organs. There is also one major mechanism which is capable of detecting and compensating for whole-body motion. This is the postural reflex, and the associated sensory organs such as the tendon organs and the inner ear's vestibular apparatus.

The most obvious effect of vibration is the response shown by the body's postural control system. It is quite apparent that the vibratory motion at lower energy intensities is accepted as a tolerable environment. At higher intensities, and low frequencies (0.1–20 Hz), which implies large displacement motion, many people feel unstable and have difficulty in maintaining a steady posture. There are also natural body motions, shown by sway and tremor, whose magnitude alters with the arousal state. Over-stimulation of the body's postural reflexes by vibration is characterised by the inability to maintain posture, and often also by more dramatic effects such as the sensation of motion sickness. In normal motion the spatial senses of the vestibular system and the proprioceptive sensors, the visual and aural senses, combine to convey a balanced information input. If this balance is disturbed then a novel situation is perceived which may be very disconcerting. It has been argued that it is this disturbance of these sensory mechanisms, and the apparently novel and conflicting sensory stimulations, which are the probable cause of motion sickness (Reason, 1978). It is certain that motion having low frequencies (less than 0.5 Hz) and high acceleration levels ($0.1g$ to $1g$) trigger the mechanisms causing the sickness effect, and that this effect is mainly transduced through the inner ear's vestibular apparatus.

Pathological effect. Much less is known with certainty about the health-damaging effects of vibration in man. Certain aspects of vibration damage have received much attention, the two prime examples being noise-induced hearing loss and the so-called white finger syndrome (Raynaud's phenomenon) which is associated with the use of vibrating hand tools.

There is clinical evidence that repeated exposure to the displacements caused by high levels of vibration constitutes a health hazard, in which effects are manifest in spinal and gastrointestinal damage. However, it is also important to consider the environment, including the posture and work pattern, of the human operator exposed to these large displacement vibrations. This complex problem is typified in considering the drivers of agricultural tractors, and other cross-country vehicles, some aspects of which are described by Troup (1978).

The pathological effects of high-level accelerations are usually injuries or even death, and arise especially in environments where there is a rapid onset of acceleration or velocity change. These situations are typified in accidents and escape systems. It is then debatable whether it is the initial shock pulse or the secondary impacts, caused by hitting some part of the structural surroundings, which actually cause the injuries.

Psychological effects. It is widely accepted that increased vehicle motion makes its occupant feel fatigued, whether or not he is performing a task. Vibration acts upon the human being performing tasks in two main ways: it induces mechanical interference in the sensory-motor control loop, and the sensation of fatigue. The latter effect is often compounded by simultaneous exposure to other environmental factors such as noise and heat. There is a large individual variability in the susceptibility to these so-called stresses, and indeed at some levels they may even be arousing.

Subjective assessments of vibration effects are essentially dependent upon an individual's susceptibility, and his expectations. The high vibration level in a novel and interesting transport, e.g. a hovercraft, may be considered acceptable, whereas a small but similarly safe wind-induced motion detected in a building may arouse alarm. It is because of this variability in the subjective assessment of vibration that simple and acceptable criteria for exposure to vibration have not yet been universally established. Most criteria which seek to limit vibration exposure have been generated from data based mainly on studies using sinusoidal vibration. These studies have aimed to produce empirical curves describing 'equal comfort contours'. The method used is based on exposing subjects to alternating levels of a known 'control' level vibration and a test vibration. The subject then adjusts the level of the test vibration until both vibrations have equal sensations, often described as 'uncomfortable'. The vibrations are usually sinusoidal and of different frequencies. This technique results in curves of the type shown in Fig. 4.9. For more intensive discussion see Oborne (1978*a, b*).

These laboratory studies, together with some general real-life experience, have shown the definite variation of subjective response with respect to the frequency of stimulus. The human sensitivity to vibration shows a peak at frequencies about 5 Hz, the area of the major whole-body resonances previously described.

Experiments to determine the tolerance of man to high acceleration forces have, of necessity, been restricted to voluntary tolerance, which is usually below an immediate injurious level (see Fig. 4.10). However, these tolerance curves are more usually based on a physiological rather than a psychological end-point, as shown in the studies of high acceleration levels acting for long durations.

The tolerance to duration of vibration that has an overall low acceleration is less well understood. In general most research has produced little experimental evidence to support the concept of a time-dependent decrement in performance. On the contrary moderate exposures to vibration seem to have an arousing effect.

Fig. 4.9. Comparison of intensity matching curves (dashed lines) with derived equal subjective magnitude curves (continuous lines). (After Shoenberger and Harris, 1971.)

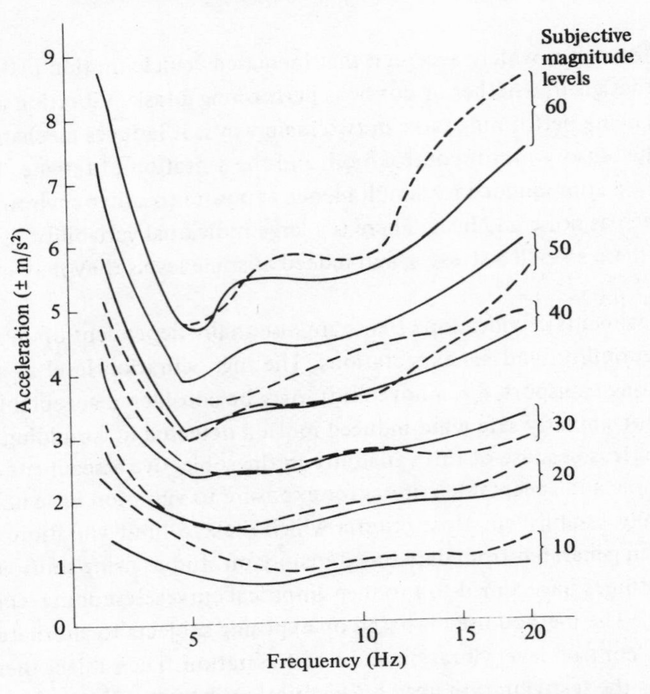

Fig. 4.10. Tolerance to whole-body impact in the attitudes and restraints illustrated. The vertical arrow indicates the inertial force vector perpendicular to the earth's surface. (After Glaister, 1978.)

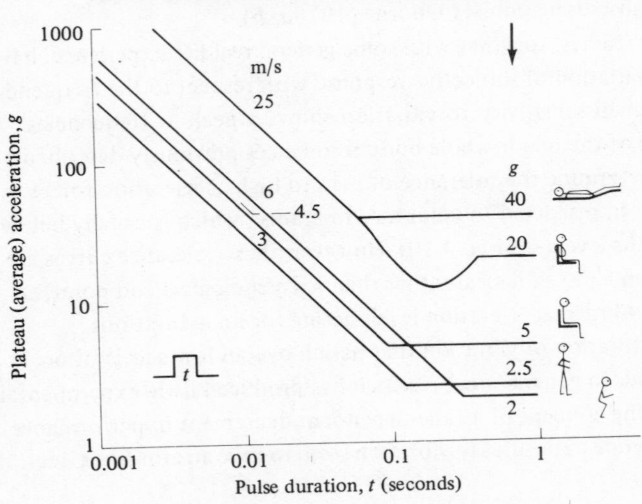

Criteria and limits of exposure to vibration

In deriving specifications to prevent breakdown in a mechanical system, both existing data and experimentation are used. This process has also been used to derive criteria for limits to vibration exposure. However, in the case of man it has been difficult to establish just what these criteria should describe, i.e. whether they should be defined in terms of effects on comfort, or as a performance decrement, or as a health risk. The limits associated with these criteria imply that if the limit is exceeded then there is a significant increase in the probability of the effect occurring.

The present International Standard, accepted as providing guidelines on vibration exposure (ISO, 1974), describes these three criteria. It gives duration limits of exposure for vibrations, both periodic and non-periodic, and some shock excitations, transmitted from a solid surface to the human body. The very similar British Standard (BSI, 1974) closely follows the limits laid down in the ISO document, together with qualifications for their use. These two documents, together with an updated German Standard, cover the vibration frequency range 1–80 Hz. In the future this range is to be extended down to 0.1 Hz to accommodate the motion sickness region. An outline of the limits is shown in Fig. 4.11.

These documents also describe curves for the variation of the exposure limits with time. This topic has aroused much controversy and is being further investigated in current research programmes. One drawback of these documents is their lack of adequate limits for shock or impact exposure, either single or repeated. The shock waveform is characterised by having a very high crest factor (the ratio of the waveform's peak to r.m.s. levels), greatly in excess of the factor of three quoted as a maximum for application of the ISO and BSI documents. One technique at present being investigated for use with shocks is based on a Dynamic Response Index (DRI). This uses an active single degree of freedom model of man's vertebral column to correlate an acceleration waveform's maximum equivalent displacement value (the DRI) with the probability of spinal injury.

A specific area of vibration exposure which has been the subject of a British Standards Institution document is that of the human hand–arm system (BSI, 1975). This document is particularly aimed at workers who use power tools or hold vibrating material and is an attempt to reduce the occurrence of 'white finger syndrome'.

Finally, one area of particular concern is that of safety for human subjects. Experimentation in the field of vibration and impact is fraught with difficulty when neither short-term nor long-term effects are fully understood. Risks arise from both personnel and equipment failures. The need for advice for engineers, biologists and others has been recognised by the British Standards Institution, who have produced a document (BSI, 1973) to provide such advice as is currently available on these safety aspects of experimentation.

4.2 Techniques

Mechanical measures

Vibration is an oscillation, and so its quantifiable measurement inevitably involves detecting the relative motion between two bodies or surfaces. In studying vibration in real-life and in experimental situations, the techniques and methods developed for its measurement have been based upon determining the

Fig. 4.11. Suggested limits for exposure to vibration. a_x, a_y and a_z are acceleration in the direction of the *x*, *y* or *z* axis respectively, where the *x* axis is back to chest, *y* right to left side and *z* foot (or buttocks) to head. (i) a_x a_y 1 minute exposure (ex) boundary; (ii) a_x a_y 1 minute fatigue decreased proficiency (FDP) boundary; (iii) a_z 1 minute ex boundary; (iv) a_x a_y 1 hour FDP boundary; (v) a_z 1 minute FDP boundary; (vi) a_z 1 hour FDP boundary; (vii) a_{xy} 8 hour FDP boundary; (viii) a_z 8 hour FDP boundary. (After BSI, 1974).

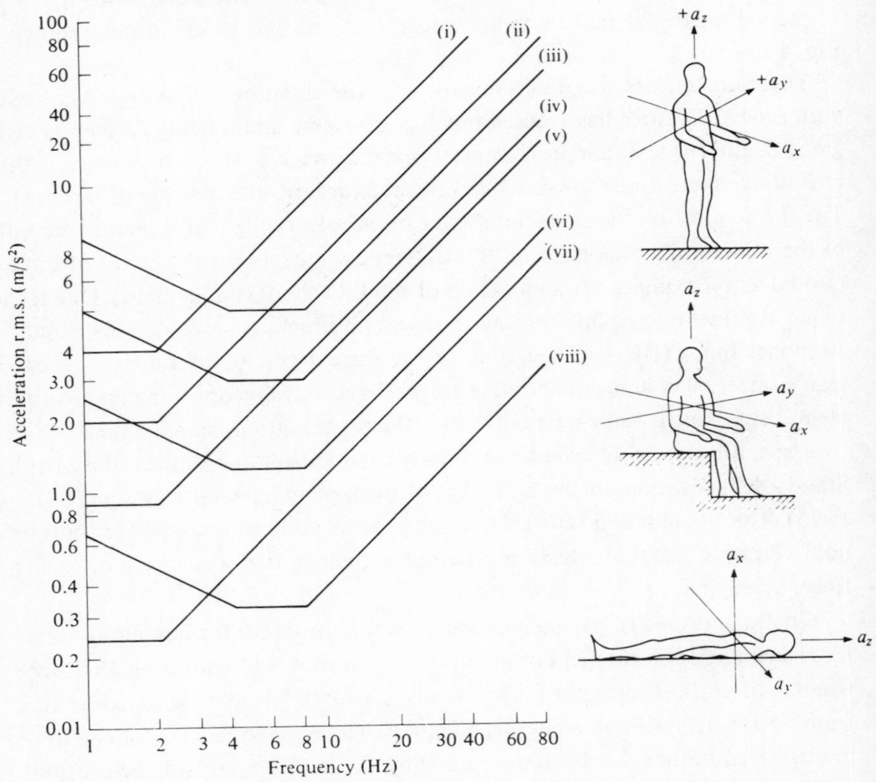

physical parameters of the motion. By inference, the basic parameter is that of the displacement between the moving body and its fixed surroundings. Alternatives to the displacement measure are those of acceleration and velocity, which are related to it by time. There are, however, good practical reasons for selecting a particular measurement technique. Displacement transducers are commonly used in equipment and machines where their mechanical requirements can be met, and where their dimensions are not prohibitive. Velocity transducers tend to be rather large and difficult to incorporate into a measuring system. Recent advances in technology have led increasingly to the use of the accelerometer, an acceleration-sensitive transducer. These can now be made very small, sensitive and capable of operating uniformly over a wide frequency range. Similar improvements in electronic circuitry allow the relevant integrations with respect to time to be made, thus producing the velocity and displacement values. The reduced size of accelerometers frees the subject system from unnecessary transducer loading factors, and allows almost 'point measurements' to be made.

Transducers

Displacement transducers are of a single type and construction. They are, in effect, linear electronic potentiometers. Thus they must move within closely constrained paths and must be connected to both moving and stationary surfaces. This affects both the maximum relative displacement and the overall sensitivity of the displacement measurement.

Accelerometers are transducers which produce a voltage signal directly proportional to the acceleration value of their motion. There are several different types, depending upon how the force acting on the transducer's internal mass is detected. Diagrams illustrating these are shown in Fig. 4.12. Three techniques are involved. First there is the variable inductance type where the motion of a ferromagnetic seismic mass core within an electrically energised coil directly influences the induced voltage in the coil. Secondly there is the piezo-electric type, which depends on the mechano-electric qualities of certain crystal structures, for instance quartz and some ceramics. These structures when compressed by a seismic mass generate an electric charge which is directly proportional to the force exerted on the structure and hence the overall acceleration. Thirdly, and most commonly used in the context of human vibration research, is the piezo-resistive type in which the seismic mass is connected to a cantilever beam whose deformation or bending is detected by strain gauges. These gauges have the property of varying their ohmic resistance in response to stretching or compression. Most linear transducers, which are sensitive to acceleration along a single 'active' axis, are of one of these three types. Accelerometers sensitive to angular motion are not very common, and are usually based upon the inductance-mass transducer system.

All transducers, of whatever detecting system, require an amplification of their output voltage signals if these signals are to be of further use. The type of amplifier used depends upon the type of transducer. For instance, the charge effect of the piezo-electric type requires a very high input impedance amplifier or 'charge amplifier' if the charge produced by the transducer is not to be lost. This effectively converts the charge to a normal voltage signal. Piezo-resistive

Fig. 4.12. The three main types of accelerometer design, and their circuitry. (*a*) An electromagnetic type. The mass *m* is on a pendulum, whose variation of position relative to the sensor is detected and fed back to the torque rotor to maintain the mass's mid position. This feedback signal is output as a signal proportional to acceleration. This system is also used in angular acceleration measurement. (*b*) A piezo-electric type. *f* and *r*, feedback capacitance and resistance (time constant parameters); *c*, cable capacitance; *m*, mass of piezo-electric material P_e. (*c*) A piezo-resistive type. Black rectangles, strain gauges; *m*, mass on a cantilever beam.

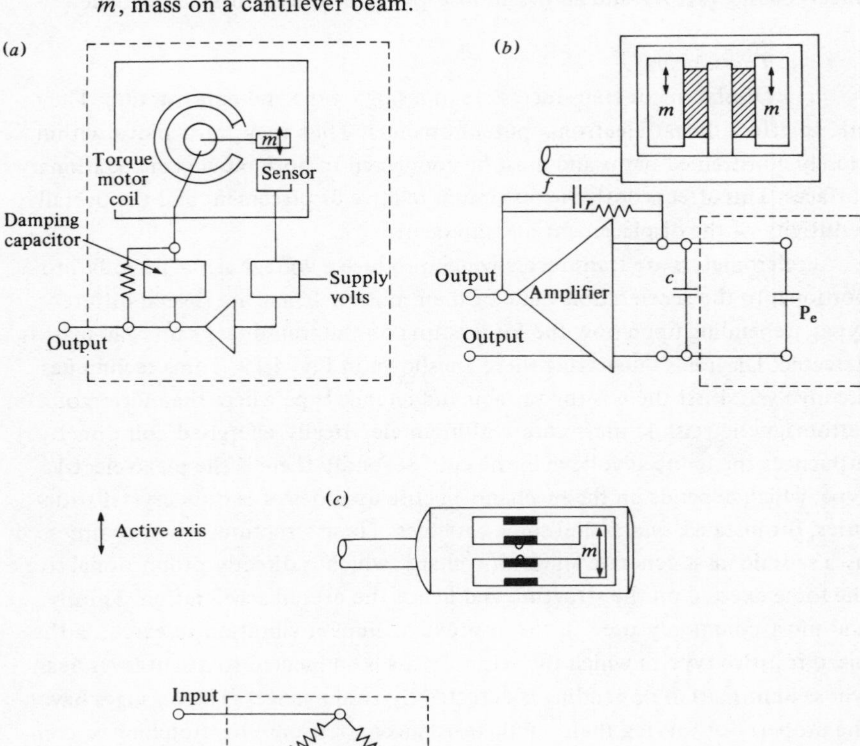

transducers require only a normal voltage gain amplifier of the type commonly found in many electronic circuits.

Selection of a particular type of accelerometer for a particular measurement system will depend on the characteristics required. In general the piezo-electric types are very robust and have a wide operating frequency range of about 0.2 Hz to several kilohertz; however, they are relatively large. The piezo-resistive type, although they can be very small and very light-weight, are not so robust. They do not operate up to such a high frequency but they will measure a 0 Hz acceleration level. The inductance types are usually fairly large, they have a low maximum frequency, and are quite delicate. However, they can be made to be extremely sensitive.

Recording

Once a vibration measurement signal has been obtained, it is usually necessary to record it. This may be for several reasons including subsequent analysis, experimental safety, portability of equipment, and lack of space in the vibrating environment. Since the frequency range of interest from the human factors point of view is 0 Hz to about 100 Hz, any recording system must typically be able to cope with this range of frequencies. In magnetic tape-recorders it is the lowest frequencies which are the most difficult to handle, whereas in paper and chart-pen recorders it is the higher frequencies which can prove awkward. Methods for overcoming these difficulties are now well established, and in the future are likely to become less of a problem due to advances in instrumentation technology. Magnetic tape-recorders used for this type of 'instrumentation' recording are based on a frequency modulated (FM) system. This system records the analogue vibration signal summed together with a high-frequency carrier signal onto the tape. This overcomes the inherent lack of low-frequency resolution found in a magnetic recording system. High-frequency paper and chart recorders use a photographic system, usually based on an ultra-violet light source, or alternatively a type of digitally controlled spark-paper erosion technique.

Future recording systems will probably be based on digital recording techniques which are, at the moment, limited by the physical dimensions of the electronic memory. These future systems will typically be able to produce recording accuracy of about 0.02% or better, compared with an FM instrumentation recorder which allows an accuracy of usually less than about 0.1% (60 dB) of a full-scale deflection signal. The advantage of an electronic data storage system is the availability of the recorded signal for retrieval and computerised analysis, which allows the use of sophisticated analysis techniques. At the moment the only feasible bulk store for electronic data is analogue magnetic tape.

Current instrumentation recorders are capable of recording from 1 to 42 channels of analogue information continuously for up to 12 hours. This advantage of parallel time recording can only be achieved digitally by a fairly sophisticated electronic buffer system together with the single analogue–digital converter (ADC). In the more usual multiplexed digitiser system, the input signals are fed serially to the ADC. This introduces a finite time delay between input data channels and creates erroneous phase information if the data are used for a comparative frequency analysis. A future solid-state recording system will exhibit several other advantages over a magnetic tape system. These include increased robustness and the elimination of media variations. These variations are manifested in magnetic tape-recorders as so-called wow and flutter. They can, to a large extent, be compensated for by suitable recording and instrumentation handling techniques.

No large electronic data storage system actually displays the data to an operator. Suitable visual devices, such as oscilloscopes and ultra-violet paper recorders, can be connected to the storage system in a 'read after write' manner, thus allowing the input data to be visually verified.

Analysis

The detailed analysis of vibration waveforms and of the very large quantities of data typically generated in a vibration experiment has only been made possible by the advent of the computer, mainly the digital rather than the analogue type. In general, data of the waveform type can be analysed in either the time or the frequency domain.

Time domain analysis may be carried out on either analogue or digital computers. This usually results in either a histogram display of signal amplitude versus signal peak occurrence, or as some overall amplitude measurement, e.g. the r.m.s. level. Overall measurements often afford some sort of 'single-figure' answers, whose usefulness is in relating one vibration signal to another. Often these signals are frequency weighted, in a fashion similar to the techniques for measuring noise. The analysis can be displayed as a correlation function, i.e. a measure of the extent to which two time domain signals correlate as a function of the time displacement between them.

Frequency domain analysis is usually carried out using digital techniques, and has some inherent disadvantages. The two most important factors to be considered are those of actually acquiring the digital data, and the necessity of creating store (memory) blocks of data points, in effect taking samples of the time domain signal. Acquisition of digital data requires some form of analogue–digital conversion, converting a continuously varying signal into a series of numerical values each separated by a finite time interval. Since frequency is a reciprocal function of time, it is evident that the interval duration (Δt) between data points must be sufficiently small to allow the frequency spectrum to contain valid data. The effect where high frequencies appear as low-frequency data is called aliasing. To remove this effect it is necessary to low-pass filter the input analogue signal to remove any high-frequency components, and to ensure that the digitiser's sampling frequency is greater than twice the highest desired frequency. This effect is illustrated in Fig. 4.13. Creating store blocks of digital

Fig. 4.13. (*a*) and (*b*) The sampling precautions necessary when digitising analogue data. Dashed line, digitised signal; f_m, maximum required frequency; f_s, sampling frequency, one sample Δt apart. Digitised signal errors: (*a*) $f_m = f_s$ and (*b*) $f_m > f_s$. Minimum sampling rate: $f_s = 2f_m$ (Nyquist frequency); $t = ½T$. (*c*) The aliasing error of digitised signals. The alias error is added to the actual signal, creating an apparently increased high-frequency signal component.

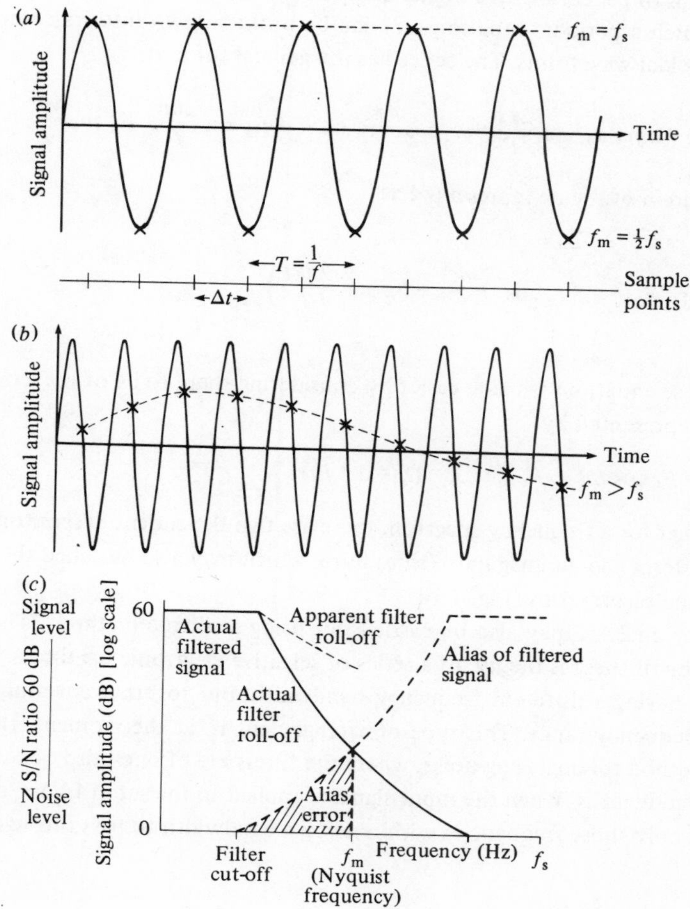

data points is an inherent effect in a digital computer system, and in using the methods for calculating the Fourier transform. This effect generates a series of equivalent time domain samples whose frequency characteristics can be calculated. This series implies that the statistical properties of any one block are the same as for any other block. This may not always be the case, as previously discussed with reference to non-stationary random waveforms, and so any averaging process must be used with caution.

Having generated the valid digital data, further computation may involve both time domain and frequency domain techniques. A commonly used technique is to create the frequency spectrum of the time domain data block. Nowadays this involves generating the Fourier transform, usually by using the so-called Fast Fourier Transform (FFT) algorithm. This algorithm is a fast, efficient means of generating in a digital computer a trigonometric series (Fourier series) by which a periodic function $x(t)$ can be broken down into its component sinusoidal waveforms. The series has the general form of

$$x(t) = a_0 + a_1 \cos \frac{2\pi t}{T} + a_2 \cos \frac{4\pi t}{T} + \ldots + b_1 \sin \frac{2\pi t}{T} + b_2 \sin \frac{4\pi t}{T} + \ldots$$

This may more usefully be represented as

$$x(t) = a_0 + \sum_{k=1}^{k=\infty} \left(a_k \cos \frac{2\pi kt}{T} + b_k \sin \frac{2\pi kt}{T} \right)$$

Handling these equations is made easier by considering them to be of the 'complex' type, represented by

$$F = F_0 \cos \omega t + jF_0 \sin \omega t \quad (F = a + jb)$$

This infers that for a frequency spectrum there are two dependent descriptors, a real (cosine) term and an imaginary (sine) term, which may also be called the magnitude and phase terms (Fig. 4.14).

Frequency analysis may also be carried out using analogue methods. The most common of these is the use of a series of selective electronic band-pass filters, each having a different frequency bandwidth, but together covering a very wide frequency range. This type of arrangement is, at the moment, the preferred method for analysing noise, where the filters are of one-third octave frequency bandwidths. When the input signal is applied to the set of filters, each filter allows only those frequencies within its own bandwidth to pass through.

Fig. 4.14. The use of j, the complex operator. (*a*) Cartesian and polar representations of a rotating vector F. F is fixed by values of x and y and also by F_0 and ϕ. (*b*) Vector representation in the complex plane. In the vector diagram the description to fix F relative to the two axes is the vector sum of the values on the axes: $F = x + jy$. To incorporate the phase ϕ, by applying trigonometry: $x = F_0 \cos \phi$ and $jy = F_0 \sin \phi$, thus $F = F_0 \cos \phi + jF_0 \sin \phi$.

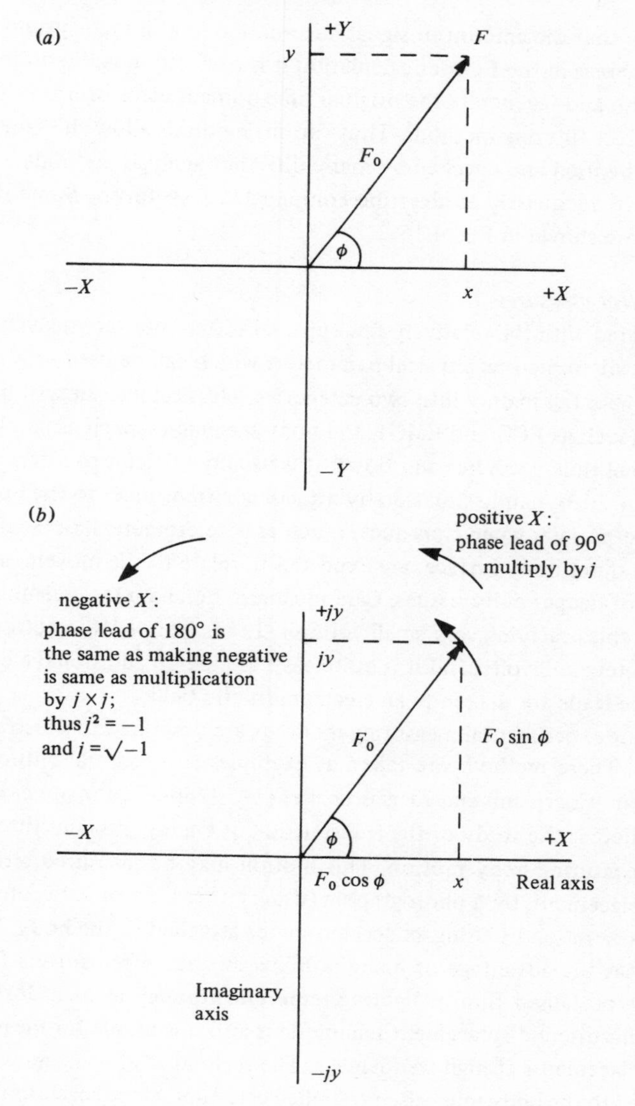

Integrating circuits then measure this passed energy and display the value relative to the values from all the other filters. This type of analysis does not require sampling, it can work in 'real time', although of course the characteristics of the integrating networks must be controlled. Electronic techniques can now incorporate digital filter networks to replace the rather bulky equivalent analogue circuits.

It can be seen that the amount of signal information in a filtered system may be degraded. Whereas in the Fourier calculation it is perfectly possible to inverse Fourier transform and regenerate the original time domain waveform this is not possible with direct filtering methods. Thus various methods allow the same time original data to be used and checked by many different analysis methods, which may be necessary adequately to describe complicated waveforms. Some illustrations of this are shown in Fig. 4.15.

Biological measures

Compared with the relatively few types of different direct mechanical measures, there are numerous physical parameters which can be used on the human body. These fall mainly into two categories: physical measures of body electric signals (such as ECG and EMG), and body mechanics, such as acceleration and internal fluid pressures and flows. It is usually difficult to attempt to take measures in the dynamic situation by attaching a transducer to the body. Problems arise especially when transducers such as accelerometers, or electrodes attached to the outer body surface, are expected to relate to the movements and properties of deeper body tissues. Care must also be taken in screening and routing signal cables carrying very small voltages. Leads from EMG electrodes only conduct a few millivolts, and it is quite easy to pick up additional electrical signals when the leads are shaken in an electromagnetic field.

Two different experimental measuring methods are described to illustrate the difficulties. These methods are taken as examples from several authors including Roman, Coermann and Ziegenruecker (1959) and Coermann *et al.* (1960). First there is the study of the transmission of energy into and through the body by measuring body motion. This motion may be measured either directly as displacement, by a photographic (using either light or X-ray) technique, or as acceleration by using an accelerometer attached to the body. The first of these has the advantage of being non-invasive but often suffers from needing either specialised film or lighting requirements such as high doses of X-radiation. This direct displacement technique is also unsuitable for measuring very small displacements at high frequencies. The technique of using an accelerometer attached to the body must also be applied carefully. Since accelerometers are relatively very sensitive to small displacements at high frequencies, their attachment to the body must be very secure, and the relative movement between skin surface and bone must be reduced. The transducer should be small and light enough not to impose a dynamic load on its substrate surface.

Secondly there is the study of measures of biological changes in the dynamically moving body-system. The measuring and analysis techniques are similar to

Fig. 4.15. (*a*) Different analogue waveforms capable of creating frequency spectra having similar 'flat' low-frequency characteristics. A, random motion; B, a swept sinusoidal motion; C, a haversine pulse; D, a square wave.

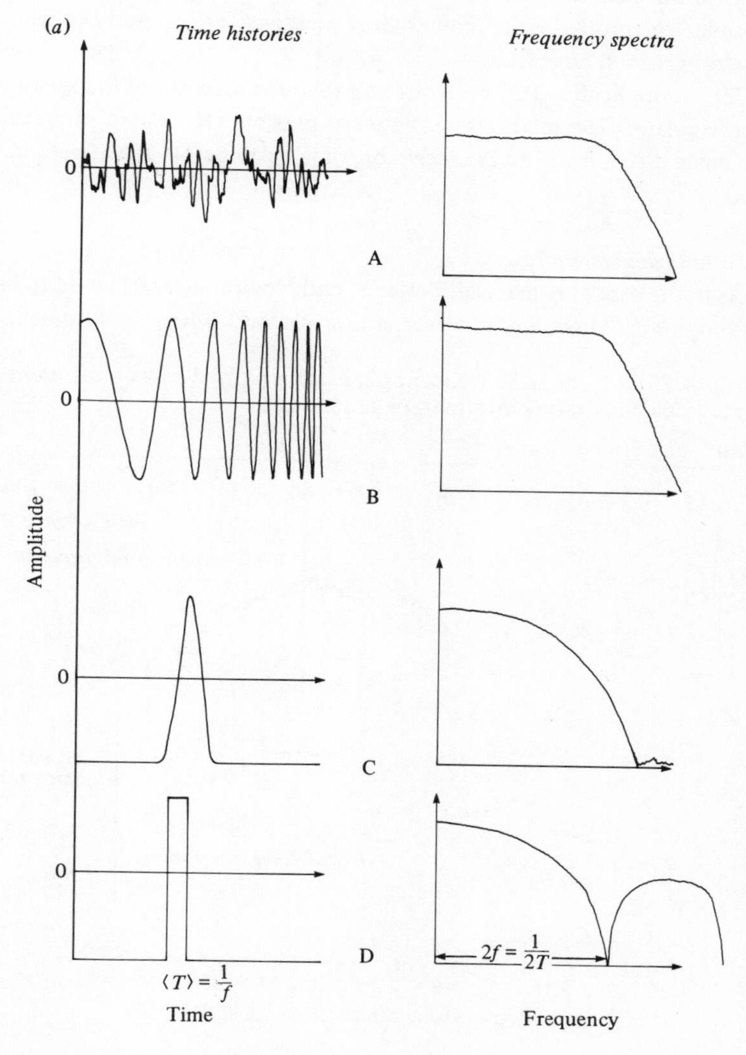

those used in the static mechanical environment, but problems arise which are mainly due to the motion induced in the body. Breathing studies must consider the apparent gas flows, measured in a flow transducer positioned at the mouth, that are generated by the mechanical pumping action of the lungs. Chemical changes in body fluids are not only relatively rather slow-acting in response to fast mechanical changes in the environment (for example the integrating effect of urine sampling), but some samples are sometimes rather difficult to obtain (for example from a blood catheter) whilst the body might be moving. Chemical techniques for the analysis of blood are now quite sensitive and sophisticated, with the advent of protein radioimmunoassay methods for the study of hormones and other blood chemicals.

In the long term, health effects are difficult to ascertain without the use of population statistics. The relationships between possible effects and probable causes are made difficult due to the inevitable time delay before the effects are manifested.

Pathological measures

As in the case of some health effects, pathological measures tend to be relatively long term. There are, however, a few instances where fairly quick

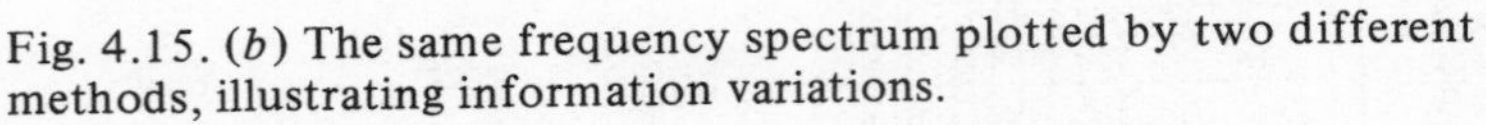

Fig. 4.15. (*b*) The same frequency spectrum plotted by two different methods, illustrating information variations.

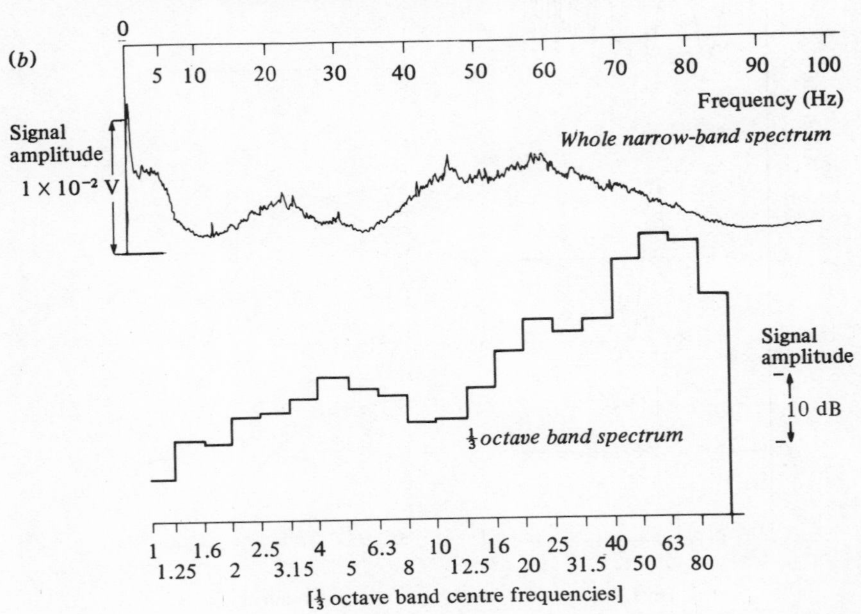

effects can be observed. Two industrial situations which have been documented are those of health risks to agricultural tractor drivers, who suffer an increased risk of gastrointestinal disorders and spinal damage (see Rosegger and Rosegger, 1960; also summarised in Matthews, 1964); and those to some workers who use vibrating hand tools, especially forestry workers handling chain saws, who tend to suffer a marked increased incidence of the so-called white finger syndrome or Raynaud's phenomenon. This area of industrial health has been reviewed by Griffin (1980).

Most pathological damage to tissues is associated with their absorption of energy. This may be manifested in two ways: first as the vascular damage in Raynaud's phenomenon, or secondly as the gross relative movement between large organs having different resonant frequencies, causing rupture of the surrounding tissues.

One further source of information on the effect of transient vibration and shock is the study of accident and suicide statistics. These have provided important information in the past, especially about the vertebral column and its strengths and weaknesses. Some of these studies have been described by Snyder (1971).

Subjective measures

Many methods of subjective measurement are well known to experimental psychologists and are described, for example, by Guilford (1954). The methods mainly fall into two groups: (*a*) rating, either by a line with or without bounds or use of descriptors; and (*b*) comparison, either within experimental conditions or referred to a known control condition. Threshold measures may also be made as an alternative, although rather specialised method.

Obtaining a verbal description can be very difficult, and so most experiments use some form of comparison where no such semantic problems exist. It is unfortunate that methods such as that of paired comparisons are tedious to perform, and present unknown time exposure factors to the subject.

Most experiments generating some form of equal-exposure or equal-annoyance contours have used the vibration intensity matching technique, with quite a high degree of success. The technique most frequently used for the estimation of comfort has mainly used rating scales, usually with some type of bounds, and end descriptors. These techniques are reviewed in Oborne (1978*b*).

Real-life versus laboratory experiments

In the study of man in his mechanical environment it is always important to bear in mind the relationship between laboratory studies, which may generate criteria and their related exposure limits, and the actual experience of the real-life situation. A good example of the anomalous situation which can arise has been the generation of the present International Standards Organisation vibration exposure limits, the data for which have been drawn mainly from laboratory experiments which used sinusoidal waveforms. It is quite apparent that, with the exception of some helicopter and ship motions, sinusoidal waveforms are very rarely encountered in real-life. In addition, the laboratory situation has very few uncontrolled variables, whereas the real-life environment abounds with them. It is certain that there are many stresses apart from vibration which are impressed upon man: vibration as a stressor rarely exists alone and is found together with noise and quite often with heat. Whether or not these individual stressors combine, and how they may combine, will not be known until much more experimental work has been completed. Some attempts at answering these questions have been started, and one by Grether *et al.* (1971) describes one such multi-stress study. This whole area of simulation of the mechanical environment has been reviewed by Floyd and Sandover (1972).

The study of the long-term health effects of vibration is fraught with difficulty caused by the time span of the study. The relationships between cause and effect are difficult to quantify, and are often only investigated using methods which involve population statistics studies.

In the laboratory, simulation of a motion recorded in real-life can be very difficult. The machines used for creating the simulated motion are restricted in their displacements and possibly in their axes of motion. Whereas this is not too much of a problem for vertical motion, it is quite a problem for lateral and even more so for fore-and-aft motion. It is generally possible to re-create a similar frequency spectrum for the motion, but the related time domain motion is usually much more difficult to achieve.

It is thus important to realise the limitations of laboratory studies, but to appreciate that they can be used to examine certain fundamental aspects of the problem, while their major asset is in allowing many experimental parameters to be controlled to some extent. Much of the emphasis on 'real-life' exposure to higher levels of vibration has been of a military, especially aerospace, nature. This is shown by the major contribution made by military research organisations and may be illustrated by reference to some reviews in, for example, conference proceedings (e.g. AGARD, 1975) and other works already cited.

Analogies

A similar discussion to that of the value of laboratory versus real-life experiments is of the value of using non-human subjects in experiments to study likely human reactions. The main alternatives to using man as a subject in possibly risky and dangerous experiments are to use either other animals (alive or dead), or cadavers, or mechanical or electrical analogue models. In all cases the use of non-human subjects, sometimes necessary for ethical reasons, poses difficult questions of the relevance of the results obtained to the equivalent human situation.

Animals have different tissue distribution, weights and centres of gravity, as well as having physiological differences when compared with man. Cadavers have been used in some mechanical environment experiments. However, although they have most of the human characteristics they do not show muscle tone, an important facet in postural control, nor do they show the tissue fluid balance especially associated with the blood circulation.

Mechanical and electrical circuit analogies, the latter usually based in a computer, can be made to model a reaction shown as a human response. These models are usually rather specific to their particular occasion. The difficulties in their use lie in the quality of data used to generate the model, consideration of all of the relevant variables and, finally, awareness of the linearity of the human response. This last point is of great importance if the model is to be used in predictive studies, where a human subject might be injured if exposed to a higher energy input parameter. The use of models has been reviewed (Kornhauser, 1970), and Table 4.1 shows some of the main factors requiring consideration.

The studies in which models have been shown to be most useful have been those investigating the effects of mechanical energy, either short or long term (shock or vibration), as the input to a system containing a man. The philosophy behind the use of models in the context of vibration research has been twofold. First there is the requirement to study environments which may be potentially too dangerous for any human subject. Examples of these environments are many and varied, but are often associated with accident or emergency situations: for example in car-crash testing, or in shock impact testing as seen in aircraft ejector seats. Models used in these studies include anthropomorphic dummies (used in seat belt tests), and computer analogues, for example the Dynamic Response Index (DRI), an 8-Hz resonance simulation of the spine (used in biodynamic studies of the human response to impact accelerations). More complex models of the human biodynamic response also have been described, as shown in Fig. 4.16. Animal subjects have also been used in early biodynamic and space research. The second main use of models has been the attempt to quantify the human subjective response to vibration and ride environments. If it proves to be possible to model and simulate these responses, then it will be possible to derive ride-meters and other analytical and control systems for use in everyday, and testing, situations.

Table 4.1. *Relative advantages and disadvantages of each approach to obtain useful data*

Method	Advantages	Limitations
Man: experimentation, results of accidents	Realistic material properties, neuromuscular reactions and injury modes	Experimentation only to point of discomfort; accident details hard to reconstruct; accident data statistical
Animals: experimentation	Tissue properties similar, neuromuscular reactions and injury modes similar to man	Structures different from man; organic differences; size scaling necessary
Computer models	Ease of 'experimentation'; programs reproducible and permanent; model may be improved to any degree of fidelity	Parameters must be adjusted per experiment results; need good set of subsystem (organs etc.) models for overall model
Physical models	Superb visual aid; inexpensive way of observing kinematic effects	Expensive to model internal organs adequately; serious problems of friction damping, material properties; injury hard to predict

From Kornhauser (1970).

4.3 Cases

Most vibration studies have been carried out in the context of the transport industry. These have been aimed mainly at achieving improved performance of a vehicle crew or improved comfort and safety for the vehicle's passengers (e.g. Oborne, 1977; Griffin, 1978).

Crew performance can be affected by vibration. Mention has already been made of the visual and manual control effects of vibration. There is also one further area where the vibration effect can be improved or worsened, and that is in the design of a support or seat. Having a reclined seat will support the body, but the seat-back will allow the vibration to by-pass the natural damping charac-

Fig. 4.16. Mechanical model of the human body exposed to G_z vibration or impact and to external pressure loads. Resonant frequencies, f_0, of body segments: a (head), ~ 30 Hz; b (chest wall, stiff diaphragm), ~ 60 Hz; c (vertebral column), ~ 8 Hz; d (abdominal mass) ~ 4–8 Hz; e (input impedance, sitting subjects), ~ 4–6 Hz. F, input force. (After von Gierke, 1964.)

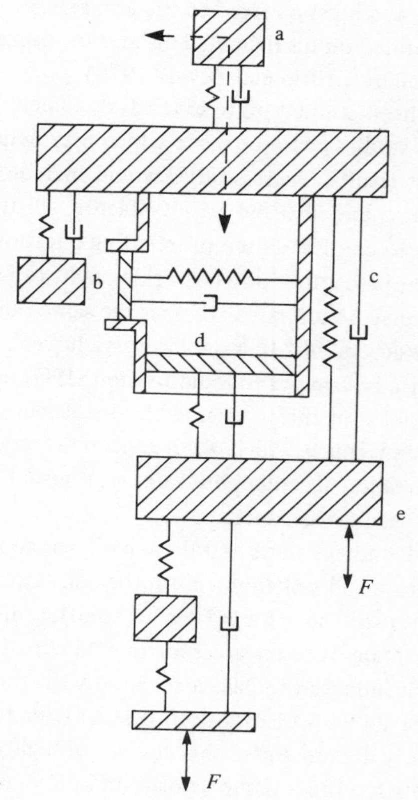

teristics of the vertebral column and thus increase the vibration level reaching the head. Sensitivity to the vibration is changed by varying the angle of the seat's recline. For a given vertical seat motion the G_z vibration component is reduced and the G_x vibration is increased if the recline angle is increased. As has been discussed previously, this will increase the human's sensitivity to the vibration.

One other context in which seating has been considered is in the use of vibration-absorbing seats. This type of seat represents a mechanical single degree of freedom resonance system which is tuned to have a resonance peak below the resonance of the man-control system sitting on it. As shown in Fig. 4.6, where the ratio of frequencies is greater than $\sqrt{2}$, the input energy is attenuated. This is achieved by allowing the relative displacement movement, between seat pan and floor, of a flexible seat support. It can be seen, therefore, that this relative motion, whilst reducing the vibration energy reaching the man, may actually cause him greater difficulties as he seeks to use control systems which are still connected to the floor – for example the steering wheel and pedals in a driving task. To some extent these difficulties can be reduced by having the control systems attached either to the seat, as a joystick fixed to the arm rest, or to the man, such as a head up display mounted on his helmet. This area of human performance has been reviewed in detail by Griffin and Lewis (1978).

Vehicle passenger comfort has been studied quite extensively. There are, however, some areas where further studies could be made which may help explain some of the variation in past results. These studies would include, for instance, motivation and expectancy, and also the novelty factor. All these human traits can be seen in answer to comfort–discomfort rating questions. In road travel people will tolerate their perception of what a ride ought to feel like: a bus-ride may be quite rough but may be tolerated whereas the same ride in a high-quality saloon car may not be acceptable. In a similar way a hovercraft, according to the International Organisation for Standardisation (ISO) criteria and limits, produces a very rough ride, but this is accepted by its passengers. People pay to travel on fair-ground machines which often generate several *g* in their motion. Expected motion is readily accepted, but safe but unusual motion (for example in a building) can cause panic.

Modern hand tools often include moving parts and there has been an intensive study of white finger syndrome (Raynaud's phenomenon), the condition of the hand and fingers associated with the use of such tools. Griffin (1980) studied it in forestry workers using chain saws. Many forestry areas are in cold climates, and it is this temperature effect which induces the pain associated with the condition. The population as a whole shows a distribution of sensitivity to the phenomenon, but it has been clearly demonstrated that the use of high-powered tools accelerates its onset, probably by direct tissue damage and the induction of chronic disorders of the peripheral blood vessels and nerves. The response to chilling the affected limbs is an abnormal vasoconstriction in which the fingers go white and numb.

4.4 Overlaps: vibration and its implications

It has already been suggested that vibration should be considered together with other environmental and human parameters. For example, the human's subjective opinion of vibration will be greatly affected by his attitude, expectation and experience. Many sports, such as horse-riding, flying and driving, expose their voluntary participants to levels of vibration which they may not tolerate in other environmental contexts. Also, vibration is only one stressor in an environment which often has other additional stressors. The relative intensities of these stressors may affect their relative perceptions. For instance, a particular level of vibration may be considered to be uncomfortable to a person also exposed to a low noise level. If the noise level is increased then the same vibration level may be perceived to be less uncomfortable. It is not yet known precisely how particular stressors act, but their action is unlikely to be linear, and is often illustrated by a U-shaped curve, a shape seen, for example, in the exposure criteria (BSI, 1974). If this is the case then stressors are most unlikely to react, or sum, in a linear manner and their combined effect may be complex; they may interact either synergistically or antagonistically. Low frequencies of vibration may be either disorientating or cause drowsiness; higher frequencies may actually be arousing before becoming uncomfortable.

The mechanical and performance impairment effects of vibration have also been mentioned. Such an effect may be caused either directly through the movement of limbs, or indirectly through the disturbance of vision or other sensory pathways. It has been emphasised that it is important to consider both the man and his task equipment within the confines of the vibration environment. The three areas of prime importance are (1) those of the task or operating system's sensitivity to vibration (for instance a wheel control is less sensitive to disturbance than is a thumb-controlled joystick), which merges with (2) the man–machine interface, which in turn overlaps with (3) man's sensitivity as an individual to the vibration.

References

AGARD (1975) *Conference on Vibration and Combined Stresses in Advanced Systems,* ed. H.E. von Gierke. Paris: Advisory Group for Aerospace Research and Development.

BSI (1973) *Guide to the Safety Aspects of Human Vibration Experiments.* BSI Draft for Development DD23: 1973.

BSI (1974) *Guide to the Evaluation of Human Exposure to Whole Body Vibration.* BSI Draft for Development DD32: 1974. London: British Standards Institution.

BSI (1975) *Guide to the Evaluation of Exposure of the Human Hand–Arm System to Vibration.* BSI Draft for Development DD43: 1975. London: British Standards Institution.

Coermann, R.R., Ziegenruecker, G.H., Wittwer, A.L. and von Gierke, H.E. (1960) The passive dynamic mechanical properties of the human thorax-abdomen system and of the whole body system. *Aerospace Medicine*, **31**, 443–55.

Floyd, W.F. and Sandover, J. (1972) Problems associated with the application of the results of vibration research to practical situations. In *Society of Environmental Engineers: Vibration and Man Symposium.*

Glaister, D.H. (1978) Human tolerance to impact acceleration. *Injury*, **9**, 191–8.

Grether, W.F., Harris, C.S., Mohr, G.C., Nixon, C.W., Ohlbaum, M., Sommer, H.C., Thaler, V.H. and Veghte, J.H. (1971) Effects of combined heat, noise and vibration stress on human performance and physiological functions. *Aerospace Medicine*, **42**, 1092–7.

Grether, W.F., Harris, C.S., Ohlbaum, M., Sampson, P.A. and Guignard, J.C. (1972) Further study of combined heat, noise and vibration stress. *Aerospace Medicine*, **43**, 641–5.

Griffin, M.J. (1978) The evaluation of vehicle vibration and seats. *Applied Ergonomics*, **9**, 15–21.

Griffin, M.J. (1980) Vibration injuries of the hand and arm; their occurrence and the evolution of standards and limits. *Health and Safety Executive Research Paper 9.*

Griffin, M.J. and Lewis, C.H. (1978) A review of the effects of vibration on visual acuity and continuous manual control. I: Visual acuity. II: Continuous manual control. *Journal of Sound and Vibration*, **56**, 383–457.

Guignard, J.C. and King, P.F. (1972) *Aeromedical Aspects of Vibration and Noise.* Paris: Advisory Group for Aerospace Research and Development. AGARD-AG-151, Nov.

Guilford, J.P. (1954) *Psychometric Methods*. New York: McGraw-Hill.

Harris, C.M. and Crede, C.E. (eds.) (1961) *Shock and Vibration Handbook*, 3 vols. New York: McGraw-Hill.

ISO (1974) *Guide for the Evaluation of Human Exposure to Whole Body Vibration.* Report ISO 2631-1974 (E). International Organisation for Standardisation.

Kornhauser, M. (1970) Biodynamic modelling and scaling; anthropometric dummies, animals and man. In *AMRL Symposium on Biodynamic Models and their Applications*. Report AMRL-RT-71-29, Dec. Ohio: Wright-Patterson Air Force Base.

Matthews, J. (1964) Ride comfort for tractor operators. *Journal of Agricultural Engineering Research*, **9**, 3–31.

Oborne, D.J. (1977) Vibration and passenger comfort. *Applied Ergonomics*, **8**, 97–101.

Oborne, D.J. (1978*a*) The stability of equal sensation contours for whole body vibration. *Ergonomics*, **21**, 651–8.

Oborne, D.J. (1978*b*) Vibration and passenger comfort: can data from subjects be used to predict passenger comfort? *Applied Ergonomics*, **9**, 155–61.

Reason, J. (1978) Motion sickness: some theoretical and practical considerations. *Applied Ergonomics*, **9**, 163–7.

Roman, J.A., Coermann, R.R. and Ziegenruecker, G.H. (1959) Vibration, buffeting and impact research. *Journal of Aviation Medicine*, **30**, 118–25.

Rosegger, R. and Rosegger, S. (1960) Health effects of tractor driving. *Journal of Agricultural Engineering Research*, **5**, 241–75.

Shoenberger, R.W. and Harris, C.S. (1971) Psychophysical assessment of whole body vibration. *Human Factors*, **13**, 41–50.

Snyder, R.G. (1971) Man's survivability of extreme forces in free-fall impact. In *AGARD Conference Proceedings on Linear Acceleration of the Impact Type*. Report AGARD-CP-88-71, June.

Troup, J.D.G. (1978) Drivers' back pain and its prevention. A review of the postural, vibratory and muscular factors, together with the problem of transmitted road-shock. *Applied Ergonomics*, **9**, 207–14.

von Gierke, H.E. (1964) Biodynamic response of the human body. *Applied Mechanics Review*, **17**, 951–8.

5 EFFECTS OF CLIMATE

David McK. Kerslake

5.1 Physiological principles

The ergonomist does not need to know much climatic physiology. If the workings of the body were better understood by physiologists he would doubtless benefit from a knowledge of them, but present attempts to describe thermoregulation in terms of current engineering practice are only partly successful, and it is not always clear whether they are intended to be any more than quantitative parables (Hammel, 1968; Hardy, Gagge and Stolwijk, 1970; Bligh and Moore, 1972; Cabanac, 1975). Body temperature alone is a very poor guide to the state of a human subject: nobody yet knows the proper way to stir together such things as rates of heat transfer and temperature distributions, and the opinion of the subject, at least as to how near he is to collapse, is astonishingly unreliable.

If we do not understand what goes on inside the body we can turn attention to the skin surface, where physiology meets physics. Fortunately events here conform with simple concepts of the Ohm's law type. The thick books on heat transfer deal with aspects of non-steady-state conduction and the theory of heat transfer in fluids which rarely arise in the study of human heat exchange. The keen ergonomist may wish to pursue them, and would not lose thereby, but it would be a work of supererogation. The chef does not need to be a chemist.

Heat balance

The living body produces heat at all times. This must eventually be transferred to the environment, but rates of heat production and heat loss need not always be exactly equal and indeed are rarely so. They are usually expressed per unit body surface area (the average area of adult men is about 1.8m^2). The relevance of this to heat loss through the skin is obvious, and it so happens that the resting heat production of different-sized people is more closely related to surface area than to height or weight. The rate of heat production is the difference between the total rate of energy production, M, which can be found from the rate of oxygen consumption, and W, the rate at which external work (force times distance) is being performed. The heat is disposed of in various ways :

1. Radiation between the skin or clothing surface and surrounding surfaces (R).
2. Convection, which is really a form of conduction to the surrounding air (C).
3. Evaporation of water either through the outer layers of skin (insensible perspiration) or from the skin surface when it is wetted by sweat or external agency (E).
4. Warming and wetting air which is drawn into the lungs and then breathed out (L).
5. Conduction to surfaces in direct contact with the skin or clothing (K).

The equation for heat balance can be written:

$$M - W = R + C + E + L + K + S \tag{5.1}$$

The last term, S, is the rate of storage of heat in the body. Heat balance exists when S is zero, and while this is the state which the body can be said to seek, it does so without much enthusiasm. If the rate of heat production is increased in a fixed and comfortable environment, the body heat content increases slowly and would level off after about 2 hours, except that by this time the next meal, or even bedtime, is approaching. Diurnal and digestive cycles affect the body's temperature preferênces, which occasion changes in heat content (Mills, 1966). Heat balance can be considered the norm, but it is rarely attained or maintained in practice.

Thermoregulation

It is convenient to think of the body as composed of a central core in which temperature is controlled within a narrow band, and a surrounding shell where greater variations in temperature are tolerated. The thermoregulatory mechanisms behave much as though there were a sensitive thermostatic control of core temperature, set at a level which depends on the subject's work rate (Leithead and Lind, 1964; Hammel, 1968). (This concept has to be greatly elaborated to account for the time relations of responses to changes in work rate (Cabanac, 1975).) In general it is roughly true that core temperature depends on work rate and is independent of environment, whereas skin temperature depends on the environment and is independent of work rate (Newburgh, 1949; Nielsen and Nielsen, 1965). The main mechanisms for controlling heat exchange are as follows (Wyndham, 1973):

1. *Vasomotor.* The skin needs some blood supply to keep it alive, but skin blood flow can be increased to many times this basic level. Increasing it raises skin temperature and increases heat transfer to the environment. Decreasing it

cools the shell tissues and conserves heat in the body core. Changes in skin blood flow are most marked in the extremities of the limbs, and less so in the trunk and head.

2. *Sweating.* As skin temperature approaches core temperature it becomes difficult to transfer heat from core to skin. Skin temperature can be held down in hot environments by evaporation of sweat.

3. *Shivering.* Even when skin blood flow is minimal there may be excessive loss of heat from the core by conduction through the shell tissues. Core temperature can then be maintained by increasing the rate of heat production. Shivering is a disorganised muscle activity having this effect.

Over a narrow band of conditions near the state of thermal comfort, vasomotor changes provide an adequate control of heat loss. If it is at all warm the sweating mechanism comes into play, but this does not necessarily imply much discomfort. People will pay large sums to lie on hot beaches. Humid heat, when the sweat can be felt on the skin, is usually regarded as uncomfortable. Nobody likes shivering.

Acclimatisation

Thomas Mann (1924) has said that acclimatisation consists of two processes: getting used to it, and getting used to not getting used to it. This is probably more useful than the plethora of words such as acclimation and accustomisation which have been coined and variously defined in recent years. People do get used to hot and cold environments. There are physiological changes which make things more tolerable, changes in behaviour as the subject learns to comport himself better, and a merciful tendency to come to prefer things as they are. This last was conspicuous in the late 1940s, when the British had become accustomed to thick clothing and cold rooms. On visiting the United States they were struck by the heat and stuffiness of the buildings. Changing to the local conventions of dress made it less hot but did not remove the oppressiveness, because they had learnt to prefer a cool head and warm body.

In hot climates physiological adaptations are such as to keep the subject cooler (Robinson *et al.*, 1943). This is not as obvious as it sounds, since it involves a greater water intake. (The camel has a mechanism for conserving water at the expense of keeping cool.) The most important changes are in the sweat glands themselves, which seem to become trained in much the same way as muscles can be trained by use (Fox *et al.*, 1964). As with muscles, full development requires frequent severe activity (Williams and Wyndham, 1968). Mad dogs and Englishmen have the message. Merely living in a hot climate, avoiding the heat where possible and taking little exercise may be ergonomic acclimatisation, but it will not improve sweating capability. There are also circulatory changes which promote the transport of heat to the skin surface (Fox and Edholm, 1963). Blood volume increases to cater for the extra capacity of dilated vessels.

Some of these changes also occur in athletic training, but there is little point in separating the causes since heat acclimatisation is best produced by doing physical work in hot environments. As the subject becomes more acclimatised the conditions should be made more severe so that his resources are still stretched. When groups of subjects are to be treated one can regulate the severity of the exposure individually by changing the rate of working. The aim is to maintain a precarious heat balance at as high a core temperature as possible, consistent with safety. The best results are obtained with exposures lasting at least an hour, and these should continue daily or on alternate days for two or three weeks. Such an acclimatisation regime produces a dramatic improvement in the ability to regulate core temperature in severe environments and a marked reduction in distress in more moderate conditions. When exposure to heat is discontinued the subject reverts to the unacclimatised state in a few weeks (Williams, Wyndham and Morrison, 1967). A high level of acclimatisation can be maintained by booster exposures every week or so (Strydom, *et al.,* 1975). The expense of the necessary climatic rooms, skilled supervisory staff and medical cover may be justifiable in some military and industrial contexts, e.g. for mines rescue personnel, but other forms of protection against heat are usually more cost-effective and more acceptable.

It is less obvious what form acclimatisation to cold should take (Edholm and Bacharach, 1965; Itoh, Ogata and Yoshimura, 1972). Should shivering be increased so as to increase heat production? Should the body shell be allowed to cool further so as to maintain core temperature? Or should the temperature of the extremities be raised so that the hands do not become numb and the danger of frostbite is reduced? The first of these does not seem to occur, but the others do and, surprisingly, can co-exist. The body core contracts, so that more of the tissues are included in the shell. The contracted core is therefore better insulated than before and less heat is needed to maintain its temperature. There is a tendency not to use the hands very much in the cold, but certain people who use them a great deal develop the ability to keep them warm. This necessarily involves increased heat loss, but without it the manual activity would be impossible. The herring filleters on the North Sea coast used to work in the open, their hands either immersed in near-freezing water or exposed, wet, to the wind. In the unacclimatised this rapidly produces severe pain, but the filleters could work all day with little discomfort (Nelms and Soper, 1962). The changes take months or years to develop, and artificial acclimatisation to cold is not feasible in normal ergonomic practice.

5.2 Measuring the environment

For a fuller treatment of the subject of measuring the environment see McIntyre, (1980).

The mercury thermometer should not be despised. Its failures are of the all-or-none type and easily recognised. The same cannot always be said of more sophisticated devices, and it is disappointing to return from a thermal safari to find that the calibration has shifted.

Air temperature

The temperature of the ambient air, T_a, better called the dry bulb temperature to distinguish it from the wet bulb temperature (see below), can be measured by any type of temperature transducer exposed to the air and screened from radiation exchange with surfaces which are not at air temperature. The screen can be dispensed with if a moderate current of air is maintained over the instrument. Since there are usually small fluctuations in air temperature from place to place and from moment to moment, which should be ironed out, there is no need for a rapid response. The mercury thermometer is adequate in this respect, and air movement over the bulb can be supplied either by mounting the thermometer in a sling, rather like a football rattle, which is whirled round, or by means of a fan which draws a current of air over the bulb, in this case surrounded by a silvered screen (Assman psychrometer). When reading the temperature it is tempting to take hold of the end containing the thermometer bulb in order to catch the light on the mercury column. Equally disastrous is over-close scrutiny of the scale, accompanied by heavy breathing over the bulb.

Humidity

The wetness of the air can be expressed in various ways (Fig. 5.1). The importance of humidity in human heat exchange lies in its effect on evaporation, and the quantity directly concerned is the partial pressure of water vapour in the air, p_a. This is sometimes expressed in terms of dew point, T_{dp}, which is the temperature at which the saturated water vapour pressure is p_a. In meteorology it is common to use relative humidity, r.h., the ratio of p_a to saturated water vapour pressure at the prevailing air temperature. It is important to realise that r.h. by itself gives no clues to the value of p_a. One must know the air temperature as well. Climates are frequently reported in terms of daily maxima and minima. If the water content of the air (and therefore p_a) remains constant throughout 24 hours, r.h. may swing from a low value in the heat of the day to near 100% in the middle of the night. Maximum daily values of temperature and r.h. may be wrongly combined to give a gross overestimate of heat stress.

Air-conditioning engineers prefer to use the humidity ratio (also called the mixing ratio), which is the mass of water vapour per unit mass of dry air. The advantage of this is that in most air-conditioning processes the mass flow of dry air is the same throughout most of the system, regardless of temperature and pressure. The humidity ratio is constant unless water vapour is added or removed.

Humidity is commonly measured by means of wet and dry bulb thermometers, usually mounted together in a sling or Assman psychrometer (see above). The bulb of the wet bulb thermometer is covered with thin muslin which is wetted with distilled water. Evaporation of the water cools the thermometer bulb and heat passes to it from the surrounding air. An equilibrium is reached when the rates of heat loss and gain are equal. The thermometer then reads the wet bulb temperature, T_{wb}. A good current of air must be maintained over the bulb in order that the convective and evaporative heat exchanges shall be large compared with radiant heat exchange and stem conduction. So long as this is so the air velocity is not critical. In the Assman psychrometer the bulb is usually wetted by means of a pipette immediately before a reading is taken. The fan is then

Fig. 5.1. Various ways of expressing the humidity of the environment indicated by the temperature and water vapour pressure at the point, *A*. The curved line shows saturated water vapour pressure. Dew point is the temperature at which the prevailing water vapour pressure would be equal to the saturated water vapour pressure. Relative humidity is the ratio *AC*/*BC*, usually expressed as a percentage. Wet bulb temperature is the temperature reached if water is evaporated adiabatically into the air until saturation is reached. (From Kerslake, 1972.)

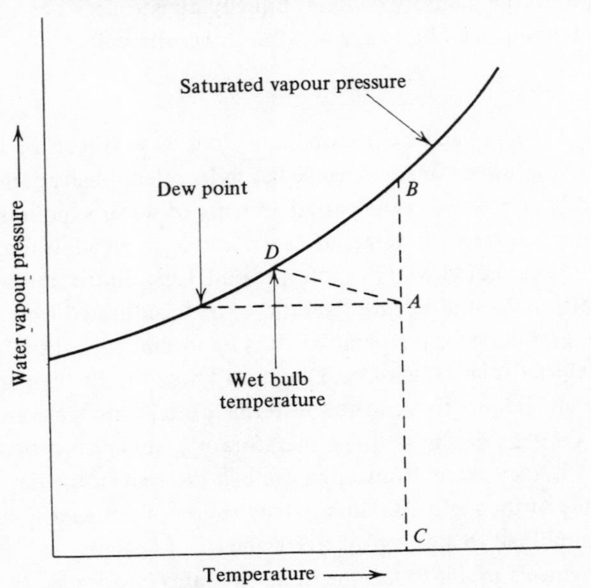

started and the thermometer should reach a steady reading after a minute or so. This is T_{wb}. Eventually, as the muslin dries out, the temperature will begin to rise again. In very dry air, or at high altitude, evaporation may be so rapid that the bulb begins to dry out before the plateau level is reached. Re-wetting with water near T_{wb} may overcome this difficulty. Sling psychrometers and some fan-ventilated instruments have a water reservoir into which is dipped a wick leading to the wet bulb. If the rate of evaporation is limited by the characteristics of the wick rather than by the conditions around the bulb an erroneous but steady reading may be obtained. This is less easy to recognise and correct. In using the sling psychrometer it is important to read the wet bulb thermometer immediately the whirling action is stopped, i.e. before the bulb has begun to warm up as a result of the reduced air movement over it.

The relation between T_a, T_{wb}, T_{dp}, r.h. and p_a is shown in Fig. 5.2. The line for 100 % r.h. indicates saturated water vapour pressure. In saturated air T_{wb} equals T_a and T_{dp}. The wet bulb lines are nearly straight and parallel. Fig. 5.2 is for normal atmospheric pressure. At reduced pressure water evaporates more readily and heat passes from the air to the wet bulb less readily. Thus at given

Fig. 5.2. The psychrometric chart, showing the relation between dry bulb temperature (T_a), water vapour pressure (p_a), relative humidity (r.h.) and wet bulb temperature (T_{wb}).

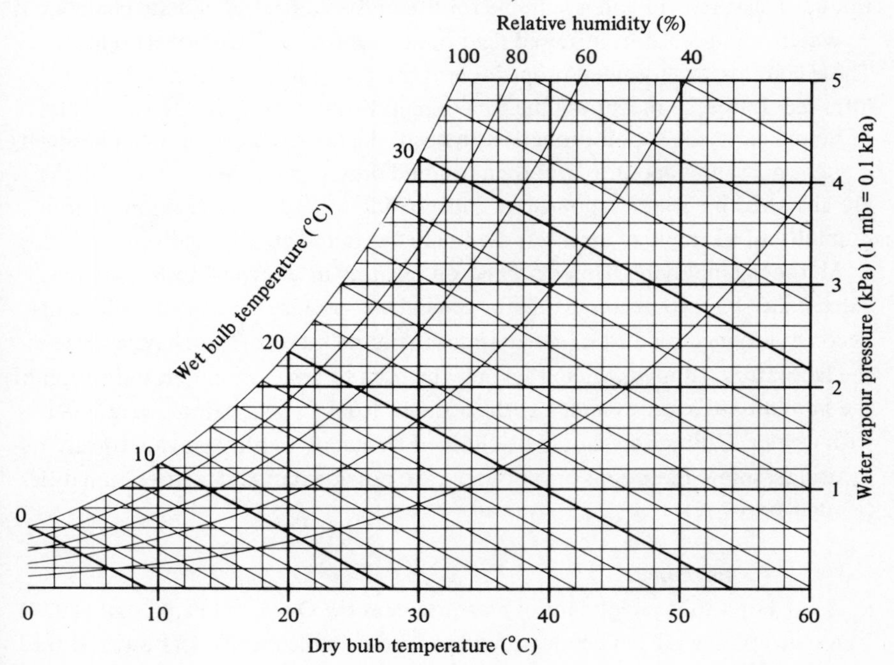

T_a and p_a the wet bulb temperature is lower and the wet bulb lines on the chart are less steep. Charts and tables are available for different altitudes. For sea level the following equation is good enough for most purposes (Hodgman, 1965):

$$p_a = p_{wb} - 0.68\ (T_a - T_{wb}) \tag{5.2}$$

Here p_{wb} is the saturated water vapour pressure at T_{wb}.

In some applications the wet bulb is not forcibly ventilated. The advantage of this is that readings can be obtained at any time as a wick and reservoir can be relied upon to keep the bulb wet, because the rate of evaporation is low. Special tables are available for such wet bulb thermometers, which should normally be provided with wind and radiation screens. An exception is the instrument used for the Wet Bulb Globe Temperature index, which is screened from neither wind nor radiation. Its readings are incorporated directly into the index and no tables or charts are required.

Another method of measuring humidity is to cool the air until the water vapour begins to condense. This will occur at the dew point if adequate precautions are taken to prevent supercooling. The onset of condensation can be detected by electrical or optical means. The apparatus is rather complicated and cumbersome, but as the end-point is a physical change in the air the calibration is stable and the readings reliable. The method can be arranged to give a continuous measurement and is suitable for use in the control of climate chambers.

Water vapour absorbs infra-red light of certain well-defined wavelengths. These comprise a very small proportion of the total radiation from a typical infra-red source, so measuring the total amount transmitted through a sample of humid air is not very informative. Instead, the infra-red detector can be a cell containing water vapour. It will then respond only to those wavebands which are absorbed by water vapour, and can provide a sensitive indication of the humidity of a sample of air interposed between it and an infra-red source.

Methods employing hairs or based on changes in electrical resistance or capacitance tend to be unreliable for general use because the sensitive elements become poisoned by dust or other atmospheric pollutants. They have a place in the laboratory, though, where they may be used to give continuous indication of the humidity of a gas stream; e.g. the effluent from a sweat capsule ventilated with dry air. Calibration should be checked frequently, either gravimetrically or by using saturated solutions of various salts to provide standard relative humidities (Hodgman, 1965).

Air movement

For a fuller treatment of air movement see Ower and Pankhurst (1977). The speed of a wind can be measured in various simple mechanical ways. If the direction is steady the vane anemometer, a miniature windmill, works well down to perhaps 0.2 m/s. The speed of rotation is proportional to the wind speed down to some minimum at which rotation ceases. If the direction is variable but the plane of the air movement is constant, e.g. horizontal, the cup anemometer is appropriate. Large examples are familiar on the tops of poles, and smaller hand-held versions are available. They are less sensitive than vane anemometers but well suited to outdoor measurements. Exploration of small winds indoors, e.g. in the vicinity of air duct outlets, can be done with a draught gauge, consisting of an open-ended tube connected to a sensitive manometer. The probe has to be pointed into the wind and is correctly directed when the reading is maximal. The method thus provides values for both speed and direction. Very low wind speeds can be measured by timing the movement of smoke.

In most circumstances, particularly indoors, air movement does not have the simple translational character of a wind. Eddies of varying direction are present which would be missed or incompletely registered by the above methods. In the present context air movement is measured in order that inferences may be drawn about heat transfer for the human subject. Properly, the velocity sensor should be of the same order of size as the subject, but useful information can be provided by instruments of more manageable proportions.

The katathermometer (Bedford, 1964) is an alcohol thermometer with a very large bulb which is silvered on the outside. There are only two graduation marks on the stem. The bulb is heated until the alcohol is well past both graduations. It is then dried if necessary and the thermometer hung up where the measurement of air movement is required. The time taken to cool from the upper to the lower graduation is measured. A nomogram entered with this time and the ambient temperature provides a statement of the wind speed (in a wind tunnel) which would provide the same cooling coefficient. This nomogram is appropriate for a bulb of a certain size and shape, cooling between two defined temperatures. In order that cooling may be neither rudely swift nor yet untimely slow, katathermometers are provided in sets appropriate to several ranges of air temperature.

The same principle of measuring the cooling power of the air is embodied in the more convenient hot wire anemometer. Here electrical energy is used to heat a fine wire, the temperature of which depends on the wattage dissipated, the air movement and the ambient temperature. The temperature of the wire can be measured by an attached thermistor or by using the wire itself as a resistance thermometer. Alternatively the power can be dissipated in the thermistor. These devices have rapid responses which are not required in normal environmental measurements. Indeed it is often necessary to ‘slug’ the output considerably in order to obtain a useful indication. A small heated sphere is in many ways a more appropriate device. There is no need to silver it if the temperature is measured with reference to a similar unheated sphere (Gagge *et al.*, 1968).

Instruments which depend on the cooling power of the air inevitably tend to generate their own natural convection currents. The ion anemometer does not have this defect and can be used at very low air movements without causing disturbance. It consists of a spherical wire framework at the centre of which is a radioactive source. An electrical potential difference is maintained between the two. Particles emitted by the source leave ionised tracks in the air, and the ions move towards the electrodes. The resulting electric current is reduced if the ions are swept out of the frame by the air current before they have time to complete their journey.

Thermal radiation

The temperatures of surfaces which bound the thermal environment (including the sky!) can be measured with a radiometer. Where access is possible, contact thermometers of various sorts can be used, but the ideal of perfect contact with no interference with heat exchange may be hard to approach.

The most useful single statement about thermal radiation is the mean radiant temperature, $\overline{T}_r$ (Vernon, 1932). If all the surroundings were black surfaces at $\overline{T}_r$, radiant heat exchange would be the same as it is in the real environment. Skin and most clothing surfaces behave as black bodies for infra-red radiation but not for sunlight, and it is useful to distinguish these.

Indoors, in the absence of direct sunlight, $\overline{T}_r$ can be measured by means of the globe thermometer. The standard instrument is a blackened hollow metal sphere 15 cm in diameter with a temperature sensor at the centre. The surface exchanges heat by radiation with surrounding surfaces and by convection with the air, reaching equilibrium at the globe temperature, T_g. $\overline{T}_r$ can be deduced if air temperature and air movement are known, but in many circumstances it is unnecessary to make this computation, since T_g is itself an adequate statement of environmental temperature (see p. 272).

A standard globe with mercury thermometer takes 20 minutes or so to reach equilibrium. The metal of the globe equilibrates in about half this time, and the response can be speeded up somewhat by using a smaller sensor (thermocouple or thermistor) instead of the mercury thermometer (Hellon and Crockford, 1959). If the material of the globe itself is used as a resistance thermometer, the response time can be reduced further (Gagge *et al.*, 1968).

The most convenient measurement of sunlight is provided by the horizontal pyrheliometer, which measures solar intensity on a horizontal surface. The solar radiation falling on a subject is the product of this intensity and the area of the subject's shadow, measured on a horizontal surface (Chrenko and Pugh, 1961).

Sunlight reflected from the sky and terrain can be measured with a conventional radiometer, using filters to eliminate infra-red radiation (Lee, 1964).

5.3 Physiological measurements

Core temperature

Temperature varies a little from place to place in the body core. The main influence is the temperature of the blood leaving the heart and this is what one would like to be able to measure. Elsewhere the temperature is modified by local heat production, by the proximity of veins carrying cooler or warmer blood and by direct conduction of heat to adjacent tissues. Temperatures measured in readily accessible sites therefore differ from one another, but the differences tend to be systematic and the temperatures at different sites go up and down in the same general way and to much the same extent (Cooper and Kenyon, 1957; Piironen, 1970). Choice of a site is frequently dominated by practical expediency, but reliability of the site as an indicator of central temperature, speed of response and social acceptability must be considered.

Mouth

Mouth temperature should be taken under the tongue, where there is a profuse blood flow. Mouth breathing must be prohibited. The clinical thermometer is usually marked with a time of the order of 30 seconds. This is the time it takes to reach equilibrium in a stirred water bath. It takes much longer – 5 minutes or more – in the mouth because the tissues around the bulb share in the temperature changes as equilibrium is approached. The response is faster if the subject is warm, and the error smaller if the thermometer is warmed before insertion. (As it is a maximum reading device beware of overwarming.) A thermistor or other small probe causes less thermal disturbance, but mouth breathing, talking or cool drinks will cool the sublingual tissues, and the mouth must be closed for several minutes before a reliable reading will be obtained. Mouth temperature is affected by heat loss from the head, because cool blood returning from the face and scalp passes close to the arteries supplying the floor of the mouth and tongue. It responds to change in central arterial temperature more rapidly than most other sites (Piironen, 1970).

Rectum

Rectal temperature should be measured at a depth of at least 10 cm to avoid local cooling effects. The temperature reflects that of the central arteries but is modified by the proximity of veins carrying blood from the legs (Eichna *et al.*, 1951). Thus leg work tends to have a greater effect on rectal than on other core temperatures. The blood flow in the rectal wall is not as high as that in the mouth and the response is slower. Rectal thermometry is well tolerated by practised subjects but is unpopular with the naive. An unpublished study has shown that the mouth temperatures of resting naive subjects were significantly raised if rectal temperature was also measured. Mere threat of this was sufficient to produce the effect.

Stomach and intestine

The middle part of the gut can be reached by radio pills which transmit temperature information after being swallowed (Fox, Goldsmith and Wolff, 1961). The pill moves through the gut along with the other contents, so its position is constantly changing. Response is slow because of the mass of the pill and surrounding gut contents, but the method has the great merit of allowing the subject complete freedom of movement.

Oesophagus

Probably the best indirect measure of central arterial temperature is given by a properly positioned sensor in the oesophagus (Saltin and Hermansen, 1966). In part of its course through the chest the oesophagus is in close contact with the aorta, and a sensor in this region registers aortic temperature unless it is also close to one of the breathing passages. The sensor can be positioned under X-ray control, but it is simpler and safer merely to lower it slowly while observing the indicated temperature. This will show a high plateau when the sensor is in the best region. An electrode incorporated in the tip allows the electrocardiogram potentials to indicate the correct position. It takes most people some practice to swallow a tube and hold it in the oesophagus for long periods. Swallowing causes transient cooling of the sensor, particularly if there is mouth breathing. These episodes can be recognised but introduce gaps into the core temperature information. Despite the low vascularity of the oesophageal wall this site has a rapid response.

Ear canal

The external ear canal has become a popular site since the original use of the ear drum by Benzinger (1959). He used a thermocouple in contact with the drum, insisting that unless there was some pain the reading would not be valid. Probably the painful contact causes some mild inflammation of the drum, improving its performance as an indicator of core temperature. Others have ignored the requirement for contact with the drum. The response is then sluggish and the indicated temperature very dependent on heat exchange from the head. The ear canal should always be plugged and a large insulating pad applied over the outer ear and the side of the head (Cooper, Cranston and Snell, 1964).

Skin sites

Skin sites such as the axilla and groin are sometimes used medically, but are poor indicators of core temperature. It is also possible to use the skin of the chest by placing over it a heat-flow sensor which in turn is covered by a heater and insulating pad (Fox *et al.*, 1973). The heater is controlled so that there is no flow of heat through the sensor. Since there is no heat loss, the skin temperature under the device rises to equal the local arterial temperature, which will be near to core temperature if there is not much heat loss from the chest and arms. This is a handy clinical instrument, which is comfortable and unobjectionable. It has not so far been used much outside hospital but would seem suited to some field applications (Smith, Davies and Christie, 1980). The adiabatic principle of this device can also be used to improve the performance of other thermal probes.

Skin temperature

The temperature of exposed skin can be measured by radiometric techniques which have a negligible effect on heat exchange and are rapid and accurate. They have the further advantage of indicating the mean temperature of a few square centimetres or more of skin, rather than the very small region sensed by a thermistor or thermocouple. However, in most ergonomic situations the subject is clothed and the skin temperature at most sites must be measured by transducers held in contact with the skin. Ideally their presence would have no effect on sensible or evaporative heat transfer at the skin surface, and the indicated temperature would not be directly influenced by air or clothing temperature. To ensure good contact the device must be pressed onto the skin, but the pressure must not be great enough to produce significant changes in blood flow or heat transfer within the skin (Molnar and Rosenbaum, 1963). Mounting the transducer on a copper disc or a small piece of gauze is inadmissible if sweating is present. It may instead be mounted on a wire which bridges a shallow ring that can be fixed to the skin with tape or tied on with elastic threads. If applied to a site where the surface is convex, contact with the skin will be good and interference with heat exchange minimal. When thermocouples are used they are best made end-to-end and tied round the body with elastic thread. The joint between wire and thread should be a few centimetres from the thermojunction so as to minimise errors from heat conduction along the wire.

If it is not to interfere with heat exchange the sensor must be small. It therefore senses the temperature of only a small region of skin in contact with it. (Although it impedes heat transfer in the region of immediate contact, conduction from neighbouring skin will prevent a significant temperature difference developing so long as the region of contact is small.) Skin temperature differs from place to place over the body surface. Differences of a degree or so can occur over short distances in the vicinity of a large cutaneous vein. Differences of several degrees may exist in the same limb segment because of differences in external heat transfer conditions (e.g. apposition of the inner side of the arm to the side of the chest) or in local anatomy (e.g. the kneecap). Differences between large regions such as limb segments are greatest when mean skin temperature is low, and diminish as skin temperature approaches core temperature. A rise in mean skin temperature from 35 to 36 °C requires an enormous increase in skin blood flow if heat transfer is to be maintained from the body core at 37 °C, whereas a rise from 25 to 26 °C would require only a trivial change. The accuracy with which one needs to know skin temperature is thus conveniently greatest when the variations from place to place are least.

In practice the mean skin temperature must be derived from measurements at only a few sites. Each must be as representative as possible of some anatomical region, and the means must be weighted according to the areas of these regions, which should involve the whole body surface. A scheme using 15 sites is shown in Fig. 5.3. This is based on the breakdown of total surface area shown in Table 5.1 (Hardy and DuBois, 1934).

With a total of only 15 sites the whole head must make do with one, so that the mean temperature of this 7% of the total area is not known with much precision. However, the wiring and fixing arrangements for even 15 channels begin to affect the hang of clothing and to discourage free movement. More would rarely be feasible and fewer often desirable. In planning schemes using fewer sites (8 to 11 are often used) it is best to adhere to the general regional breakdown, merely reducing the number of sites on the larger regions. In the

Fig. 5.3. A suggested scheme for siting fifteen skin temperature sensors. (From Clifford, Kerslake & Waddell, 1959.)

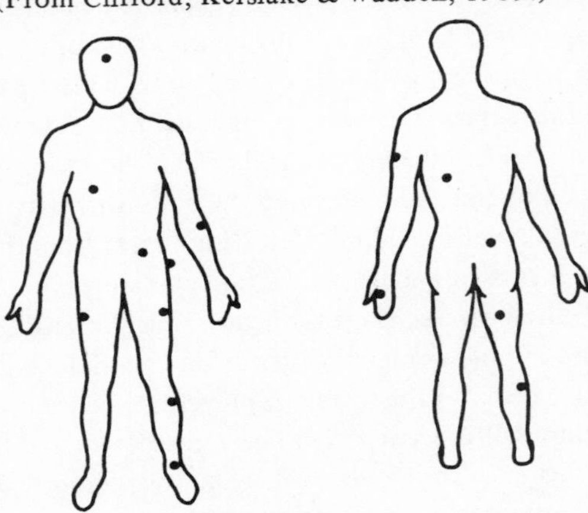

case of clothed subjects it may be possible to combine some regions, e.g. legs and feet.

Claims that as few as one site will give some indication of mean skin temperature are, of course, valid, but the precision of the estimate depends in large part on the range of conditions involved. Thus for a single subject, with clothing and activity standardised and in conditions near comfort, there is likely to be a strong correlation between skin temperatures at different sites. The mean can be predicted with some confidence from any of the individual values. Remove any of these restrictions and the confidence is reduced.

Heat storage

Heat storage is very difficult to measure directly (Winslow, Gagge and Herrington, 1940). In principle it can be done by measuring all the other quantities in the heat balance equation (equation 5.1), but this is a formidable task even under laboratory conditions. The total amount of heat stored over a period must be reflected in temperature changes somewhere in the body, but the number of accessible sites is rather small. Even so, some sort of estimate of changes in stored heat can be based on changes in mean skin ($\bar{T}_{sk}$) and core (T_c) temperatures. In warm or comfortable conditions the mean body temperature is usually taken as $\frac{2}{3} T_c + \frac{1}{3}\bar{T}_{sk}$. (This does not imply that the core is twice as big as the shell, since whereas T_c may be assumed to be the mean temperature of the core, $\bar{T}_{sk}$ is that of the coolest part of the shell. If we assume that the mean temperature of the shell is half-way between T_c and $\bar{T}_{sk}$, the weightings for mean body temperature suggest that the shell is twice as big as the core.) For immersion in cool water the weighting $0.8T_c + 0.2\bar{T}_{sk}$ is more appropriate (Stolwijk and Hardy, (1966). In severe cold exposure no estimate based on T_c and $\bar{T}_{sk}$ is much use (Webb, 1973).

Table 5.1. *Approximate regional surface areas for adults with suggested distribution of 15 temperature sensors*

Region	% total area	No. of sites
Trunk	35	5
Thighs	19	3
Lower legs	13	2
Feet	7	1
Upper arms	7	1
Forearms	7	1
Hands	5	1
Head	7	1

Changes in mean body temperature are converted into changes in body heat content on the basis of a specific heat of about 3.5 kJ/kg·°C. This is close to 1.0 W·h/kg·°C. The watt hour is convenient here because rates of heat exchange are usually reckoned in watts, but exposure times in hours rather than seconds.

Sweating

The rate of loss of water through the skin and lungs can readily be measured by weighing the subject at intervals. Allowance must be made for food and water intake and for excreta. There is also a small loss of weight due to the difference between the masses of carbon dioxide lost and oxygen absorbed. This amounts to about 10 $\dot{V}_{O_2}$ g/h, where $\dot{V}_{O_2}$ is the rate of oxygen consumption in l/min at STP. For a resting subject this is only about 4 g/h and is usually neglected since the accuracy of the weighing is unlikely to be as high, and the evaporation of water at this rate would give a heat loss of less than 2 W/m^2.

For most purposes an accuracy of ± 10 g/h is adequate. This can be provided by a conventional platform balance of good quality if weighings are at intervals of half an hour or so. Evaporation of 1.5g water per hour takes up heat at the rate of 1 W, so for a subject of area 1.8 m^2, 10 g/h represents an evaporative heat loss of 3.7 W/m^2.

If sweat does not drip onto the floor or accumulate in the clothing the rate of weight loss equals the rate of evaporation. It includes water loss from the lungs and by diffusion through the skin, and allowance for these processes can be made if an accurate measure of sweat production is required (Kerslake, 1972).

$$\text{Respiratory water loss} = (2.0 - 0.36\,p_a)\,\dot{V}/A_D \qquad \text{g/m}^2\cdot\text{h} \qquad (5.3)$$

$$\text{Diffusional water loss} = 1.75\,(1-\phi_{sk})\,p_{sk} \qquad \text{g/m}^2\cdot\text{h} \qquad (5.4)$$

where p_a is the ambient water vapour pressure (kPa), $\dot{V}$ the respiratory minute volume (l/min), A_D the area of the subject (m^2), p_{sk} saturated water vapour pressure at mean skin temperature (kPa) and ϕ_{sk} the relative humidity of the skin (see p. 263). In the absence of thermal sweating $\phi_{sk} \cdot p_{sk}$ is close to p_a and the diffusional water loss is 1.75 ($p_{sk} - p_a$) g/m^2 · h. Non-thermal sweating from the palms, axillae, etc., amounts to about 6 g/m^2·h at rest. This increases with physical or mental activity.

Difficulties arise when sweat is lost by dripping or taken up in important amounts by clothing. If there is no dripping, or if the drips can be collected, intermittent weighing will provide a measure of the rate of evaporation. Weighing the clothing at the beginning and end of the exposure will allow the total sweat loss to be found. This may give a useful general guide to the mean condition of the subject over the period of a working shift, but the maximum rate attained during the shift is a better indicator of approach to tolerance limits, and this cannot be deduced from the total sweat loss alone. Valuable additional information can be gained from a ventilated sweat capsule applied to a few square centimetres of skin on the trunk or arm. The local sweat rate can be measured continuously by this means and is roughly proportional to total sweat rate (Custance, 1965; Brebner and Kerslake, 1969).

Pulse rate

The pulse rate, or heart rate, can be measured by feeling the pulse on the thumb side of the wrist. If this can be clearly felt it can be counted quite easily. Some people have difficulty counting high rates, but when the pulse rate is high it is also regular, and one can count every second beat. There will be an error of a second or so in the time measurement, so an error of plus or minus a beat will be acceptable. The pulse can also be felt at other sites but usually less easily.

The heart can be heard with a stethoscope if the subject is still. It makes a double sound for each heart beat, easily recognised at low rates but a possible source of confusion at high.

The electrical activity of the heart (electrocardiogram: ECG or EKG) can be recorded graphically for later counting or fed into rate meters or counters. In exercising subjects it is hard to avoid occasional extraneous signals from muscle activity and movement of electrodes. These can be recognised in a trace, but not so readily by a rate meter. No doubt these difficulties will diminish as the discernment of electronic devices continues to improve.

Heat production

Methods for measuring metabolic rate are described in Chapter 2. External work, e.g. walking downhill, is often difficult to estimate, but tasks other than bicycle ergometer riding are usually rather inefficient and the estimate of heat production may remain quite good.

5.4 Judging physiological state

Measurements of core temperature, skin temperature, sweat rate and pulse rate can provide useful information about the general thermal state of a subject, how near he is to his tolerance limits, whether his performance is likely to be adversely affected by the heat or cold, and so on. The significance of each of these quantities is considered below.

Core temperature

There is widespread belief that core temperature is so closely controlled that the slightest departure from the 'normal' value of 37.0 °C is a sign of illness.

In fact, the healthy subject has a diurnal swing of some ± 0.3°C about this average level (Reinberg *et al.*, 1980). This is partly due to the daily pattern of exercise, meals and sleep, but a basic cycle remains if these effects are removed (see *Ergonomics*, vol 21 no. 10, 1978). At rest the core temperature rises and falls a little with external heat or cold. Residents of Singapore show the same daily temperature cycle as those of Britain, but their temperatures run a point or two higher (Macpherson, 1960). This seems to be a consequence of habitat rather than race or habituation. The effect on core temperature diminishes at higher rates of heat production.

Physical exercise is normally attended by a considerable rise in core temperature. Except at very high work rates the increase is roughly linearly related to metabolic rate (Davies, 1979), more precisely to oxygen consumption expressed as a proportion of the maximum of which the subject is capable (Saltin and Hermansen, 1966). At a level equivalent to half the maximum oxygen uptake the core temperature T_c rises to about 38.0 °C. The relation is roughly

$$T_c = 36.5 + 3.0\ \dot{V}_{O_2}/\dot{V}_{O_2\ max} \qquad (5.5)$$

Departures from this 'preferred' level of core temperature are strongly resisted by the body. Exposure to cold or heat has very little effect on core temperature until the heat conservation or loss mechanism is overwhelmed (Fig. 5.4). Beyond this point the body loses control of heat exchange, and though heat balance may

Fig. 5.4. Equilibrium levels of core temperature at three work rates in different environments. Core temperature is almost independent of environmental conditions over a wide range known as the prescriptive zone, but regulation fails when conditions are very severe. (After Leithead and Lind, 1964.)

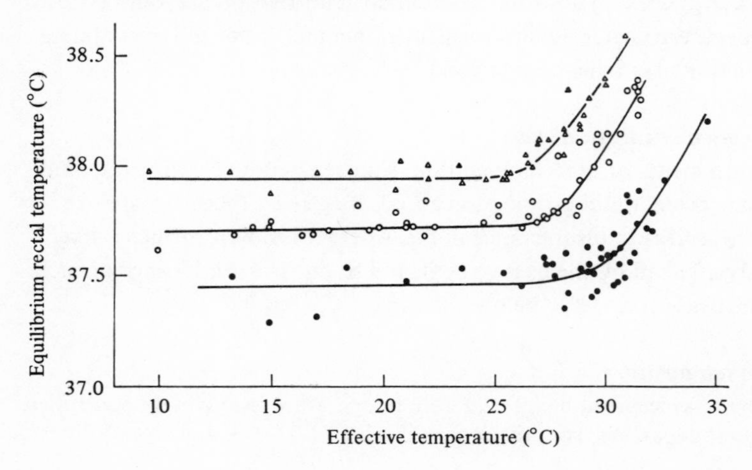

(rarely) be attained it is precarious. The thermally safe zone of working conditions is that in which the core temperature does not depart from its preferred level. At the cold end of this zone the body can choose between allowing core temperature to fall below its preferred level or maintaining it by increased heat production (shivering). These are both failures to maintain the preferred relation between heat production and core temperature. The prescriptive zone so defined depends on the clothing, the work rate and the physiological characteristics of the subject. The last may not be administratively convenient but should not be neglected.

Determination of whether an individual is operating within his prescriptive zone is not as simple a matter as it might appear. Equation (5.5) is not a sufficiently accurate yardstick, and the preferred level for the individual in question must be found by experiment in an innocuous environment. Day-to-day variations of perhaps ±0.2 °C considerably blur the ends of the prescriptive zone (Fig. 5.4).

When work rate changes it usually takes about 45 minutes for the oesophageal temperature to level off, and 60 minutes or more for the rectal (Nielsen and Nielsen 1965). If work rate is steady and the environment suddenly changes there is often a transient paradoxical change in core temperature. On entering a warm environment from a comfortable one the core temperature at first falls, returning to its previous level after 10 to 20 minutes (Kerslake and Cooper, 1950). This is due to an increase in blood flow through the body shell. The thermal contents of the body are stirred, core temperature falling and shell temperature rising. The converse effects may occur if the environment becomes cooler (Burton and Edholm, 1955).

Departures from the prescriptive zone are usually attended by massive physiological responses which may have adverse effects on task performance. These effects do not seem to be a direct consequence of the level of core temperature. Thus if a resting man is exposed to conditions which raise his core temperature to 38.5 °C he will be very distressed, unable to concentrate, there may be some visual disturbance and he may be in danger of fainting. The same core temperature might be produced by exercise in the prescriptive zone, when the only distress would be that directly attributable to the activity. The symptoms of overheating are relieved as soon as heat loss increases and before the core temperature has fallen to its prescriptive zone level.

Core temperature may be raised by infection or other illnesses, when it is regulated at the new level much as it is in exercise (Macpherson, 1960). However, the subject almost always feels ill, so this cause is unlikely to be overlooked. Except in severe fever the malaise does not appear to be directly due to the rise in temperature. Anxiety, excitement and anticipation of activity can also produce increases in resting core temperature.

Direct effects of core temperature are apparent beyond the range 35.5 °C to 39.5 °C, and are summarised below.

42 °C	Usually fatal
41 °C	Thermoregulatory responses cease. Temperature tends to rise further. Coma. Convulsions
39.5 °C	Upper acceptable limit. Drowsiness
35.5 °C	Lower acceptable limit. Onset of physical and mental slowing
34.5 °C	Shivering diminishes. Temperature tends to fall further. Extreme physical and mental slowing
33 °C	Coma
Lower	Deep coma. Heart becomes weaker but also irritable with danger of fibrillation and death. At about 27 °C the heart stops and will probably fibrillate and fail during rewarming

Skin temperature

The surface of the skin is the environment of the rest of the body (lungs excepted) and its temperature should be an excellent indicator of thermal stress. It is probably the difficulty of making the necessary measurements which has discouraged more extensive use of skin temperature as a definitive measure of thermal state.

A subject in thermal balance at rest is most comfortable with a mean skin temperature of 33–34 °C (Fanger, 1970). The preferred skin temperature decreases with increasing heat production to about 30 °C at a heat production of 200 W/m^2. Preferences during work are less critical than at rest, and subjects can be trained to prefer levels so low that sweating is absent (Webb, Troutman and Annis, 1970). This is advantageous in the special case of liquid-cooled suits which otherwise tend to become wet inside from condensation.

It would be hard to argue that any departure from thermal comfort is unattended by some deterioration in task performance. Thermal comfort is therefore a most important state, but is better indicated by the subject's opinion than by his skin temperature.

If the mean skin temperature is 36 °C or more, it is likely that the subject is near or beyond the upper limit of the prescriptive zone. This critical skin temperature varies somewhat with work rate and the fitness of the subject, and probably reflects the difficulty of transporting heat from core to skin. Core temperature increases with work rate, easing the problem of transporting greater quantities of heat to the surfaces. The critical skin temperature is therefore less dependent on work rate than it would otherwise be.

The lower end of the prescriptive zone is more difficult to define than the

upper and fewer data are available. Low skin temperature can readily be produced by immersion in water, but the distribution of skin temperature then differs from that produced by exposure to cold air with or without clothing. The consequence is that although at rest in air a mean skin temperature of 33 °C is comfortable and associated with heat balance, in water 35 °C is preferred, 33 °C marking the lower end of the prescriptive zone. Heat is then lost faster than it is produced, until mild shivering redresses the balance.
The effects of *local* skin temperature are summarised below:

Over 45 °C	Rapid tissue damage: burns
42 °C	Pain
40 °C	Uncomfortably hot
25 °C	Uncomfortably cold
5 °C	Numbness
Below 0 °C	Frostbite

Low skin temperatures are associated with pain, but the temperature at which this occurs depends on the region and extent of the cooling. Prolonged cooling to temperatures above 0 °C (as high as 25 °C) can produce skin and nerve damage (chilblains, immersion foot).

The effects of local cold are most noticeable in the hands, where loss of dexterity appears at about 15 °C, increasing at lower temperatures. Sense of position of joints is first impaired, and the joints themselves become stiff. Below about 5 °C the sense of touch becomes grossly impaired, but pain, which is usually present, is reduced. Near freezing point the hands become numb. Loss of dexterity may occur rather suddenly as the hand cools, coinciding with constriction of the blood vessels.

Frostbite occurs when the skin freezes (Burton and Edholm, 1955). The freezing point for moist tissues is −0.6 °C. Normal 'dry' skin can often be supercooled a few degrees, but freezing then occurs very suddenly. Frostbite can occur within a minute of exposure to cold air (e.g. −20 °C, 20 m/s) and is almost instantaneous on contact with cold metal. The frozen skin sticks to the metal, so damage is likely to be severe, subcutaneous tissues and even muscle being killed.

Minor frostbite, evidenced by the appearance of a numb white patch often on the nose or cheek usually recovers rapidly if covered by a warm hand, but the extent of the damage cannot be assessed until thawing is complete. Nearby tissues not actually frozen will suffer if their blood supply is destroyed. First aid treatment consists of removal to a warm environment where feasible, and rewarming of the frozen region as rapidly as possible, keeping the temperature below 45 °C.

Sweat rate

Some sweat is produced all the time from regions such as the palms and axillae, but over most of the body surface sweating normally occurs only in response to heat stress (Newburgh, 1949). Such thermal sweating in a resting subject suggests that he is too hot. During physical work most people prefer conditions which demand some sweating. Fanger (1970) has examined the preferences of a large number of subjects of both sexes, and recommends that sweating should account for about 36% of the excess heat production (above resting level). Call the excess heat production H_X W/m^2. Evaporation accounting for $0.36H_X$ W/m^2 requires sweating at the rate of $0.54\,H_X$ g/m^2·h. For an average male subject of area 1.8 m^2 this would be $0.97H_X$ g/h. Thus Fanger's recommendation can be conveniently summarised for a subject of average size as follows: the sweat rate in g/h should be equal to the excess heat production in W/m^2.

As a subject becomes hotter his body calls for more sweat production, and the glands usually respond to this. However, there is one serious snag which precludes identification of sweat rate with the degree of stress on the subject. This is hidromeiosis (Kerslake, 1972). In humid conditions, when the skin is running with sweat, the glands slowly become blocked and sweat rate decreases progressively. In exposure to severe humid heat the sweat rate typically reaches a peak at about the end of the first hour, diminishing thereafter despite rising core and skin temperatures. If the stimulus to sweat production is kept constant and the whole of the body surface is kept wet, sweat production decreases exponentially towards zero with a time constant of about 2 hours. Sweating recovers rapidly on exposure to conditions which eliminate frank wetting of the skin surface. Although hidromeiosis develops rather slowly, its importance over a period of hours may be considerable. If the skin were wet throughout an exposure lasting 4 hours, the total sweat loss would be less than half that in the absence of hidromeiosis and the final sweat rate less than one-seventh. Complete wetting of the skin only occurs in very severe conditions. In most circumstances hidromeiosis is confined to local skin areas where ventilation is poor.

When all the sweat evaporates, differences in sweat production between different subjects are smaller than one might suppose. Subjects who sweat readily have skin temperatures a degree or so lower than poor sweaters. This may make an enormous difference to skin blood flow requirements, allowing good sweaters to tolerate the conditions much better. For subjects in heat balance the difference of 1 °C in skin temperature will increase evaporative demand in the good sweaters by perhaps 12 W/m^2 (depending on clothing and wind speed), equivalent to only about 30 g/h. In dry environments sweat rate is thus a good indicator of the state of the environment but a poor indicator of the state of the subject.

Pulse rate

The heart has to supply blood to the brain and viscera, the muscles and the skin. An increase in heart output usually involves increases in both stroke volume (output per beat) and pulse rate, and the latter is a fair indicator of the demand being made on the circulation. In the resting, comfortable subject heart rate shows a general relation with physical fitness, athletes having rates commonly below 60/min, others up to 80 or so. The resting heart rate varies cyclically (heart rate variability: HRV) and close study of these variations casts light on such matters as state of arousal; but the variations disappear as pulse rate is increased (Luczak, 1979; Hitchen, Brodie and Harness, 1980).

At rest in the heat, pulse rate is increased in parallel with the increased blood flow to the skin. Skin blood flow must also be increased in exercise because of the need for greater heat loss, and the working muscles make a further demand. The result is that the effects of work and of heat on pulse rate are roughly additive (Sengupta *et al.,* 1979). Although variation between subjects in the same situation is large, unadjusted pulse rate is a good indicator of a subject's state. The athlete finds resting as well as working less of a strain, and the upper limits of pulse rate in fit and unfit subjects are similar, although the associated rates of working differ greatly.

5.5 Heat exchange with the environment

Radiation

All surfaces emit radiant heat at the rate $\epsilon \cdot \sigma \cdot T^4$ W/m^2, where ϵ is the emittance, a property of the material, having a value between 0 and 1; σ is the Stefan-Boltzmann constant (5.67×10^{-8} W/m$^2 \cdot$K^4); and T is the absolute temperature in K(= °C + 273). If ϵ is 1.0 the surface will emit radiation at the maximum rate consistent with its temperature. It will also absorb all radiation which falls on it, and is said to behave as a black body.

Emittance depends to some extent on the wavelength of the radiation, which in turn depends on the temperature of the surface producing it, so it is possible for a surface to behave as a black body for infra-red wavelengths, e.g. for its own radiation at a temperature of 40°C, but to reflect a large proportion of the sunlight which falls on it (proportion reflected = $1 - \epsilon$). White-painted surfaces have this property, and so run cooler in the sun than both dark-painted surfaces, which absorb more sunlight, and polished metal surfaces, which, while reflecting most of the sunlight, have a low emittance in the infra-red and do not lose their heat so readily. Nearly all non-metallic surfaces – skin, clothing, paint, etc. – regardless of visible colour are black for infra-red radiation. Emittance depends on the molecules very near the surface, and ϵ for polished metal, perhaps 0.05 when clean, is greatly increased by finger marks, other traces of grease or dirt, or the thin layers of transparent plastic with which the shiningness of aluminium foil is often protected.

Two black surfaces facing each other each emit radiation at the rate $\sigma \cdot T^4$ and each absorbs all the radiation falling on it from the other. The net exchange of heat is therefore $\sigma\,(T_1^4 - T_2^4)$ W/m^2. This is true only if each surface completely fills the radiant environment of the other. Where this is not so a geometrical factor must be introduced indicating the proportion of the radiant environment which is involved, and this need not be the same for both surfaces. One can be considered as 'subject', losing radiant heat at the rate $\sigma \cdot T_1^4$, no matter whither, and the other as part of the environment, contributing heat to the subject at the rate $F \cdot \sigma \cdot T_2^4$, where F is the geometrical factor.

Any point on the clothing surface can be regarded as 'seeing' a radiant environment composed of various surfaces having different temperatures and different orientations and lying at different distances and directions. The importance of each depends on the solid angle subtended at the point in question and on the angle at which the radiation strikes the clothing surface. Distance is unimportant except in so far as it affects the solid angle subtended. The total radiation exchange between subject and environment is the integral of the exchanges at each point on the subject's surface. Fortunately it is not necessary to resort to tedious calculations from first principles, as angle factors for rectangular slabs of radiant environment in relation to seated and standing subjects have been computed by Fanger (1970) and are reproduced by McIntyre (1980).

Detailed analysis by angle factors is of particular value when a building is in the design stage, or when constructional changes are contemplated. Where the environment already exists it is often sufficient to work in terms of mean radiant temperature, which can be measured with the globe thermometer. For indoor conditions the range of temperature is usually small, and it is permissible to avoid the cumbersome fourth-power relation, using instead a first-power relation of the form

$$R = h_r\,(T_s - \overline{T}_r) \qquad (5.6)$$

where R is the rate of heat loss by radiation (W/m^2 body surface), T_s the mean temperature of the clothing surface and exposed skin (°C), $\overline{T}_r$ the mean radiant temperature (°C), and h_r a somewhat bogus coefficient, of value about 4.0 W/m$^2 \cdot$°C for a seated subject at room temperature, which makes R come out right.

The coefficient h_r really depends on the two temperatures and includes an allowance for the fact that the area of the subject available for radiant heat exchange (radiant area) is less than his total surface area because some parts of the body 'see' or are in contact with others. This radiant area factor is about 0.70 for seated subjects. Estimates for standing subjects range from 0.72 to 0.78. To calculate h_r for a particular case one multiplies the radiant area factor by $4\sigma \cdot T^3$, where T is the average of T_s and $\overline{T}_r$ (K). At about 20°C h_r increases by about 1% per degree increase in temperature.

Sunlight

At its temperature of nearly 6000 K the sun emits intense radiation in the visible wavebands. Some is scattered and absorbed by the atmosphere and some reflected by the terrain and other surfaces (Breckenridge and Goldman, 1971). Sunlight thus reaches a subject either directly or after reflection from the sky or elsewhere. The relative magnitudes in four types of terrain are shown in Table 5.2. Light reflected from the sky and from the terrain together fill the whole radiant environment, whereas direct sunlight, coming from an object of negligible solid angle, falls only on the silhouette of the subject. Despite this, direct sunlight is the biggest factor on a clear day (Lee, 1964).

Projected area depends on posture and on the angle between the sun and the subject usually expressed as solar altitude and azimuth (Ward and Underwood, 1967; Breckenridge and Goldman, 1971). Azimuth here means the angle between the direction in which the subject is facing and the horizontal direction of the sun. In Fig. 5.5 the projected area for standing nude subjects is expressed in terms of total skin area. The ratio for clothed subjects is similar, taking outer clothing area instead of skin area (Chrenko and Pugh, 1961). The effect of solar altitude on heat load is not so large as Fig. 5.5 may suggest, because when the sun is low in the sky its intensity is less. Values for standing and sitting subjects, clothed and nude, male and female, are given by Fanger (1970).

Table 5.2. *Solar radiation flux in different terrains*

	Solar radiation flux (W/m^2)		
Terrain	Direct	Reflected from sky	Reflected from terrain
Desert	1000	188	133
Rain forest	840	180	46
Tropical steppe	1020	114	105
Subarctic	950	143	64

From Breckenridge and Goldman (1971).

Sky temperature

As well as reflected sunlight, some infra-red radiation reaches the ground from the atmosphere, in effect from the sky. The strength of this can be expressed as the temperature of a black surface covering the sky which would provide the same radiation intensity. If the sky is clear its temperature is about 20 °C below that of the air near the ground; if cloudy the temperature approaches that of the air (Monteith, 1973).

Convection

Convection means the direct exchange of heat between the skin or clothing surface and the surrounding air. It may be thought of as taking place

Fig. 5.5. Projected area, expressed as a proportion of total surface area, of male (continuous lines) and female (dashed lines) subjects as a function of solar altitude. Curves are shown for three angles of azimuth (0° corresponds to facing the sun). (After Underwood and Ward, 1966.)

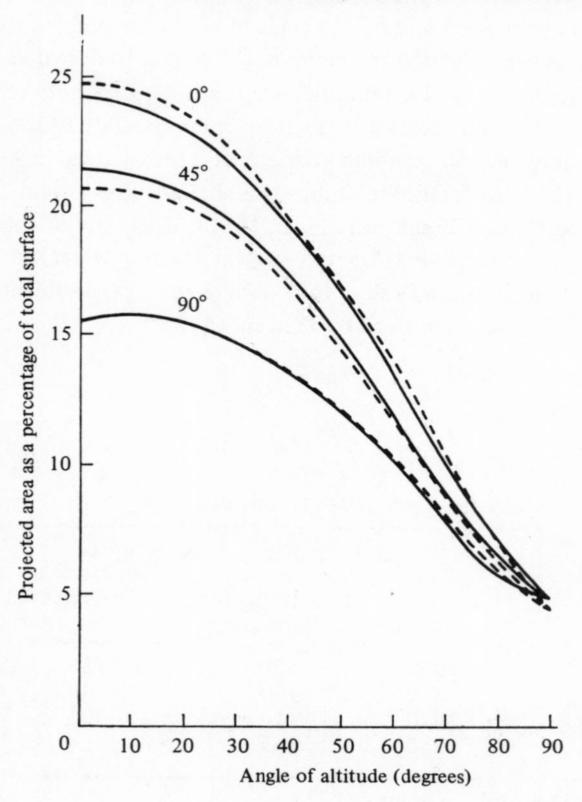

across a boundary layer of still air, the outer part of which is readily stripped off by wind, reducing the thermal resistance. The basic equation for heat exchange by convection is

$$C = h_c (T_s - T_a) \tag{5.7}$$

where C is the rate of heat exchange (W/m²), T_s and T_a the temperatures of the surface and the air (°C), and h_c the convection coefficient (W/m²·°C), in effect the conductance of the boundary layer. It is sometimes more convenient to use the thermal resistance of the layer, the inverse of its conductance. The equation is then

$$C = (T_s - T_a)/I \tag{5.8}$$

where I is the thermal resistance or insulation of the boundary air layer (°C·m²/W). (Beware of confusing this I with the ambient insulation I_a, used in the clo system. I_a combines convection and radiation.)

The convection coefficient depends in a rather complicated way on the properties of the air and the way it is flowing, and it might be thought that a knowledge of the theory of fluid flow and its relation to heat transfer would be of value here. Fortunately for the ergonomist this is hardly so. Papers concerned with convective heat transfer in man and animals frequently begin with a *précis* of the relevant theory, but this is rarely used in the remainder of the piece. This is because the theory deals essentially in similarities. It can compare cases of large and small objects of the same shape immersed in fluids (which includes liquids and gases) of different sorts. It can predict the effects of pressure and atmospheric composition, which is useful in the cases of deep mines, diving bells and so on, but what it cannot do is to predict the magnitude of the convection coefficient, h_c, for different postures and wind speeds. It does not even predict the effect of wind speed (V). All it says is that for a given posture h_c will be some function of V. This function is often expressed in the form $a \cdot V^n$, where a and n are empirical parameters, although this form has no theoretical justification. Different values of a and n may be required for different ranges of V. To discover them we must measure h_c at various values of V. There is no short cut. However, once we know h_c for one type of atmosphere we can calculate what it will be in another.

Measurements made in various laboratories are in general agreement that at wind speeds between 0.5 and 4.0 m/s h_c varies approximately as $V^{0.5}$. Results of individual investigations are often better fitted by an expression containing a somewhat higher fractional power of V, but as the variation between laboratories is about ± 10% the more convenient square root relation is adequate. The figures in Table 5.3 are rendered down from various sources and are probably close to consensus values (Kerslake, 1972; Gagge and Nishi, 1977).

This dependence of h_c on V applies in the regime of forced convection, when the pattern of air movement over the subject is dominated by the imposed wind velocity. If this is less than about 0.2 m/s, convection currents generated by changes in density of the air as it exchanges heat with the skin become dominant. This process of free convection depends on the temperature difference between the skin or clothing surface and the air. It is adequately described for both seated and standing subjects by

$$h_c = 2.4\,(T_s - T_a)^{0.25} \tag{5.9}$$

There are situations in which both imposed and generated air movements combine to determine h_c, and elaborate methods for treating such cases have been proposed. In practice it is usual, and probably adequate in view of the rather low accuracy of predictions of h_c, to make separate calculations for forced and free convection and to use the larger of the two (Fanger 1970).

The coefficients referred to above are mean values for the whole body. If the surface temperature differs from place to place when a determination is made, that value will only apply, strictly, to that particular surface temperature distribution (Kerslake, 1963). Determinations on nude subjects are generally made in warm conditions when skin temperature is fairly uniform. In the cold the subject reduces the blood flow to the limbs, which consequently contribute less to the total heat loss. The cold subject behaves as if he were composed more of trunk and less of limbs.

Evaporation

When sweat evaporates it takes up large amounts of heat (about 2.4 kJ/g). The process of evaporation from the skin is similar to that of convective heat exchange except that the driving force is water vapour pressure instead of temperature.

Table 5.3. *Values for the convection coefficient, h_c, in man (V in m/s)*

Posture	h_c in air at sea level ($W/m^2 \cdot {}^\circ C$)
Seated	$11.6\sqrt{V}$
Standing	$8.3\sqrt{V}$
Walking at V_w into wind at V_a	$8.7\sqrt{V_w} + 1.9\,V_a$

$$E_{max} = h_e \, (p_{sk} - p_a) \, , \tag{5.10}$$

where E_{max} is the rate of evaporative heat loss from skin covered by a film of water (W/m^2), h_e the evaporation coefficient ($W/m^2 \cdot kPa$), p_{sk} saturated water vapour pressure at skin temperature (kPa), and p_a ambient water vapour pressure (kPa). Like h_c, h_e depends on the thickness of the boundary layer, and the two are closely related. In air at normal atmospheric pressure h_e is about $15h_c$. (The theoretical value is said to be $16.5h_c$ (Gagge and Nishi, 1977), but experiments lead to the value 15, perhaps because the skin is never properly wetted.)

Equation 5.10 shows the rate at which evaporation will proceed if the skin is completely covered with a film of water. If the rate of sweat production is insufficient to maintain this film evaporation must be determined by the sweat rate instead of by physical factors. The form of equation 5.10 can be retained by introducing a term, W, the wettedness of the skin (Gagge, 1937). This is the ratio of the actual rate of evaporation to the maximum defined by equation 5.10.

$$E = W \cdot h_e \, (p_{sk} - p_a) \tag{5.11}$$

The skin surface may be thought of as a mosaic of wet and dry patches, the wet ones following equation 5.10, the dry ones in vapour pressure equilibrium with the ambient air.

Instead of using W to modify the area from which evaporation is taking place, we can modify the vapour pressure at the skin surface (Mole, 1948). The relative humidity of the skin, ϕ_{sk}, is the ratio of the effective vapour pressure to p_{sk}.

$$E = h_e \, (\phi_{sk} \cdot p_{sk} - p_a) \tag{5.12}$$

There is a good deal of evidence that dampness of the skin is associated with discomfort. Some have sought to equate this dimension of discomfort with W (Winslow, Herrington and Gagge, 1937; Gagge and Gonzalez, 1974), others with ϕ_{sk} (Kerslake, 1972). The distinction is unimportant when E is near E_{max}, and both are near unity, but at low rates of sweating in humid environments the difference may be quite large. For example, zero evaporation requires W to be zero, but ϕ_{sk} would equal p_a/p_{sk}.

Sweat rate and skin ventilation vary from place to place on the body, so that in some places evaporation may be governed by sweat rate while elsewhere it is limited by environmental factors. Four evaporative regimes may be distinguished for the body as a whole (Sibbons, 1966):

1. Full wetness. Evaporation follows equation 5.10, W and ϕ_{sk} are unity.
2. Restricted evaporation. Some parts of the body are fully wet while in others evaporation of sweat is complete. Sweat may run from the

former to the latter and some may be lost by dripping or taken up by clothing.

3. Full evaporation. All sweat is evaporated from the skin surface.
4. Diffusion. In comfortable and cool conditions when there is no thermal sweating, evaporation from most of the body surface is by diffusion through the skin. There is a little residual sweating.

Respiratory tract

When air is drawn into the lungs it rapidly reaches temperature equilibrium with the blood and becomes saturated with water vapour. However, expired air does not leave the body saturated at lung temperature. This is because the inspired air is warmed and wetted by the upper respiratory passages, thereby somewhat cooling and drying them. Air leaving the lungs passes over these surfaces, warming and wetting them, and is itself cooled and dried. The condition of the expired air depends mainly on the state of the inspired air (wet bulb temperature is an adequate definition for this purpose: McCutchan and Taylor, 1950), and little on the temperature of the blood in the lungs, simply because this does not vary much. The heat lost with each breath is the difference in heat content between the expired and inspired air, and depends on the volume of air breathed. The volume of air breathed per minute is proportional to metabolic rate (with considerable individual variation) and in effect ambient wet bulb temperature determines the proportion of energy production which is lost by respiration (Fig. 5.6).

Conduction

Heat exchange between the skin and contiguous surfaces (excluding clothing) has received scant attention from physiologists, who usually prefer to keep the area of contact small. Their subjects sit on bicycle saddles or recline on nets. In normal sitting or lying the area of contact may be 10% or more of the body surface, but the larger the area of contact, in general the softer and more insulating will be the contacting material. Although heat transfer may not be large, seating or bedding is important in that it reduces the area of skin available for heat exchange by other channels.

Conduction is of particular importance when contact is with metals or other good conductors of heat at high (45 °C) or low (0 °C) temperatures. Skin rapidly approaches the temperature of such materials on contact, causing pain and possibly burns or frostbite. Reflex withdrawal of a limb may have serious consequences unrelated to the thermal injury. Transient problems of ergonomic significance can arise when sun-soaked seats are occupied or tea is being poured, but in terms of body heat balance conduction is rarely of major importance.

Clothing

The clo unit

Development of clothing for military purposes during the Second World War was at first hampered by the different systems of units used by workers in the various disciplines concerned. To remedy this, and to make their pronouncements more readily comprehensible to non-technical people, a group of scientists actively concerned in clothing research introduced a system of units centred on a new unit of clothing insulation catchily called the 'clo' (Gagge, Burton and Bazett, 1941). Now that the general standardisation to the MKS system has been agreed the first of these objectives is obsolete. So, indeed, should be the clo because it is not an MKS unit. Outside clothing technology it is almost unknown and it actually hinders communication with scientists unfamiliar with this field. It survives largely because it has a name, which the MKS unit of thermal insulation, $°C \cdot m^2/W$ does not. This is a great convenience in conversation.

The system is described in detail by Burton and Edholm (1955). The thermal insulation surrounding the body is considered to consist of two layers: the clothing insulation, I_c, next to the skin, and the ambient insulation, I_a, between the clothing surface and the environment (Fig. 5.7). Although often called the 'air insulation', I_a applies to combined heat transfer by radiation and convection. If these processes are described by the conductances h_r and h_c they can be

Fig. 5.6. Relation between respiratory heat loss, expressed as a proportion of metabolic rate, and wet bulb temperature of the inspired air (T_{wb}).

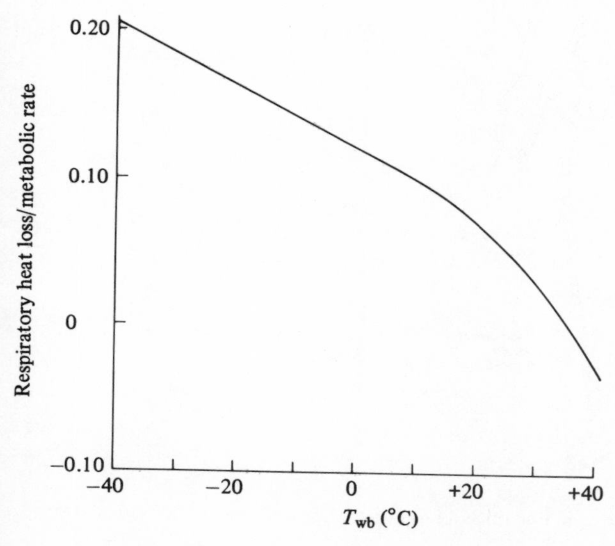

added since they are in parallel with one another. However, it is more convenient to express their joint effect as a resistance, since it is in series with I_c. Resistances in series can be added. The formal definition of I_a is*

$$I_a = 1 / (0.61\,(T/298)^3 + 1.9V^{1/2}\,(298/T)^{1/2}) \text{ clo units} \qquad (5.13)$$

The terms in the denominator will be recognised as h_r and h_c. Each is based on its value at 25 °C (298 K) and varies with temperature in the manner indicated. T is in K and V in m/s. In principle, equation 5.13 requires that air and mean radiant temperatures shall be equal, but when they are not, the globe temperature will serve as a suitably weighted mean.

The clo unit was chosen so that 1 clo would be the thermal resistance of the normal indoor clothing of the period. (People nowadays tend to wear less than 1 clo.) More precisely, '1 clo will maintain a sitting-resting man, whose metabolism is 50 kcal/m^2·h (58 W/m^2) indefinitely comfortable in an environment of 21 °C, relative humidity less than 50%, and air movement 20 ft/min (0.010 m/s)'. This definition is illustrative rather than practically applicable, and the clo is also

Fig. 5.7. The various thermal insulations, and their range of values, that make up the total thermal insulation of Man. (From Burton and Edholm, 1955.)

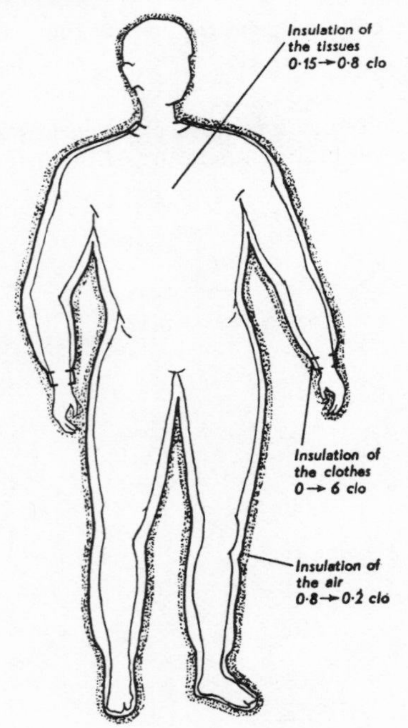

* The final square root is omitted in Burton and Edholm (1955) but used in their calculations.

defined as a mean thermal insulation of 0.18 °C·m²·h/kcal (0.155 °C·m²/W). Equations linking clo units with heat transfer contain one of these constants or its inverse. The constant used may be a useful clue to which heat units the author is using. Confusion is worse confounded by the practice of mixing insulations expressed in clo with conductances in MKS units (instead of reciprocal clo, presumably olc) (cf. equation 5.15).

This rather simple model of heat transfer through the clothing to the environment can be used to calculate clothing requirements in different situations. It is assumed that the skin temperature for comfort is 33 °C and that 25% of the heat production is lost by evaporation, the remainder passing through the clothing as sensible heat. Fig. 5.8 shows 'what 1 clo is good for'. The indoor case is correct by definition, but one suspects that clothing suitable for a desk job indoors would not be adequate outside at –20 °C no matter how hard one ran. In fact, the calculations and the suspicions are both correct. One clo of clothing insulation would be adequate for this condition, but the uniform which provides 1 clo in the still air of the office will not do so in the wind outside. The wind

Fig. 5.8. Pictograms to illustrate the importance of metabolic rate in determining the clothing required in different environments. The Met is the unit of metabolic rate in the clo system, equal to 50 kcal/m² · h (58 W/m²). (From Burton and Edholm, 1955.)

will penetrate the fabric. Also, physical movement pumps air in and out of the space beneath and within the clothing, reducing its insulating power. Another difficulty is that the uniform could not provide comfort at –20°C because it gives no protection to the face. Mean skin temperature is not an adequate definition of comfort.

These limitations of the simple I_c and I_a model were appreciated by those who devised the clo system, but are sometimes ignored by their successors. In fact, the effects of wind and movement on clothing insulation turn out to be so large that the concept of a fixed I_c can be thoroughly misleading (Fig. 5.9). This is so even for assemblies with windbreak outer layers. It has not been found possible to express these effects other than empirically, and if I_c varies with wind speed there might seem to be no virtue in separating it from I_a. However, the distinction remains useful because radiant heat transfer, including the important solar component, takes place at or below the clothing surface.

In office and domestic situations the air movement is normally low and the subject sedentary. It is usually assumed that I_c is independent of small changes

Fig. 5.9. Effect of wind speed on the insulation of a flying overall. Total insulation ($I_c + I_a$) was measured on a copper man. Clothing insulation, I_c, is the difference between this and the ambient insulation, I_a, measured on the unclothed manikin.

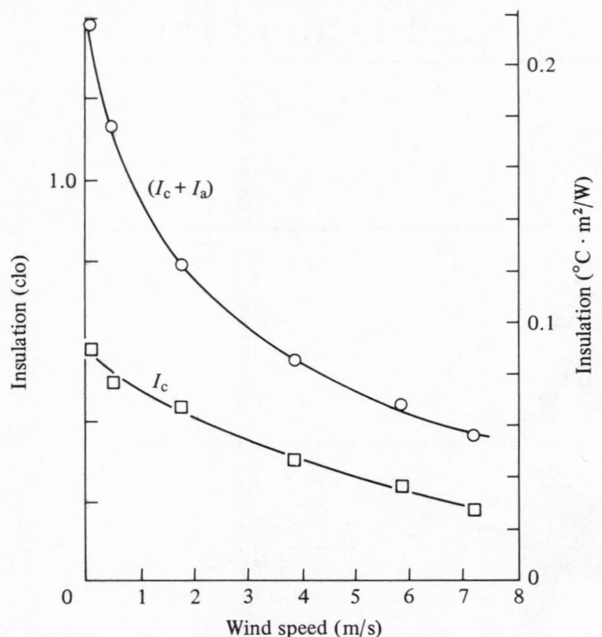

in air movement and activity, but precisely because the air movement is low, clothing intended for indoor use is frequently of open weave and loose fit, so that this assumption may not be justified. Walking about generates a relative air movement of 1 m/s or so and it is conceivable that the materials and design of indoor clothing have evolved in such a way that activity produces about the right drop in clothing insulation to maintain comfort.

Measurement of I_c

Thick fabrics designed for thermal insulation have a thermal resistance of about 0.25°C·m²/W (1.6 clo) per centimetre thickness. Unlike the boundary air insulation this is independent of atmospheric pressure, although it does depend on the composition of the gas trapped in the fabric. The air layers beneath and within the clothing have a similar thermal resistance so that the insulation of a complete assembly (in MKS units), assuming no wind penetration and no pumping action due to bodily movement, is about 0.25 times the thickness in centimetres. The thickness can be derived from measurements of the circumference of the clothing and of the body beneath it. Thickness will vary from place to place and the mean should be weighted according to the areas of the various body segments. Although this is not a strictly correct way of obtaining mean insulation it is sufficiently accurate for many purposes. Mean insulation is an awkward concept, particularly if some parts of the skin have no clothing. In practice it is that quantity which will correctly link mean sensible heat loss and mean temperature difference.

Clothing insulation can be measured directly in the laboratory by means of a heated manikin, commonly known as a copper man (Newburgh, 1949). This is dressed in the assembly and allowed to equilibrate in constant conditions. The different body segments are usually controlled independently and results will depend to some extent on how the surface temperature distribution of the manikin is determined: whether it is held at a uniform 33°C, given the distribution associated with thermal comfort, or given a realistic distribution for the conditions of the test. Instead of controlling temperature, the heat output of the segments may be pre-set or a physiologically appropriate relation between skin temperature and heat loss may be imposed. The experiment leads directly to the total insulation, $(I_c + I_a)$, and can usually provide this for each segment – a useful indication of where improvements may be needed.

The copper man does not move. He does not even breathe, and under thick assemblies slight movements may produce perceptible stirring of the air between the layers and perhaps some entry of outside air. Thick and stiff clothing on the torso can act as an effective counter-lung with disastrous effects on its insulation value. Clothing intended for human subjects should really be assessed on a device which mimics the body movements to be expected in use. The human subject is the obvious solution, but there are problems about estimating heat storage and evaporative heat loss. The best way is to choose conditions which will maintain the subject in a thermally neutral state. If he becomes cold there will be increased uncertainty about heat storage. If he becomes warm sensible heat transfer will diminish, reducing the accuracy of the estimate of insulation, and evaporative heat loss will increase, leading to problems about how to treat water which evaporates from the body and is taken up by the clothing. In some experiments the environmental temperature has been steered to the level preferred by the subject, but one cannot assume that he is then in heat balance. Unusual distribution of skin temperature can disturb the subject's response and sensations.

Water vapour transfer

Clothing offers a resistance to the passage of water vapour, and so hinders heat loss by evaporation from the skin. The transfer of water vapour through the clothing and thence to the environment has much in common with the transfer of sensible heat, but an important difference is that there is no vapour transfer channel analogous to thermal radiation. This has proved a serious obstacle to attempts to provide a unified description of heat transfer for the clothed subject.

Woodcock (1962) has linked evaporative and sensible heat transfer by his permeability index, i_m. Denoting vapour resistance by R (expressed in heat units, kPa·m²/W) and thermal resistance by I, the ratio I/R for still air is taken as 16.5 °C/kPa. This is the yardstick for i_m. If the ratio I/R for a sample of clothing material is, say, 12, its i_m value is 12/16.5, or 0.73. The permeability index can be applied to the whole thermal enclosure:

$$i_m = (I_c + I_a) / 16.5 (R_c + R_a) \tag{5.14}$$

Table 5.4. *Values of total insulation and permeability index,* i_m, *for US army fatigues covered with various lengths of plastic raincoat*

Clothing	$(I_c + I_a)$ clo	$(I_c + I_a)$ °C·m²/W	i_m
Fatigues only	1.3	0.20	0.45
Plus raincoat			
Quarter length	1.4	0.22	0.26
Half length	1.5	0.23	0.21
Three-quarter length	1.6	0.25	0.18
Full length	1.7	0.26	0.15

From Breckenridge and Goldman (1977).

Here the subscript 'c' refers to clothing and 'a' to ambient. I_a includes radiation transfer and is not equal to $16.5R_a$. The value of i_m lies between zero, for a clothing assembly impermeable to water vapour, and unity. It depends on air movement even if there is no wind penetration of the clothing, because of the radiation factor included in I_a. For the nude subject it does not have the pleasing value of 1.0, but tends to this at high wind speeds as the relative importance of radiation diminishes. Measurements of total insulation and i_m are shown in Table 5.4 for various lengths of plastic raincoat worn over standard US army fatigues (Breckenridge and Goldman, 1977). Air movement was 0.3 m/s.

A different formulation, the permeation efficiency factor, F_{pcl}, has been introduced by Nishi and Gagge (1970). This is analogous to a sensible heat transfer factor, F_{cl}, derived from Burton's earlier work. It applies to the whole thermal enclosure and can be defined as $R_a / (R_c + R_a)$. 'Normal porous clothing' transmits latent and sensible heat like a layer of stationary air ($i_m = 0.93$)* and for such assemblies the permeation efficiency factor can be expressed in terms of the convection coefficient, h_c, and the clothing insulation, I_{clo}:

$$F_{pcl} = 1 / (1 + 0.143\, h_c \cdot I_{clo}) \qquad (5.15)$$

where h_c is in $W/m^2 \cdot °C$ and I_{clo} in clo units. The factor 0.143 is 0.93×0.155. (1 clo is $0.155\ °C \cdot m^2 / W$). Like i_m, F_{pcl} depends on air movement, here indicated by h_c, but unlike i_m it does have the value 1.0 for the nude subject.

5.6 Judging the environment

In order to make generalisations about the thermal environment, e.g. to lay down acceptable limits of heat or cold, it is desirable to reduce the number of environmental factors by establishing equivalents where possible. Thus, convective heat exchange depends on air temperature and wind speed, and it may be possible to combine these two factors into a single number indicative of the hotness or coldness of the wind. Ideally the whole environment could be summed up by a single figure. The way the factors are combined may depend on the properties of the subject, the clothing, the work rate and upon what criteria are chosen to define equivalence.

Operative Temperature

Since the principles of heat exchange are fairly well understood, we can start by distinguishing situations in which certain aspects of heat exchange are equivalent. The simplest example, and one of the most important, is the combi-

* Nishi and Gagge (1970) derive the value 15.3 for I/R for still air. Woodcock (1962) uses the value 16.5, appropriate for boundary air.

nation of air temperature, T_a, and mean radiant temperature, $\overline{T}_r$, to yield Operative Temperature, T_o (Herrington, Winslow and Gagge, 1937). Calling the subject's surface temperature (clothing or skin) T_s, the convective heat exchange is given by

$$C = h_c (T_s - T_a)$$

and the radiant heat exchange by the first-power approximation

$$R = h_r (T_s - \overline{T}_r)$$

The total sensible heat exchange is found by addition:

$$(C + R) = h_c (T_s - T_a) + h_r (T_s - \overline{T}_r) \quad (5.16)$$

A standard environment can be imagined in which air and mean radiant temperatures are equal (isothermal enclosure) and in which the total sensible heat exchange is the same as in the real environment. The temperature of this standard environment is the Operative Temperature T_o, and this is associated with a heat transfer coefficient, h_o, such that

$$(C + R) = h_o (T_s - T_o) \quad (5.17)$$

By equating the right sides of equations 5.16 and 5.17, expressions for T_o and h_o can be derived:

$$T_o = \frac{h_c}{(h_c + h_r)} \cdot T_a + \frac{h_r}{(h_c + h_r)} \cdot \overline{T}_r \quad (5.18)$$

$$h_o = h_c + h_r \quad (5.19)$$

T_o is a weighted mean of T_a and $\overline{T}_r$. The factors applied to these temperatures add up to 1.0, as they must since when T_a equals $\overline{T}_r$ they are both equal to T_o by definition.

The properties of the subject which affect the weighting factors are size and shape, which, together with air movement, determine h_c and h_r. Inspection of equation 5.18 shows that the weightings depend on the ratio h_c/h_r rather than on the magnitudes of h_c and h_r. For the standard globe thermometer h_c is somewhat larger than it is for an adult human subject at the same air movement. The value of h_r for the globe is also larger because some parts of the body screen others from radiant heat exchange with the environment. The ratio h_c/h_r is much the same for the globe as for the human subject, so the standard globe provides a direct reading of Operative Temperature.

To what extent can differently constituted environments of equal Operative Temperature be regarded as equivalent? On the face of it they must be so since the human body has no way of distinguishing between convective and radiant heat loss. However, h_c and h_r vary from place to place over the skin or clothing surface. The ratio of their local values is not everywhere equal to the ratio of their mean values, which is used to define T_o. This means that when T_a and $\overline{T}_r$ are varied in such a way that the mean sensible heat exchange is unaltered, the distribution of skin heat loss and skin temperature will change. These changes in temperature distribution are unimportant unless T_a and $\overline{T}_r$ are very different, but they demonstrate that if a different criterion of equivalence were chosen, e.g. sensible heat exchange from the face, this could lead to different weightings in equation 5.18.

Sunlight is measured and expressed as a flux of energy, delivering, say, S W/m² at the clothing surface. (S will depend on the colour of the clothing, the posture and orientation of the subject as well as on the solar intensity.) This can be converted into an equivalent increase in T_o. The effect of sunlight is to reduce sensible heat loss by S W/m², and the same effect could be produced by increasing T_o by S/h_o °C. Thus the Operative Temperature concept will work for sunlight, and the two quantities, h_o and T_o, are sufficient to describe the environment for the purpose of calculating sensible heat exchange.

Thermal comfort

It is possible to spend many happy hours failing to define comfort. For the present purpose it is assumed that thermal comfort means an absence of discomfort, and that it is the state which people prefer. They usually make proper allowance for the nature of their activity, different conditions being demanded by a sitting subject according to whether he is working at a desk, watching a play, or dozing in an armchair. Much is due to differences in heat production, but 'comfortably warm' conditions are conducive to sleep and 'comfortably cool' are preferred for mental activity. Quite small departures from the preferred state can be shown to be associated with adverse effects on performance during long exposures (Griffiths and Boyce, 1971; Wyon, 1974; McIntyre, 1980), and there does not seem to be a band of moderate discomfort which is free from such effects. In the short term this is not necessarily so. A report on a project intended to demonstrate the hazard of conducting cockpit checks in a sun-soaked aircraft had to be entitled 'The initial stimulating effect of warmth upon perceptual efficiency' (Poulton and Kerslake, 1965).

In the original clo system, comfort is identified as a state of thermal balance in which the mean skin temperature is 33 °C and 25 % of the heat loss is by evaporation. Though inadequate in some respects, this definition works well enough with normal indoor clothing and a fairly uniform environment, and it enables comfortable conditions to be defined in terms of T_o, I_a and I_c.

In sedentary activity the heat production will be assumed to be 60 W/m², of which 45 W/m^2 is lost as sensible heat passing successively through the clothing insulation, I_c, and ambient insulation, I_a. The temperature difference between skin and environment must be 45 $(I_c + I_a)$. (The insulations are here expressed in $°C \cdot m^2/W$.) We can now define T_o for a comfortable sedentary subject.

$$T_o = 33 - 45\ (I_c + I_a) \tag{5.20}$$

If air and mean radiant temperatures are equal, this is as far as we need go. If not, T_o must be split into its components as in equation 5.18, where air movement, which affects I_a, appears again.

Suppose that I_c and I_a are determined by non-thermal factors beyond the scope of the present discussion. In the office environment this is usually so, I_c depending on custom and vanity, I_a on the ventilation requirements and the need to avoid draughts. Let I_c be 0.14 $°C \cdot m^2/W$ and the air movement 0.2 m/s, which means that h_c will be 5.2 $W/m^2 \cdot °C$. Take h_r as equal to 4.0 $W/m^2 \cdot °C$. This makes h_o 9.2 $W/m^2 \cdot °C$, and its inverse, I_a, 0.11 $°C \cdot m^2/W$. Therefore

$$T_o = 33 - 45\ (0.14 + 0.11) = 21.75\ °C$$

Conditions will vary from place to place in the office, as will people's preferences, clothing and heat production, so the required value of T_o can be rounded to 22°C.

There is still some freedom of choice between T_a and $\bar{T}_r$, defined at this air movement by

$$0.57\ T_a + 0.43\ \bar{T}_r = 22 \tag{5.21}$$

If T_a is, say, 19°C, $\bar{T}_r$ will need to be 26°C. In considering how this requirement can best be met it is sometimes useful to think in terms of effective radiant field, H_r, rather than mean radiant temperature (Gagge, Stolwijk and Hardy, 1965). If all the surfaces were at air temperature, T_o would be 19°C. We require an added radiant flux which will boost this to 22°C. Just as solar flux was converted into an increment in T_o by dividing by h_o, so this increment of 3°C can be converted into the equivalent radiant flux by multiplying by h_o. In this case H_r is 27.6 W/m^2.

H_r is the radiant heat gain that the subject would have if his surface were at air temperature. It is the sum of the contributions made by those surfaces in the environment which are not at air temperature. The contribution of a cool window is $F \cdot \sigma\ (T_w^4 - T_a^4)$, where T_w is the temperature of the window (K) and F the angle factor. In this case T_w is less than T_a so the contribution to H_r is negative. Such sources as infra-red heaters and sunlight are readily described in terms of radiant flux, and the effective radiant field method allows contributions of all types to be added.

This calculation of the conditions required for comfort makes no reference to humidity. The assumption that a constant proportion of the heat production is lost by evaporation is not strictly true, since water losses from the respiratory tract and by diffusion through the skin (p. 250) depend directly on ambient water vapour pressure, p_a. The quantities involved, however, are small. At 22°C the saturated water vapour pressure is 2.64 kPa. Table 5.5 shows the not unrealistic extreme cases of relative humidity (r.h.) 80% and 20%. Over this range of humidity the difference in heat loss is 3.2 W/m². The total insulation ($I_c + I_a$) for comfort at 22°C is 0.25 °C·m²/W, so changing the relative humidity from 20% to 80% has the same effect on heat loss as increasing T_o by 0.8°C. Experiments have confirmed that humidity has little effect on the preferred temperature in long exposures under typical office conditions (Koch, Jennings and Humphreys, 1960; Fanger, 1970).

Although the effects of humidity on heat exchange are quite small in comfortable conditions such as these, it is desirable to avoid extremes of humidity. The band 30–60% r.h. is usually recommended, though there is some evidence of an increase, at the lower end, of respiratory infections, particularly among schoolchildren (Green, 1975). The cabins of passenger aircraft are ventilated with outside air, compressed so as to produce a cabin pressure close to that at sea level. At cruising altitudes the outside air is very cold, and although its relative humidity may be high the proportion of water vapour in the air is very small. Despite pressurisation, after heating to cabin temperature the relative humidity is very low. This can give rise to justifiable complaints of dryness of the nose and throat, but Table 5.5 shows that there is no basis for the claim that low humidity causes excessive water loss and dehydration. (Lubrication of a dry throat with several Martinis is a more likely cause.) Conditions can be improved by partial recirculation of the cabin air.

This approach to the requirements for thermal comfort is perhaps deceptively simple, because the variables have been considered one at a time. In Fanger's exhaustive and meticulous study they are all combined in a 'comfort equation'

Table 5.5. *The effect of relative humidity (r.h.) on the evaporative heat loss of a resting subject*

r.h. (%)	p_a (kPa)	T_{wb} (°C)	Evaporative heat loss (W/m²)		
			Lungs	Skin diffusion	Total
80	2.11	19.6	4.2	3.4	7.6
20	0.55	11.0	5.5	5.3	10.8

Air temperature 22°C, skin temperature 33°C; p_a is the ambient water vapour pressure, T_{wb} the wet bulb temperature.

which runs to four lines of text (Fanger, 1970). It would be longer still if unrestricted evaporation of sweat were not assumed, but it is justifiable to regard this as a prerequisite for comfort. The physical principles on which the comfort equation is based are well established, and its form is clearly correct if the underlying propositions about thermal comfort are adequate. These are:

1. The subject is in heat balance.
2. The mean skin temperature for comfort is given by

$$T_{sk} = 35.7 - 0.0275\,M \qquad (5.22)$$

   where M is the metabolic rate in W/m^2.
3. The sweat rate (in W/m^2) for comfort is given by

$$S = 0.36\ (M - 58) \qquad (5.23)$$

4. The thermal resistance of clothing is independent of air movement.

Fanger (1970) does not specify the acceptable limits of skin temperature distribution. Clearly the mean value is an inadequate definition, but the comfort equation is intended for use with the range of clothing associated with near-comfortable conditions, and such clothing largely determines the distribution of skin temperature (Olesen and Fanger, 1973). Extra clothing does not fully compensate for a lower environmental temperature (McIntyre and Griffiths, 1975; Wyon *et al.*, 1975). Important departures from the preferred skin temperature distribution may occur if there are marked asymmetries in the radiant environment, temperature gradients in the air, or draughts. Underfloor heating tends to promote the preferred warm legs and cool head, whereas overhead radiant heating is less acceptable (McIntyre, 1977). Lateral asymmetries may arise from heated panels, cool windows or sunlight. They may be expressed as the 'radiation vector' (McIntyre, 1974). The various contributions to effective radiant field each have a direction as well as a magnitude. At any chosen place their vector sum will have a direction and magnitude which together indicate the asymmetry of the radiant environment. Its magnitude is the difference between the values of H_r for the two surfaces of a flat sheet disposed in such a way that the difference is maximal. It can also be expressed as the difference between the mean radiant temperatures to which the two surfaces are exposed (vector radiant temperature). Under office conditions quite small fluxes of the order of tens of watts per square metre may cause discomfort, but the limits have not as yet been fully explored (McIntyre, 1980). Sunbathing raises problems.

The precision with which the comfort equation can be applied in practice depends on the accuracy with which the preferred skin temperature, rate of heat production and clothing insulation are known. The effects of inter-individual variations in these quantities are considerable. For young adults seated in 'still'

(0.1 m/s) air and wearing descriptively similar clothing assessed at 0.6 clo (0.09 °C·m²/W), Fanger and Langkilde (1975) found the standard deviation of preferred ambient temperature to be ± 1.15°C. For heavier clothing the figure is larger. This variation is larger than the band of ambient temperature acceptable to an individual, and no environment will satisfy all subjects under specified conditions of clothing and activity. Differences due to age, sex and race are negligible when clothing and heat production are standardised, but outside the laboratory these factors may lead to significant differences in environmental preference.

Subjective Temperature

Fanger's equation has been expressed in a simplified form by McIntyre (1980). The Subjective Temperature, T_{sub}, is defined as the temperature of an isothermal enclosure with low air movement (0.1 m/s) and relative humidity 50% which will produce the same sensation of warmth as the actual environment. The Subjective Temperature for comfort depends on clothing and activity.

$$T_{sub} = 33.5 - 3I_{clo} - (0.08 + 0.05 I_{clo})\,H \qquad (5.24)$$

where I_{clo} is the clothing insulation in clo units and H the rate of heat production in W/m². This equation is not based on physical principles but is empirically fitted to Fanger's equation within an envelope of realistic practical limits. The environmental factors combined in T_{sub} are air temperature, mean radiant temperature and air movement:

$$T_{sub} = \frac{0.44\bar{T}_r + 0.56\,[5 - (5 - T_a)\,\sqrt{(10V)}]}{0.44 + 0.56\sqrt{(10V)}} \qquad (5.25)$$

This is appropriate for air velocities between 0.1 and 1.0 m/s. For 'low' air movement 0.1 m/s is used, making T_{sub} equal to $0.44\bar{T}_r + 0.56\,T_a$. For clothing insulation between 0 and 1.0 clo, and rates of heat production between 60 and 120 W/m², values derived from equation 5.25 are within 0.5°C of those derived from Fanger's equation.

Hot environments

The effects of humidity become of major importance when the subject is sweating. He is relying on evaporation of sweat to hold his skin temperature down, and if the sweat does not evaporate he has no means of regulating his body temperature. Control is effectively lost when the skin becomes completely wetted with sweat and changes in the rate of sweat production no longer influence the rate of evaporation. Heat balance may still be possible, but the subject is at the mercy of his environment and his situation is potentially dangerous.

The required rate of heat loss by evaporation, E_{req}, is given by

$$E_{req} = H_s + (T_o - T_{sk}) / I_t \quad (5.26)$$

where H_s is the required rate of heat loss from the skin and I_t is $(I_c + I_a)$. The right-hand term is the rate at which heat is gained from the environment by convection and radiation. The rate of evaporation from fully wetted skin, i.e. the maximum possible under the prevailing conditions, is given by

$$E_{max} = (p_{sk} - p_a) / R_t \quad (5.27)$$

where R_t is $(R_c + R_a)$, the sum of the vapour resistances of the clothing and the boundary air, expressed in kPa·m² / W. The saturated water vapour at the skin surface, p_{sk}, depends on T_{sk}, and the equations can be solved for the case $E_{req} = E_{max}$. If we fix the insulations and vapour resistances, T_{sk} can be expressed in terms of T_o, p_a and H_{sk}. Fig. 5.10 shows the relation between T_{sk}, T_o and p_a for a chosen value of H_{sk}. The lines are straight and parallel, with slope $-R_t/I_t$ kPa / °C. The position of each can be established by imagining an environment in which p_a is equal to p_{sk}. For T_{sk} 30°C, p_a will be 4.24 kPa. Since the rate of evaporation must be zero, all the skin heat loss must be by convection and radiation, so $(T_{sk} - T_o)$ must equal $H_{sk} \cdot I_t$, in this case 19.5°C, and the line for T_{sk} = 30 °C can be constructed from this point.

Fig. 5.10. Heat balance for wet skin (W = 1.0) under the following conditions: subject seated on a bicycle ergometer wearing light clothing (short-sleeved shirt, shorts, socks and plimsolls), T_{sk} as indicated, H_{sk} = 150 W/m², V = 0.8 m/s; h_c = 7.4, h_r = 5.2, h_o = 12.6 W/m² · °C; I_a = 0.05, I_t = 0.13°C · m²/W; R_a = 0.009, R_c = 0.003, R_t = 0.012 kPa · m²/W. The lines are constructed from the circled points which represent (unattainable) environments in which there is no heat exchange by evaporation, the coordinates of which are $p_a = p_{sk}$; $T_o = T_{sk} - H_{sk} \cdot I_t$. The slope of the lines is $-R_t/I_t$.

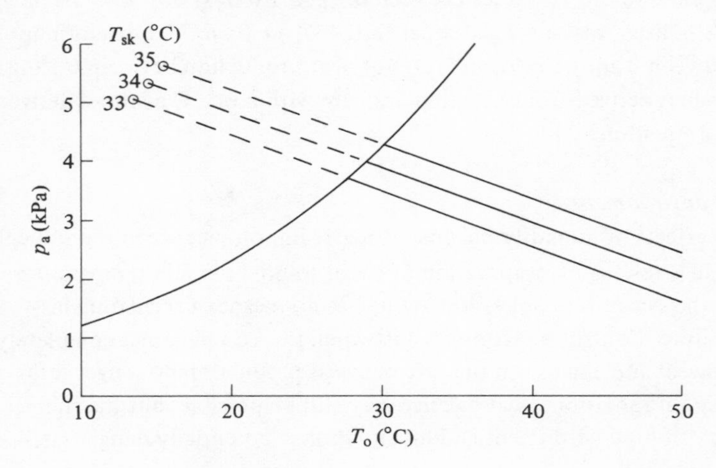

The living tissues of the body are bounded by the skin, and since this is their interface with the thermal complexities outside, it is reasonable to propose that mean skin temperature is a good single indicator of heat stress. If we neglect variation in heat loss from the lungs, the lines in Fig. 5.10 become lines of equal stress, assuming that there is sufficient sweat to keep the skin fully wet.

Sweat rate depends on work rate and skin temperature. In most subjects, at least in the category of fit young men, the physiological mechanisms are such that submaximal sweat rate can be described by

$$S = H_{sk} + K(T_{sk} - T_{sko}) \qquad (5.28)$$

where S is the sweat rate expressed in potential evaporative heat loss (W/m^2); K a constant ranging from 90 to 160 W/m^2 in subjects acclimatised to heat, lower in the unacclimatised; and T_{sko} a sort of threshold skin temperature (at which $S = H_{sk}$) ranging from about 33°C to 35°C (Hatch, 1963; Kerslake, 1972). This describes the sweat rate in the absence of hidromeiosis (p. 000). Where the skin is wet, hidromeiosis decreases the sweat rate until it is just sufficient to maintain the fully wet state, but this does not affect heat balance. Hidromeiosis will be neglected in the remainder of this discussion.

Fig. 5.10 is drawn for the case H_{sk} = 150 W/m^2, so each skin temperature line is also a line of constant sweat production (equation 5.28). Consider a subject for whom K equals 100 W/m^2 · °C, and T_{sko} 34.0°C. At T_{sk} 34.0°C his sweat rate will be 150 W/m^2, at 35.0°C 250 W/m^2, and so on. Now at T_{sk} 34°C, S equals H_{sk}, so if all the sweat evaporates heat balance will be attained at T_o equal to 34.0°C. This places an upper limit on the wet skin line for this skin temperature. At T_{sk} 35°C there is a further 100 W/m^2 evaporative heat loss available to balance heat coming in from outside. Since I_t equals 0.13°C·m^2/W, this means that when all the sweat evaporates, heat balance will occur with T_o 13°C higher than T_{sk}, i.e. at T_o equal to 48°C. The vertical lines in Fig. 5.11 are for full evaporation.

Between the regimes of full wetness and full evaporation defined in this way lies the zone of restricted evaporation, in which some parts of the body are fully wet while elsewhere there is full evaporation. By careful choice of posture and air movement this zone can be made very small, but for a nude subject standing in a wind the corner between the other two regimes is cut quite considerably, as shown by the limits of the hatched areas in Fig. 5.11. The lines for other postures and types of air movement probably run somewhere through the hatched areas.

Givoni's index of thermal stress (Givoni, 1964) and the Predicted Four-hour Sweat Rate (McArdle *et al.*, 1947) are both based on sweat rate as the criterion of equivalence, and they lead to lines of equal stress similar to those in Fig. 5.11.

Sweat rate and skin temperature are good criteria of stress in severely hot environments, but they do not correlate very highly with discomfort votes in moderate stress. It seems that wettedness (p. 263), or something akin to it, may be a more important, though not quite a definitive factor (Gonzalez, Nishi and Gagge, 1974). The criterion of equal wettedness forms the basis of the family of 'rational environmental temperature indices' developed by the Pierce laboratory (Gagge and Nishi, 1977).

Humid Operative Temperature

Humid Operative Temperature, T_{oh}, is the temperature of a saturated isothermal enclosure in which the total heat exchange and wettedness of the skin are the same as in the real environment (Nishi and Gagge, 1971). Skin temperature, clothing and air movement are assumed to be the same in the real and standard environments; sweat rate is not. Call the total thermal insulation between skin and environment I_t and the corresponding vapour resistance R_t. The total heat exchanges in the real (T_o, p_a) and in the standard (T_{oh}, p_{oh}) environments are equal:

$$(T_{sk} - T_o)/I_t + W(p_{sk} - p_a)/R_t = (T_{sk} - T_{oh})/I_t + W(p_{sk} - p_{oh})/R_t \quad (5.29)$$

This reduces to

$$(T_o - T_{oh}) / I_t = W(p_{oh} - p_a) / R_t \quad (5.30)$$

Since the standard environment is, by definition, saturated, p_{oh} is the saturated

Fig. 5.11. Heat balance of a subject whose sweat rate depends on skin temperature as described in the text. Other conditions as for Fig. 5.10. Straight, sloping lines relate to wet skin, vertical lines to conditions in which all the sweat evaporates. The hatched areas indicate the intermediate zone of restricted evaporation.

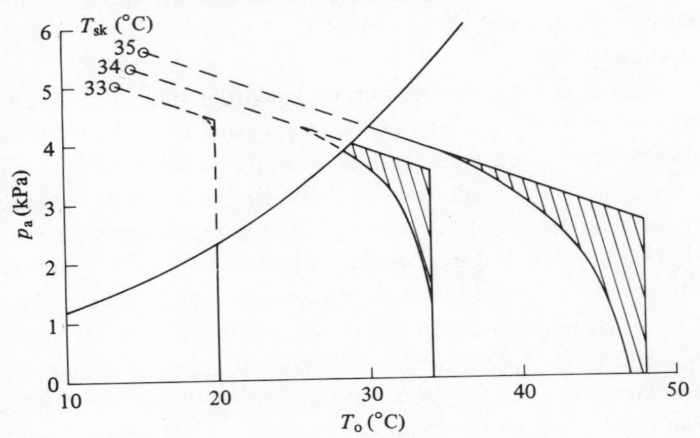

water vapour pressure at T_{oh}, and when a real environment is specified there is only one solution for T_{oh}. The family of environments sharing this value of T_{oh} will be a straight line on the psychrometric axes, of slope $-R_t / W \cdot I_t$ kPa/°C. As with the line for wet skin ($W = 1.0$) in Fig. 5.10, these lines of equal wettedness must pass through the point $p_a = p_{sk}$; $T_o = T_{sk} - H_{sk} \cdot I_t$. For given values of T_{sk}, H_{sk}, I_t and R_t they fan out from this pivot point as shown in Fig. 5.12 (Nishi and Gagge, 1971, 1974).

Unfortunately, unless the skin temperature in the real environment is known, this construction cannot be made. If the subject is known to be thermally comfortable, equation 5.22 above can provide an estimate. In other cases the estimate can be based on a mathematical model of human thermoregulation (Gagge, Stolwijk & Nishi, 1971; Nishi and Gagge, 1977).

Standard Effective Temperature

Standard Effective Temperature, SET, is an extension of these principles. SET is the temperature of an isothermal enclosure at 50% relative humidity with air movement such that h_o is 8.0 W/m²·°C (still air), in which a subject wearing standard clothing will have the same heat loss, skin temperature and skin wettedness as he has in the real environment and clothing. The insulation of the standard clothing worn in the imaginary environment is 0.6 clo for light activity, but decreases at higher levels of activity in such a way that the SET required for comfort is unchanged (about 24°C). Calculation of SET requires

Fig. 5.12. T_{oh} for the conditions as in Figs. 5.10 and 5.11 (T_{sk} = 34°C, H_{sk} = 150 W/m², I_t = 0.13°C·m²/W, R_t = 0.012 kPa·m²/W). A line of constant wettedness is also a line of constant T_{oh}. The index values are the temperatures on the saturation line.

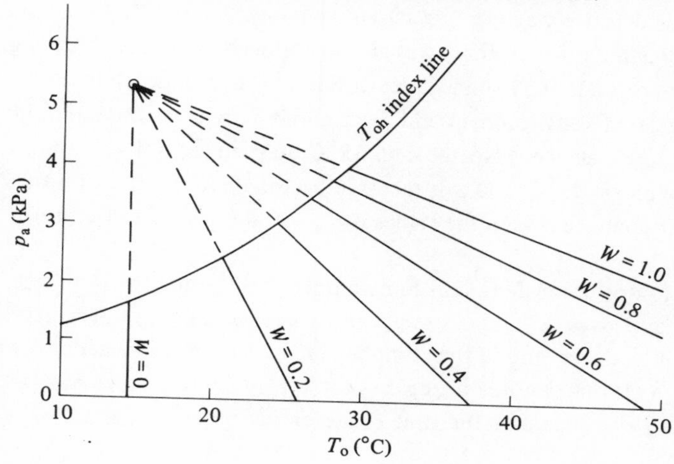

a computer, for which a Fortran program has been published (Nishi and Gagge, 1977).

New Effective Temperature

New Effective Temperature, designated ET*, is calculated in the same way as SET, but is restricted, like the old ET which it replaces (ASHRAE, 1974), to normal indoor clothing and light activity. Like SET, the index value is based on an imaginary environment at 50% relative humidity (the old ET uses 100%). In the original expositions of these indices, I_t and R_t are represented by their inverses, the net conductances for sensible and latent heat transfer between skin and environment. These are written $h \cdot F_{cl}$ and $h_e \cdot F_{pcl}$ respectively. F_{cl} is a sort of thermal efficiency factor:

$$F_{cl} = I_a / (I_a + I_c) \tag{5.31}$$

F_{pcl} is the corresponding expression for vapour resistances (p. 271). When F_{cl} is multiplied by h (which is h_o expressed per unit skin area, not clothing surface area) we get $1/I_t$. The value of F_{cl} for a given clothing assembly depends on I_a and therefore on the air movement, even if no wind penetration is assumed. Values for various assemblies are usually tabulated for the standard low air movement.

The Heat Stress Index

The Heat Stress Index, HSI, introduced by Belding and Hatch (1955), takes wettedness (W) as the sole criterion of equivalence. It uses a simple nomogram (Fig. 5.13) to find W, assuming a skin temperature of 35°C. The index value is W, expressed as a percentage. The reader is warned that the heat exchange coefficients on which the diagrams are based were changed in 1963 (Hatch, 1963) and an adjustment for clothing was also introduced (Hertig and Belding, 1963). This is appropriate for a single layer of thin clothing. Index values of more than 100 relate to environment and work combinations in which the skin temperature for heat balance is greater than 35 °C. In such cases the skin temperature will rise above 35 °C, and it is incorrect to calculate the rate of heat storage as the difference between the values of E_{req} and E_{max} as derived from the diagrams.

The assumption of a constant skin temperature makes calculation of heat exchanges easy, and gives this index its appeal. Its weakness lies in the errors this necessarily introduces, and in the complete identification of wettedness with heat stress. One feels that sensible heat gain should play a more direct part, and values of E_{req} and E_{max} are used by some in preference to the actual index value.

Fig. 5.13. The Heat Stress Index (HSI) of Belding and Hatch. For clothed subjects the upper two diagrams are entered with Operative Temperature, T_o, and ambient water vapour pressure, p_a, as shown. The HSI value is given in the right-hand middle diagram. For nude subjects the nomogram is entered through the bottom two diagrams, as indicated. If E_{max} exceeds 380 W/m^2 the final diagram is entered with this figure, as shown for the nude case, p_a = 2.0 kPa, V = 1.0 m/s. The heat exchange coefficients used in constructing this nomogram are those recommended by Hatch (1963), with the adjustment for clothing proposed by Hertig and Belding (1963). (From Kerslake, 1972.)

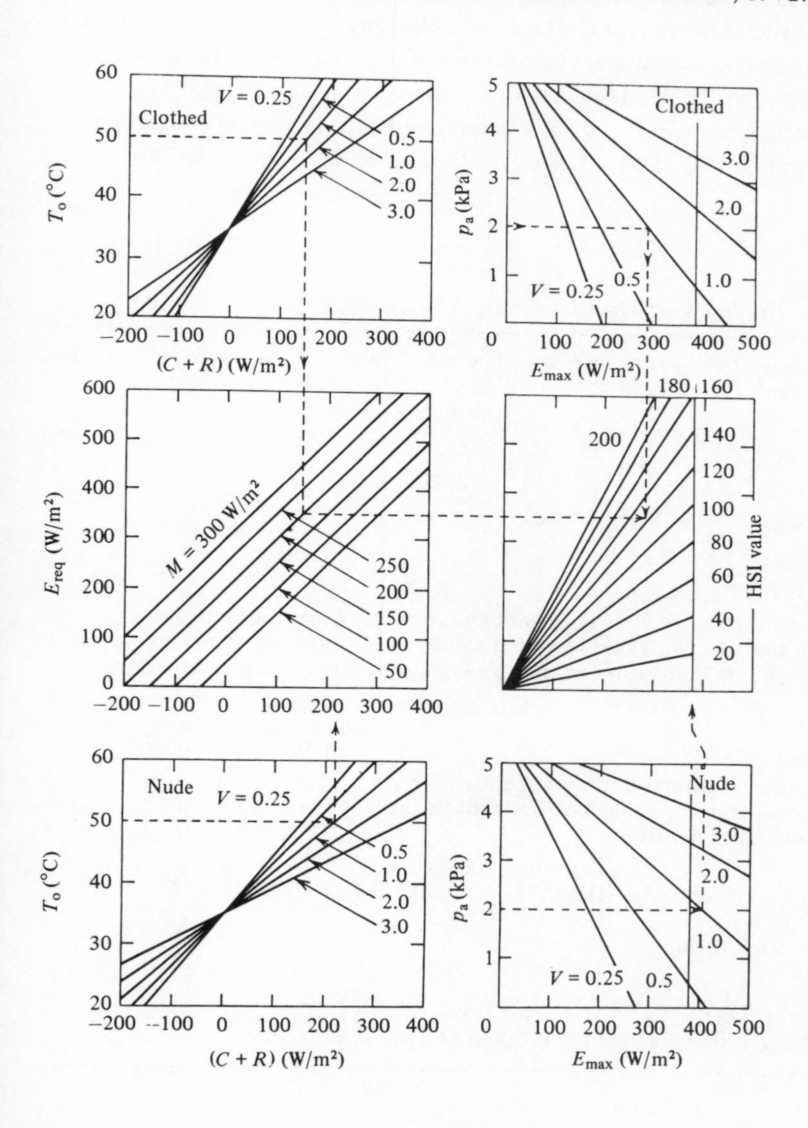

The Index of Thermal Stress

The Index of Thermal Stress, ITS, of Givoni (1964) deserves more publicity than it has had. It is intended for outdoor situations and is restricted to air movements of at least 0.5 m/s. As in the HSI a skin temperature of 35°C is assumed and E_{req} and E_{max} are calculated. Their ratio, W, which again can exceed 1.0 determines the efficiency of sweat evaporation, (evaporative heat loss/sweat rate), and this allows the required sweat rate to be found. This, expressed in g/h (not per m^2), is the index value and criterion of equivalence. The index allows for various clothing assemblies and for sunlight. It is in good agreement with more elaborate calculations of required sweat rate, as well as with the empirically based Predicted Four-hour Sweat Rate (see below). The sweating efficiency factor is a way of representing the important zone of restricted evaporation (Fig. 5.11). The calculation proceeds as shown in Table 5.6.

Table 5.6. *The Index of Thermal Stress, ITS (Givoni, 1964)*

H_o, R_s, E_{max} and E_{req} are in W/m^2. The index value, ITS, is in g/h for a subject of area 1.8m^2.

Equations

$$H_o = \alpha \cdot V^{0.3} (T_g - 35)$$
$$R_s = I_N \cdot K_{pe} \cdot K_{cl} [1 - a (V^{0.2} - 0.88)]$$
$$E_{max} = p \cdot V^{0.3} (5.62 - p_a)$$
$$E_{req} = H_p + H_o + R_s$$
$$ITS = F_\eta \cdot E_{req}$$

H_o, heat gain from the air and radiant environment excluding sunlight; R_s, heat gain from sunlight; T_g, globe temperature; I_N, solar intensity measured normal to solar flux; p_a, ambient water vapour pressure; H_p, rate of heat production.

Coefficients

Clothing	α	K_{cl}	a	p
Bathing suit and hat	10.2	1.0	0.35	153
Light summer clothing: underwear, short-sleeved cotton shirt, long cotton trousers, hat	8.4	0.5	0.52	63
Military overalls over shorts	7.5	0.4	0.52	63

Posture	*Terrain*	K_{pe}
Sitting, back to sun	Desert	0.39
	Forest	0.38
Standing, back to sun	Desert	0.31
	Forest	0.27

$$F_\eta = 2.7 \exp\{-0.6 [(E_{req}/E_{max}) - 0.12]\}$$

If E_{req}/E_{max} is less than 0.12 it is taken as 0.12; if more than 2.15 as 2.15.

Effective Temperature

Effective Temperature, ET, is one of the earliest and most successful indices of heat stress (Yaglou and Miller, 1925), though now superseded by ET*. The criterion of equivalence is the sensation of warmth on passing back and forth between two rooms. The ET is the temperature of still, saturated air which would give rise to an equivalent sensation. The normal scale, shown in Figs. 5.14 and 5.15, refers to subjects in normal American indoor clothing of the 1920s; the basic scale, not shown here, to subjects stripped to the waist. On moving from a warm damp environment to a hot dry one, water vapour which has been taken up by the clothing evaporates, with consequent transient cooling. Skin also exchanges water vapour in the same way. This is probably why ET over-

Fig. 5.14. Effective Temperature (normal scale), appropriate for subjects wearing normal indoor clothing. The dry and wet bulb temperatures (T_a and T_{wb}) are joined and the Effective Temperature read from the appropriate wind speed scale (V). Adjustments for radiation (CET and ETR) are described in the text.

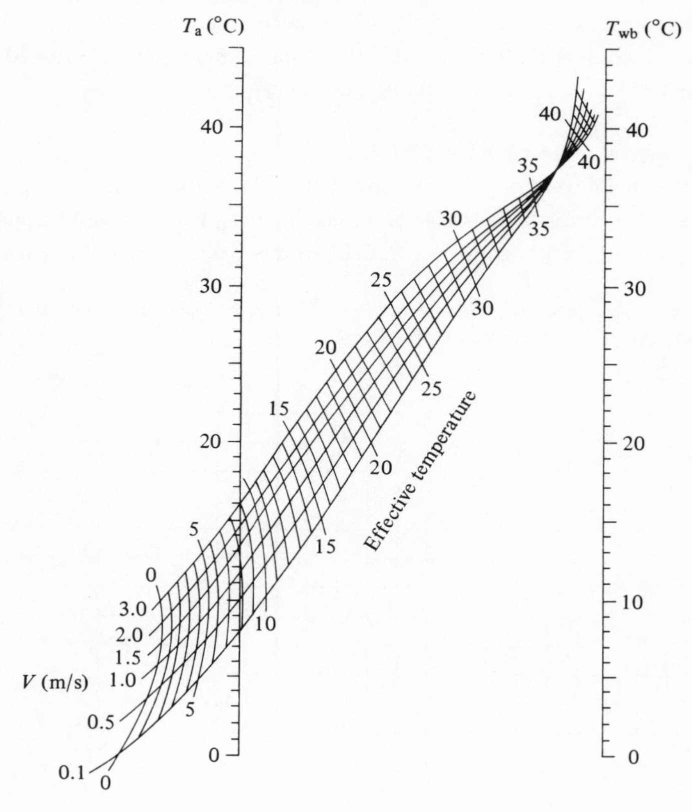

estimates the effect of humidity on steady-state warmth sensation, particularly in the regions of comfort and mild heat stress (Macpherson, 1960). It may underestimate the effect of humidity at high stress levels, possibly because the exposure times were too short Macpherson, 1960; Leithead and Lind, 1964; Wyndham *et al.*, 1967).

The ET should really be applied only to the exercise level used, namely strolling, but in the absence of a better index it has been used for many years in all sorts of working situations. If air movement is fixed, ET can be displayed on the psychrometric axes (Fig. 5.15), when it shows a general correspondence with the theoretically derived ET*.

When mean radiant temperature differs from air temperature (T_a), it is usual to enter the ET nomogram with globe temperature, T_g, instead of T_a. The resulting figure is called the Corrected Effective Temperature, CET (Bedford, 1964). This is not really a correct procedure because at constant humidity if T_a increases, wet bulb temperature, T_{wb}, will also rise. Entering the nomogram with T_g and T_{wb} instead of T_a and T_{wb} underestimates the humidity (if $T_g > T_a$). If an appropriate adjustment is made to T_{wb} the resulting figure is called the Effective Temperature including Radiation, ETR. It is approximately correct to add to T_{wb} the value 0.4 ($T_g - T_a$) (Vernon and Warner, 1932).

Predicted Four-hour Sweat Rate

Predicted Four-hour Sweat Rate, P4SR (McArdle *et al.*, 1947), takes sweat rate as its criterion of equivalence, subject to an allowance for hidromeiosis. This is built into the nomogram, which deliberately overestimates the sweat rates

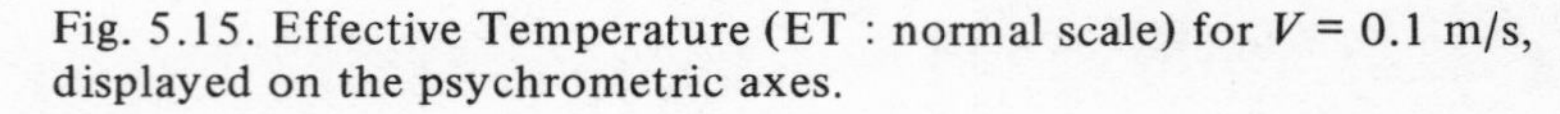

Fig. 5.15. Effective Temperature (ET : normal scale) for V = 0.1 m/s, displayed on the psychrometric axes.

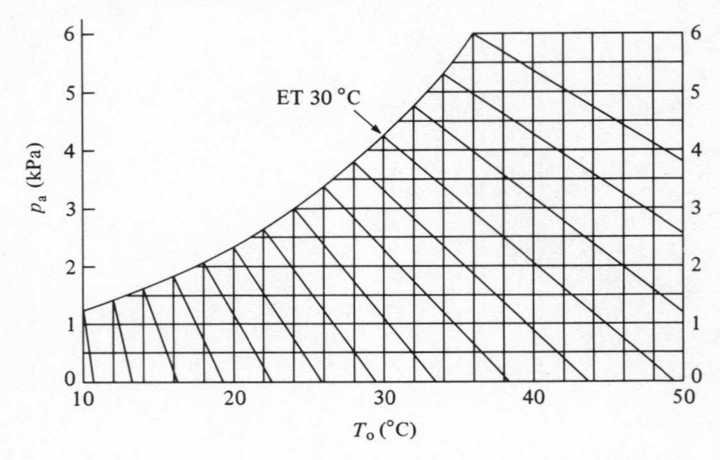

in humid environments so as to give a better indication of their general stress levels. It allows for work rates up to 230 W/m^2 and for two clothing assemblies. The calculation is somewhat tedious and tends to obscure the operation of different factors affecting heat stress, because work rate, radiant temperature and clothing all contribute adjustments to the wet bulb temperature as well as appearing directly. This complexity is justified by the close agreement between index values and the observations on which they were based. Both inside and outside the envelope of conditions actually used, the P4SR agrees well with theoretical expectation (Kerslake, 1972).

The P4SR is calculated in the following three stages:

1. *Calculation of modified wet bulb temperature*

(*a*) If T_g differs from T_a, the wet bulb temperature is increased by $0.4\,(T_g - T_a)$°C.

(*b*) If the metabolic rate (M) exceeds 63 W/m^2 the wet bulb temperature is increased by the amount indicated in the small inset chart of Fig. 5.16.

(*c*) For men wearing overalls the wet bulb temperature is increased by 1.0°C. The modifications are additive.

2. *Calculation of B4SR*

The nomogram (Fig. 5.16) is now entered with T_g and modified wet bulb temperature, using the wet bulb scale appropriate for the wind speed (V). A line drawn between these points intersects the appropriate B4SR line at the B4SR value. For wind speeds between 0.4 and 2.5 m/s the wet bulb temperature should be interpolated.

3. *Calculation of P4SR*

The P4SR is found by adding to the B4SR amounts which depend on the metabolic rate and clothing.

(a) Men sitting in shorts : P4SR = B4SR

(b) Men working in shorts : P4SR = B4SR + 0.012 (M – 63)

(c) Men sitting in overalls : P4SR = B4SR + 0.25

(d) Men working in overalls : P4SR = B4SR + 0.25 + 0.017 (M – 63)

Wet Bulb Globe Temperature

Wet Bulb Globe Temperature, WBGT (Yaglou and Minard, 1957), is a simple weighting of temperatures designed to identify conditions which are so hot that military training should be suspended. The great simplicity of the index depends on its restriction to only one situation and only one level of stress. It was not intended for use in any other context, and when so misused does not combine the environmental factors correctly.

Readings are required of air temperature, globe temperature and wet bulb temperature, measured by an unventilated thermometer kept wet by means of a wick and reservoir. No measurement of wind speed is required.

$$\text{WBGT} = 0.7\,T_{\text{wb}} + 0.2\,T_{\text{g}} + 0.1\,T_{\text{a}} \tag{5.32}$$

Fig. 5.16. The P4SR nomogram. The left-hand scale is entered with the globe temperature and the appropriate wet bulb scale with a modified wet bulb temperature (see text). The B4SR is found by joining these points and reading off from the appropriate B4SR scale. The P4SR is then calculated as described in the text. (From Kerslake, 1972.)

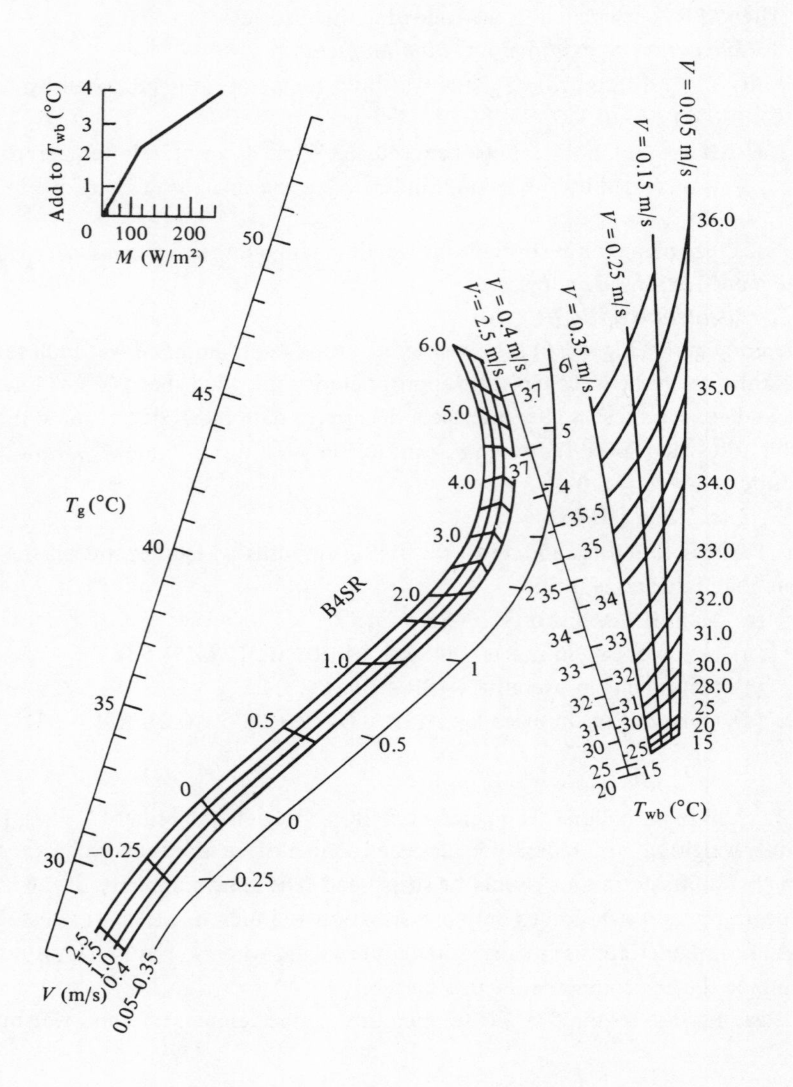

If a ventilated psychrometer (Assman or sling) is available a simpler weighting can be used.

$$\text{WBGT} = 0.7\, T_{wb} + 0.3\, T_g \qquad (5.33)$$

Oxford Index

The Oxford Index (Leithead and Lind, 1964) is another simple weighting of wet and dry bulb temperatures, in this case $0.85T_{wb} + 0.15T_a$, and is intended to apply to the upper permissible limits for men working in extreme heat. The temperature limits depend on the work rate, but this weighting of wet and dry bulb temperatures applies to all work rates when conditions are near the upper limits for sustained work. The difference between this weighting and that of the WBGT, both of which are well justified by practical experience, may lie in the fact that military trainees include a proportion who tolerate heat badly and who constitute the potential casualties. In hot industries, for which the Oxford Index is intended, the more susceptible people have been weeded out.

Temperature Humidity Index

Temperature Humidity Index, THI, is yet another combination of wet and dry bulb temperatures. These are given equal weight, but a constant is added, so different formulations are required for different temperature scales. There is no allowance for wind or sunshine.

$$\text{Fahrenheit : THI} = 0.4\, T_a + 0.4\, T_{wb} + 15 \qquad (5.34)$$
$$\text{Celsius : THI} = 0.4T_a + 0.4\, T_{wb} + 4.8 \qquad (5.35)$$

It is hard to see much merit in this index, except that it is better than applying the WBGT or Oxford weightings to inappropriate situations.

Choice of an index

The handful of heat stress indices outlined above is but the tip of the iceberg. Time has submerged most of the remainder as understanding of bio-thermal phenomena has extended, but there are still sporadic applications of regression techniques without reference to physical processes, e.g. by adding functions of temperature, humidity and wind speed. However, the general tendency has been towards rationally based indices.

The popularity of an index does not seem to depend very much on its accuracy. This is unfortunate because a widely used index carries more weight and can be used to greater effect. The ET is probably still the best-known index, and would be a natural choice for specifying indoor conditions, except that in and near the comfort zone it makes too much allowance for humidity. Its successor, ET*, corrects this.

In conditions of mild warmth, sensation is the most appropriate criterion, and this is closely related to wettedness. The indices based on wettedness are the HSI and the SET family. The latter require a computer for their determination, but solutions for particular cases of clothing, activity and air movement can be presented in chart form, as seems necessary if an index is to become popular. The possibility of precise algebraic evaluation may be useful in borderline cases where such matters as rates of pay are involved. The HSI has the merit of simplicity and a good chart. It can be described as rationally based and its principles are easy to understand. In consequence its accuracy is rather low. It is quite widely used and also misused by such practices as averaging the index value over time, or calculating heat storage and loss from the nomogram.

In severe conditions skin temperature becomes more important than wettedness in determining discomfort, fatigue and risk of collapse. Here the rationally based ITS and the empirically based P4SR are in close agreement. The Oxford Index weighting of wet and dry bulb temperatures is consistent with a skin wettedness close to unity, and hence should be appropriate for extrapolating from known limiting conditions.

The WBGT has been outstandingly successful in the specific role for which it was designed. Despite the lack of a nomogram it has achieved great popularity because of its great simplicity and because it is rather easy to build instruments which indicate the index value directly and if necessary remotely. Although its use as a general heat stress index is indefensible, such misapplication is not uncommon. It cannot be too strongly stressed that outside the band 29–32°C the WBGT has no significance, and that this index should only be applied to indicate potential hazard to physically fit young men engaged in military training or similar activity.

This restriction to fit young men applies also in principle to both the P4SR and the ITS, and it is no pedantic quibble. Not only are such subjects inherently better at sweating than other groups in the community, but they also tend to become heat-acclimatised. Their maximum sweat rates are of the order of a litre or two per hour, although in long exposure to moist heat this is reduced by hidromeiosis. Unacclimatised young men of sedentary habits have maxima of two or three hundred grams per hour, and young women perhaps only half of this. As the experiment which measures maximum sweat rate is itself an acclimatising experience, data about unacclimatised people are sparse.

In moist environments most people can produce enough sweat to wet the skin, so maximum sweat rate is not a limiting factor. It is in moderate to severe dry heat that differences between subjects become most important (the zone of full evaporation). What is a pleasant hot day to one person will be overwhelming to another whose maximum sweat rate is less than the evaporative demand, and the difference between them may not be evident in slightly cooler conditions. Clearly any heat stress index should contain a qualification about maximum sweat rate, and more should be known about this vital quantity and its distribution in the general population.

Cold environments

The factors contributing to the coldness of an environment are air temperature and air movement. Surrounding surfaces are usually near to air temperature and sunlight the only important radiant source. In dry cold regions with clear skies sunlight can greatly ameliorate conditions.

Wind-chill

Wind-chill is an index which combines air temperature and air movement (Siple and Passel, 1945). It was based on measurements of the rate of cooling of a small cylinder with a surface temperature of 33 °C. Since bare skin would not remain at this temperature in the cold conditions for which the wind-chill index is intended, the actual rates of cooling have no physiological significance and no purpose would be served by converting the original units ($kcal/m^2 \cdot h$) into MKS. They may be treated simply as index values. The results were fitted by the empirical equation

$$\text{Wind-chill} = (33 - T_a)(10\sqrt{V} + 10.45 - V) \qquad (5.36)$$

where dry bulb temperature T_a is in °C and wind speed V in m/s. Wind-chill relates more closely to the cooling of exposed skin than to general thermal balance. It has proved a valuable guide to the severity of outdoor conditions in which frostbite is the predominant hazard. Frostbite is likely when the wind-chill exceeds 1400 (e.g. −20 °C, 4.0 m/s).

Equivalent Still Air Temperature

Equivalent Still Air Temperature, ESAT, is the temperature of still air which would remove heat at the same rate as in the real environment, the clothing surface temperature being unchanged (Burton and Edholm, 1955). Using the ambient insulation, I_a, of the clo system, in the real environment

$$T_s - T_a = H \cdot I_a \qquad (5.37)$$

where T_s is the clothing surface temperature, T_a the air temperature and H the rate of sensible heat loss. In the imaginary still air environment of temperature T_{ao} and ambient insulation I_{ao}

$$T_s - T_{ao} = H \cdot I_{ao} \qquad (5.38)$$

The ESAT can be found by combining these equations to eliminate T_3:

$$T_{ao} = T_a - H(I_{ao} - I_a) \tag{5.39}$$

If insulations are measured in clo, the last term in each equation has to be multiplied by 0.155 °C·m² / W·clo. The value of I_{ao} is arbitrarily taken as 1.0 clo (0.155 °C·m² / W). Since this is greater than any real value of I_a, ESAT is always lower than air temperature. Fig. 5.17 shows the dependence of this difference on heat loss and wind speed.

ESAT has the merit of a logical basis, but is open to the objection that it disregards the important factor of wind penetration. Since this varies from one clothing assembly to another it is hard to see an alternative, but no motor cyclist would accept the trivial effect credited to wind speeds over 10 m.p.h. (4.5 m/s). The dependence of ESAT on rate of heat loss is theoretically correct, but leads to the bizarre result that ESAT is lower the harder one works. It is therefore not a very good indication of how cold one feels.

Equivalent Chill Temperature

Equivalent Chill Temperature is based on the same principles as ESAT, but uses 'calm air' (3-4 m.p.h.; 1.4-1.8 m/s) as a standard.

Fig. 5.17. The effect of wind speed (V) and rate of heat loss on Equivalent Still Air Temperature (ESAT). The ordinate is the number of degrees by which ESAT is lower than the actual air temperature (T_a). H, rate of sensible heat loss.

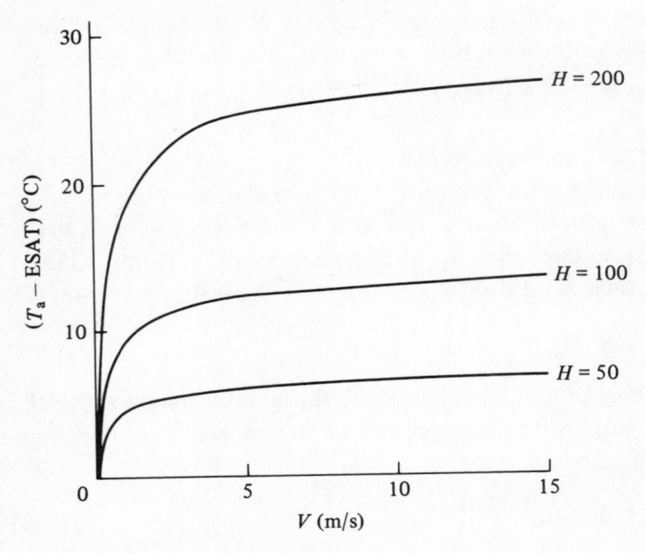

Humidity

Humidity has never been shown to have an important effect on heat transfer in cold environments. The mechanisms of evaporation described above would lead one to expect this since the water vapour pressure in cold air is always low. At 0°C saturated water vapour pressure is 0.61 kPa, at −20°C 0.10 kPa. Popular belief, on the contrary, is that damp 'raw' conditions near freezing point feel colder than 'dry' climates with far lower temperatures. Sunlight and wind are probably largely responsible for this belief, but differences in attitude to clothing and weather may contribute. Clothing which is stored at low relative humidity (not difficult if the outside air is very cold and therefore has a low water content even if saturated) becomes very dry. When it is put on and the wearer goes out into the cold air the fibres on the inside take up water vapour from the skin, and those on the outside take up water vapour from the air because they equilibrate with relative humidity rather than with vapour pressure. As the fibres take up water, heat is released which provides protection for the wearer. In a thick assembly this process can continue for several hours. It is beneficial so long as the water is taken up into the structure of the fibres. Condensation of liquid water between the fibres can also occur, particularly if the subject is sweating, and this destroys the insulating power of the clothing. Wool has a large capacity for water vapour, as have many other natural fibres. Artificial fibres in general have small capacities.

References

ASHRAE (1974) *Thermal Environmental Conditions for Human Occupancy.* ASHRAE Standard 55-74. New York : American Society of Heating, Refrigerating and Air-conditioning Engineers.

Bedford, T. (1964) *Basic Principles of Ventilation and Heating,* 2nd edn. London : H.K. Lewis.

Belding, H.S. and Hatch, T.F. (1955) Index for evaluating heat stress in terms of the resulting physiological strain. *Heating, Piping and Air Conditioning,* **27,** 129–36.

Benzinger, T.H. (1959) On physical heat regulation and the sense of temperature in Man. *Proceedings of the National Academy of Sciences, USA,* **45,** 645–59.

Bligh, J. and Moore, R.E. (ed.) (1972) *Essays on Temperature Regulation.* Amsterdam : North-Holland.

Brebner, D.F. and Kerslake, D. McK. (1969) The relation between sweat rate and weight loss when sweat is dripping off the body. *Journal of Physiology (London),* **202,** 719–35.

Breckenridge, J.R. and Goldman, R.F. (1971) Solar heat load in Man. *Journal of Applied Physiology,* **31,** 659–63.

Breckenridge, J.R. and Goldman, R.F. (1977) Effect of clothing on bodily resistance against meteorological stimuli. In *Progress in Biometeorology,* vol 1, pt II, ed. S.W. Tromp, pp. 194–208. Lisse : Swets and Zeitlinger.

Burton, A.C. and Edholm, O.G. (1955). *Man in a Cold Environment.* London : Arnold, (Facsimile, New York : Hafner, 1970.)

Cabanac, M. (1975) Temperature regulation. *Annual Review of Physiology,* **37,** 415–39.

Chrenko, F.A. and Pugh, L.G.C.E. (1961) The contribution of solar radiation to the thermal environment of Man in Antarctica. *Proceedings of the Royal Society, Series B,* **155,** 243–65.

Clifford, J., Kerslake, D.McK. and Wadell, J. (1959) The effect of wind speed on maximum evaporative capacity in Man. *Journal of Physiology (London),* **147,** 253–9.

Cooper, K.E. and Kenyon, J.R. (1957) A comparison of temperatures measured in the rectum, oesophagus and on the surface of the aorta during hypothermia in Man. *British Journal of Surgery,* **44,** 616–19.

Cooper, K.E., Cranston, W.I. and Snell, E.S. (1964) Temperature in the external auditory meatus as an index of central temperature changes. *Journal of Applied Physiology,* **19,** 1032–5.

Custance, A.C. (1965) Use of small skin surface areas for whole body sweating assessment. *Canadian Journal of Physiology and Pharmacology,* **43,** 971–7.

Davies, C.T.M. (1979) Influence of skin temperature on sweating and aerobic performance during severe work. *Journal of Applied Physiology : Respiratory, Environmental and Exercise Physiology,* **47,** 770–7.

Edholm, O.G. and Bacharach, A.L. (ed.) (1956) *The Physiology of Human Survival.* London : Academic Press.

Eichna, L.W., Berger, A.R., Rader, B. and Becker, W.H. (1951) A comparison of intracardiac and intravascular temperatures with rectal temperatures in Man. *Journal of Clinical Investigation,* **30,** 353–9.

Fanger, P.O. (1970) *Thermal Comfort.* Copenhagen : Danish Technical Press (Also published by McGraw-Hill, New York, 1972).

Fanger, P.O. and Langkilde, G. (1975) Interindividual differences in ambient temperatures preferred by seated persons. *ASHRAE Transactions,* **81**(II), 140–7.

Fox, R.H. and Edholm, O.G. (1963) Nervous control of the cutaneous circulation. *British Medical Bulletin,* **19,** 110–14.

Fox, R.H., Goldsmith, R., Hampton, I.F.G. and Lewis, H.E. (1964). The nature of the increase in sweating capacity produced by heat acclimatization. *Journal of Physiology (London),* **171,** 368–76.

Fox, R.H., Goldsmith, R. and Wolff, H.S. (1961) The use of a radio pill to measure deep body temperature. *Journal of Physiology (London),* **160,** 22*P*–23*P*.

Fox, R.H., Solman, A.J., Issacs, R., Fry, A.J. and MacDonald, I.C. (1973) A new method for monitoring deep body temperature from the skin surface. *Clinical Science,* **44,** 81–6.

Gagge, A.P. (1937) A new physiological variable associated with sensible and insensible perspiration. *American Journal of Physiology,* **120,** 277–87.

Gagge, A.P. and Gonzalez, R.R. (1974) Physiological and physical factors associated with warm discomfort in sedentary Man. *Environmental Research,* **7,** 230–42.

Gagge, A.P. and Nishi, Y. (1977) Heat exchange between human skin surface and thermal environment. *Handbook of Physiology,* sec. 9, *Reaction to Environmental Agents,* ed. D.H.K. Lee, pp. 69–92. Washington, DC : American Physiological Society.

Gagge, A.P., Burton, A.C. and Bazett, H.C. (1941) A practical system of units for the description of the heat exchange of Man with his thermal environment. *Science,* **94,** 428–30.

Gagge, A.P., Graichen, H., Stolwijk, J.A.J., Rapp, G.M. and Hardy, J.D. (1968) ASHRAE-sponsored research project RP-41 produces R-meter. *ASHRAE Journal,* **10,** 77–81.

Gagge, A.P., Stolwijk, J.A.J. and Hardy, J.D. (1965) A novel approach to measurement of Man's heat exchange with a complex radiant environment. *Aerospace Medicine,* **36,** 431–5.

Gagge, A.P., Stolwijk, J.A.J. and Nishi, Y. (1971) An effective temperature scale based on a simple model of human physiological regulatory response. *ASHRAE Transactions,* **77**(1), 247–62.

Givoni, B. (1964) A new model for evaluating industrial heat exposure and maximum permissible work load. *International Journal of Bioclimatology and Biometeorology,* **8,** 115–24.

Gonzalez, R.R., Nishi, Y. and Gagge, A.P. (1974) Experimental evaluation of Standard Effective Temperature, a new biometeorological index of Man's thermal discomfort. *International Journal of Biometeorology,* **18,** 1–15.

Green, G.H. (1975) The effect of indoor relative humidity on absenteeism and colds in schools. *ASHRAE Transactions,* **80**(2), 131–41.

Griffiths, I.D. and Boyce, P.R. (1971) Performance and thermal comfort. *Ergonomics,* **14,** 457–68.

Hammel, H.T. (1968) Regulation of internal body temperature. *Annual Review of Physiology,* **30,** 641–710.

Hardy, J.D. and DuBois, E.F. (1934) The technique of measuring radiation and convection. *Journal of Nutrition,* **15,** 461–75.

Hardy, J.D., Gagge, A.P. and Stolwijk, J.A.J. (ed.) (1970) *Physiological and Behavioral Temperature Regulation.* Springfield, Ill, : Charles C. Thomas.

Hatch, T.F. (1963) Assessment of heat stress. In *Temperature, its Measurement and Control in Science and Industry,* vol. 3, pt 3, ed. J.D. Hardy, pp. 307–18. New York : Reinhold.

Hellon, R.F. and Crockford, G.W. (1959) Improvements to the globe thermometer. *Journal of Applied Physiology,* **14,** 649–50.

Herrington, L.P., Winslow, C.-E. A. and Gagge, A.P. (1937) The relative influence of radiation and convection upon vasomotor temperature regulation. *American Journal of Physiology,* **120,** 133–43.

Hertig, B.A. and Belding, H.S. (1963) Evaluation of health hazards. In *Temperature, its Measurement and Control in Science and Industry,* vol. 3, pt 3, ed. J.D. Hardy, pp. 347–55. New York : Reinhold.

Hitchen, M., Brodie, D.A. and Harness, J.B. (1980) Cardiac responses to demanding mental load. *Ergonomics,* **23,** 379–85.

Hodgman, C.D. (1965) *Handbook of Chemistry and Physics.* Cleveland : Chemical Rubber Co.

Itoh, S., Ogata, K. and Yoshimura, H. (eds.) (1972) *Advances in Climatic Physiology.* Tokyo : Igaku Shoin.

Kerslake, D.McK. (1963) Errors arising from the use of mean heat exchange coefficients in the calculation of the heat exchanges of a cylindrical body in a transverse wind. In *Temperature, its Measurement and Control in Science and Industry,* vol. 3, pt 3, ed. J.D. Hardy, pp. 183–90. New York : Reinhold.

Kerslake, D.McK. (1972) *The Stress of Hot Environments.* Cambridge : Cambridge University Press.

Kerslake, D.McK. and Cooper, K.E. (1950) Vasodilatation in the hand in response to heating the skin elsewhere. *Clinical Science,* **9,** 31–47.

Koch, W., Jennings, B.H. and Humphreys, C.M. (1960) Sensation responses to temperature and humidity under still-air conditions in the comfort range. *ASHRAE Transactions,* **66**(2), 264–72.

Lee, D.H.K. (1964) Terrestrial animals in dry heat : Man in the desert. In *Handbook of Physiology,* sect. 4, ed. D.B. Dill, pp. 551–82. Washington, DC : American Physiological Society.

Leithead, C.S. and Lind, A.R. (1964) *Heat Stress and Heat Disorders.* London : Churchill.

Lind, A.R. and Hellon, R.F. (1957) Assessment of the physiological severity of hot climates. *Journal of Applied Physiology,* **11,** 35–40.

Luczak, H. (1979) Fractioned heart rate variability. II. Experiments on superimposition of components of stress. *Ergonomics,* **22,** 1315–23.

McArdle, B., Dunham, W., Holling, H.E., Ladell, W.S.S., Scott, J.W., Thomson, M.L. and Weiner, J.S. (1947) *The Prediction of the Physiological Effects of Warm and Hot Environments.* Report RNP 47/391. London : Medical Research Council.

McCutchan, J.W. and Taylor, C.L. (1950) *Respiratory Heat Exchange with Varying Temperature and Humidity of Inspired Air.* Air Force Technical Report 6023. Ohio : Wright-Patterson Air Force Base.

McIntyre, D.A. (1974) The thermal radiation field. *Building Science,* **9,** 247–62.

McIntyre, D.A. (1977) Sensitivity and discomfort associated with overhead thermal radiation. *Ergonomics,* **20,** 287–96.

McIntyre, D.A. (1980) *Indoor Climate.* Barking : Applied Science Publishers.

McIntyre, D.A. and Griffiths, I.D. (1975) The effects of added clothing on warmth and comfort in cool conditions. *Ergonomics,* **18,** 205–11.

Macpherson, R.K. (1960) *Physiological Responses to Hot Environments.* Medical Research Council Special Report 298. London : HMSO.

Mann, T. (1924) *Die Zauberberg,* trans. H.T. Lowe-Porter. Harmondsworth : Penguin (1960).

Mills, J.N. (1966) Human circadian rhythms. *Physiological Reviews,* **46,** 128–71.

Mole, R.H. (1948) The relative humidity of the skin. *Journal of Physiology (London)* **107,** 399–411.

Molnar, G.W. and Rosenbaum, J.C. (1963) Surface temperature measurement with thermocouples. In *Temperature, its Measurement and Control in Science and Industry,* vol. 3, pt 3, ed. J.D. Hardy, pp. 3–11. New York : Reinhold.

Monteith, J.L. (1973) *Principles of Environmental Physics.* London : Arnold.

Nelms, J.D. and Soper, D.J.G. (1962) Cold vasodilatation and cold acclimatisation in the hands of British fish filleters. *Journal of Applied Physiology,* **17,** 444–8.

Newburgh, L.H. (ed.) (1949) *Physiology of Heat Regulation and the Science of Clothing.* Philadelphia : Saunders. (Facsimile published by Hafner, New York, 1970.)

Nielsen, B. and Nielsen, M. (1965) On the regulation of sweat secretion in exercise. *Acta Physiologica Scandinavica,* **64,** 314–22.

Nishi, Y. and Gagge, A.P. (1970) Moisture permeation of clothing – a factor governing thermal equilibrium and comfort. *ASHRAE Transactions,* **76**(1), 137–45.

Nishi, Y. and Gagge, A.P. (1971) Humid operative temperature. A biophysical index of thermal sensation and discomfort. *Journal de Physiologie,* **63,** 365–8.

Nishi, Y. and Gagge, A.P. (1974) A psychrometric chart for graphical prediction of comfort and heat tolerance. *ASHRAE Transactions,* **80**(2), 115–30.

Nishi, Y. and Gagge, A.P. (1977) Effective temperature scale useful for hypo- and hyperbaric environments. *Aviation, Space and Environmental Medicine,* **48,** 97–107.

Olesen, B.W. and Fanger, P.O. (1973) The skin temperature distribution for resting Man in comfort. *Archives des Sciences Physiologiques,* **27,** A385–93.

Ower, E. and Pankhurst, R.C. (1977) *The Measurement of Air Flow.* Oxford : Pergamon Press.

Piironen, P. (1970) Sinusoidal signals in the analysis of heat transfer in the body. In *Physiological and Behavioural Temperature Regulation,* ed. J.D. Hardy, A.P. Gagge and J.A.J. Stolwijk, pp. 358–66. Springfield, Ill. : Charles C. Thomas.

Poulton, E.C. and Kerslake, D.McK. (1965) Initial stimulating effect of warmth upon perceptual efficiency. *Aerospace Medicine,* **36,** 29–32.

Reinberg, A., Andlauer, P. Guillet, P., Nicolai, A., Vieux, N. and Laporte, A. (1980). Oral temperature, circadian rhythm amplitude, ageing and tolerance to shift-work. *Ergonomics,* **23,** 55–64.

Robinson, S., Turrell, E.S., Belding, H.S. and Horvath, S.M. (1943) Rapid acclimatization to work in hot climates. *American Journal of Physiology,* **140,** 168–76.

Saltin, B. and Hermansen, L. (1966) Esophageal, rectal and muscle temperature during exercise. *Journal of Applied Physiology,* **21,** 1757–62.

Sengupta, A.K., Sarkar, D.N., Mukhopadhyay, S. and Goswami, D.C. (1979) Relationship between pulse rate and energy expenditure during graded work at different temperatures. *Ergonomics,* **22,** 1207–15.

Sibbons, J.L.H. (1966) Assessment of thermal stress from energy balance considerations. *Journal of Applied Physiology,* **21,** 1207–17.

Siple, P.A. and Passel, C.F. (1945) Measurements of dry atmospheric cooling in subfreezing temperatures. *Proceedings of the American Philosophical Society,* **89,** 177–99.

Smith, P., Davies, G. and Christie, M.J (1980) Continuous field monitoring of deep body temperature from the skin surface using subject-borne portable equipment. *Ergonomics,* **23,** 85–6.

Stolwijk, J.A.J. and Hardy, J.D. (1966) Partitional calorimetric studies of responses of Man to thermal transients. *Journal of Applied Physiology,* **21,** 967–77.

Strydom, N.B., Kok, R., Jooste, R.L. and van der Walt, W.H. (1975) Intermittent exposures to heat, and the retention of heat acclimatization. *Journal of the South African Institute of Mining and Metallurgy,* July, 315–18.

Underwood, C.R. and Ward, E.J. (1966) The solar radiation area of Man. *Ergonomics,* **9,** 155–68.

Vernon, H.M. (1932) The measurement of radiant heat in relation to human comfort. *Journal of Industrial Hygiene and Toxicology,* **14,** 95–111.

Vernon, H.M. and Warner, C.R. (1932) The influence of the humidity of the air on capacity for work at high temperatures. *Journal of Hygiene (Cambridge),* **32,** 431–63.

Ward, E.J. and Underwood, C.R. (1967) The effect of posture on the solar radiation area of Man. *Ergonomics,* **10,** 399–409.

Webb, P. (1973) Rewarming after diving in cold water. *Aerospace Medicine,* **44,** 1152–7.

Webb, P., Troutman, S.J. and Annis, J.F. (1970) Automatic cooling in water cooled space suits. *Aerospace Medicine,* **41,** 169–77.

Williams, C.G. and Wyndham, C.H. (1968) The problem of optimum acclimatization. *Internationale Zeitschrift für angewarndte Physiologie,* **26,** 298–308.

Williams, C.G., Wyndham, C.H. and Morrison, J.F. (1967) Rate of loss of acclimatization in summer and winter. *Journal of Applied Physiology,* **22,** 21–6.

Winslow, C.-E.A., Gagge, A.P. and Herrington, L.P. (1940) Heat exchange and regulation in radiant environments above and below air temperature. *American Journal of Physiology,* **131,** 79–82.

Winslow, C.-E.A., Herrington, L.P. and Gagge, A.P. (1937) Relations between atmospheric conditions, physiological reactions and sensations of pleasantness. *American Journal of Hygiene,* **26,** 103–15.

Woodcock, A.H. (1962) Moisture transfer in textile systems. *Textile Research Journal,* **32,** 628–33.

Wyndham, C.H. (1973) The physiology of exercise under heat stress. *Annual Review of Physiology,* **35,** 193–220.

Wyndham, C.H., Allan, A. McD., Bredell, G.A.G. and Andrew, R. (1967) Assessing the heat stress and establishing the limits for work in a hot mine. *British Journal of Industrial Medicine,* **24,** 255–71.

Wyon, D.P. (1974) The effects of moderate heat stress on typewriting performance. *Ergonomics,* **17,** 309–18.

Wyon, D.P., Fanger, P.O., Olesen, B.W. and Pedersen, C.J.K. (1975) The mental performance of subjects clothed for comfort at two different air temperatures. *Ergonomics,* **18,** 359-74.

Yaglou, C.P. and Miller, W.E. (1925) Equivalent conditions of temperature, humidity and air movement determined with individuals normally clothed. *ASHVE Transactions,* **31,** 59–70.

Yaglou, C.P. and Minard, D. (1957) Control of heat casualties at military training centers. *American Medical Association Archives of Industrial Health,* **16,** 302–16.

6 VISION, LIGHT AND COLOUR

P.R. Boyce

The subject of this chapter is the interaction between people and light. The emphasis is on the way in which light affects people's ability to do work. The chapter starts by summarising the operation of the visual system and defining the quantities by which light is measured. After this discussion of the components of the interaction, the performance capabilities of the visual system are presented. This is followed by a discussion of the effects of light on work and health and a consideration of various methods whereby information on the effects of light can be obtained. The final section consists of a brief description of some lighting problems of relevance to the ergonomist.

It must be emphasised that throughout this chapter the work being considered is that in factories and offices or on roads. The lighting of interest is the lighting appropriate for these locations. Decorative lighting, as in homes, and special lighting, as in aircraft cockpits, are not considered here.

6.1 The visual system

The visual system begins with the eye and ends with the brain. Very briefly, the optics of the eye form a sharp image of the object being viewed on a layer of photoreceptors. These photoreceptors act as transducers converting the incident electromagnetic quanta into electrical discharges. The signals produced by the photoreceptors are partially analysed in the associated neural network before being transmitted up the optic nerve to the brain where they are interpreted on the basis of previous experience. This 'back of a postage stamp' description does not do justice to the capabilities, complexities and mysteries of the visual system, but it does emphasise that the visual system is much more than just the eye.

Nonetheless, the obvious starting point for a more detailed examination of the visual system is the eye, a structure which can be likened to a bag of jelly, held in shape by the pressure of the fluids it contains. Fig. 6.1 shows a section through the eye. In looking at a particular object the eye is first pointed at it by the six muscles controlling the eye's position. The image of the object is then brought to focus on the retina, the photosensitive layer covering the back of the eye. To produce a focussed image, the elastic lens in the eye is rounded or flattened by the ciliary muscle thereby increasing or reducing its power. Changes in lens shape alter the focal length of the eye's optical system, but this variation occurs on top of a fixed power provided mainly by the front surface of the eye, the cornea. If the power of the complete optical system does not match the dimensions of the eye then the range of distances over which focussing can occur will be restricted. This is the basis of short- and long-sightedness. A young person with normal eyes can focus objects that occur at any distance from infinity to within about 15 cm of the cornea. The ability to focus over a considerable range is known as accommodation. The range over which focussing can occur usually reduces with increasing age, the shortest distance at which clear vision is possible becoming greater. This happens principally because as the eye ages the lens becomes larger and more rigid.

The other important optical function of the eye is to control the amount of light that reaches the retina. This control is exercised by variation in the pupil size, the pupil being the aperture formed by the iris in front of the lens. The size of the pupil is mainly controlled by the amount of light reaching the retina, although it also varies with accommodation and emotion. Feedback from the retina will reduce the pupil size if the amount of light is too great and increase it if too small. The adjustment in pupil size can vary the amount of light reaching the retina over a range of about 16 : 1. It should be noted that this covers only a small part of the total range of operation of the visual system. For a given amount of light pupil size diminishes with age.

It is these two features of the eye, the focussing of images by a lens and the adjustment of the amount of light striking a photoreceptor surface by an aperture, that have led to the common analogy between the eye and the camera. Unfortunately, the analogy breaks down as soon as anything outside these two

Fig. 6.1. A section through the eye, when adjusted for near and distant vision. (From Henderson and Marsden (1972). Reproduced by kind permission of Thorn Lighting Ltd.)

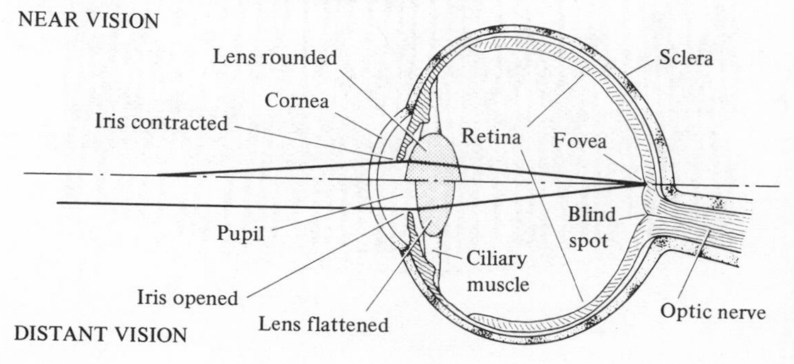

features is considered. The most obvious difference is in the comparison of the retina with the film. The retina is an exceedingly complex structure. Fig. 6.2 shows a schematic section through the retina, revealing the layers of its structure and the wealth of neural connections within and between the layers. Physiological studies have shown that there are two broad classes of photoreceptors in the retina, named rods and cones after their shape. For both types of photoreceptor the absorption of light quanta by the photopigment triggers a response which is then transmitted to the other layers of the retina. Rods and cones are

Fig. 6.2. A schematic diagram of a section through the retina. (From R.W. Gregory (1966) *Eye and Brain.* Weidenfeld and Nicholson, London. Reproduced by kind permission of the author and publishers.)

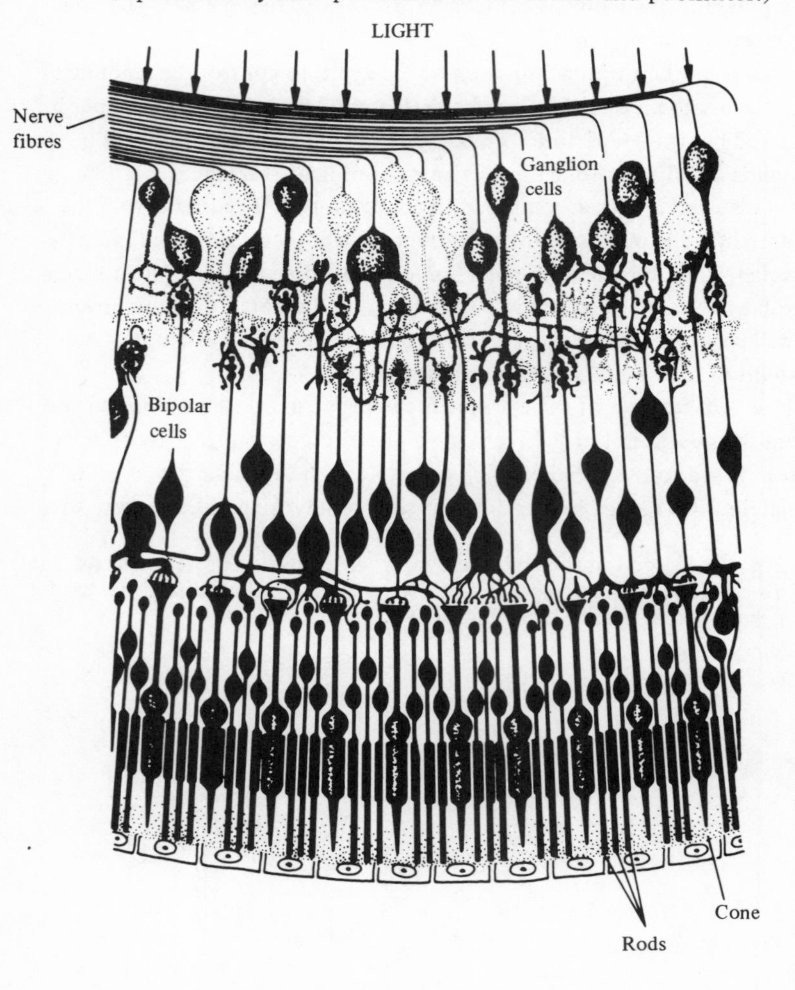

known to be sensitive to different wavelength regions which suggests they contain different photopigments. Studies of the sensitivity of the photoreceptor system to the quantity of incident light show that the rod system is much more sensitive than the cone system. Fig. 6.3 illustrates the uneven distribution of rods and cones across the retina. The central region of the retina, called the fovea, is occupied almost entirely by cones. As one moves away from the fovea into the periphery there is a domination by the rods, and a discontinuity at the 'blind spot' where the optic nerve leaves the eye (see Fig. 6.1).

Another important feature of the retina is unevenly distributed. If one considers the estimated number of photoreceptors, namely about 120 million rods and 7 million cones, all of which feed ultimately into about 1 million optic nerve fibres, it is apparent that the ganglion cells (Fig. 6.2), each of which feeds into one nerve fibre, must be receiving the output from a number of photoreceptors. The extent of this pooling of photoreceptor output is at a minimum in the fovea, where the cones are dominant, and at a maximum in the periphery, which has mainly rods. This variation in the extent of pooling, the relative sensitivity to light of the two types of photoreceptor system and their distribution across the retina imply that the fovea and periphery are suitable for different purposes. The rods, which are the more sensitive, are most dense in the periphery where the extent of pooling is highest. Thus the periphery will be very effective in detecting very low light signals but not very good for the discrimination

Fig. 6.3. The density of rods (filled circles) and cones (triangles) across the retina on a horizontal meridian. (After G. Osterburg (1935) Topography of the layer of rods and cones in the human retina. *Acta Ophthalmologica*, **13**, Supplement 6.)

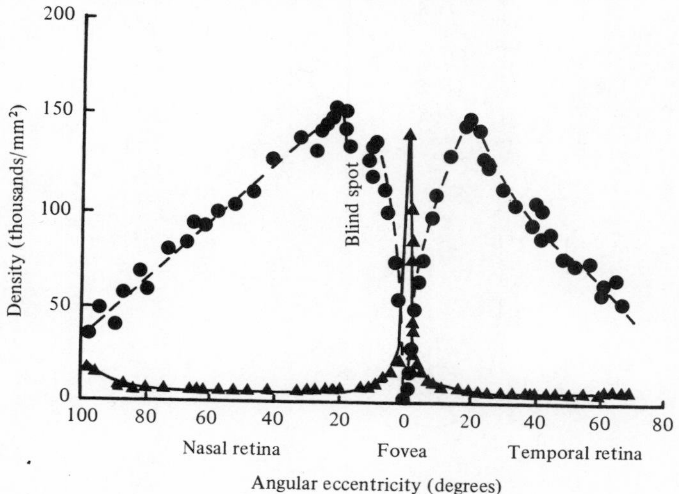

of detail. The cones are less sensitive but are concentrated in a small region where pooling is at a minimum, i.e. the fovea. Thus the fovea is the area of the retina most suitable for the discrimination of detail, provided there is sufficient light for the cones to operate. Fittingly, the fovea is the region of the retina which is used when an observer looks directly at an object.

Although pooling of responses is important it is not the only function of the complicated retinal structure. Micro-electrode examination of animals with visual mechanisms similar to man has shown that the responses to light stimuli are greatly modified at retinal level. For example, ganglion cells have been found which respond to movement in a specific direction, or to light and colour boundaries. It is the number and variety of forms of this signal modification as well as the histology which suggest that the retina may legitimately be considered as an extension of the brain.

When the signals leave the retina they pass up the optic nerve along the route shown in Fig. 6.4. The first area of interest is the optic chiasma. It is here that the fact that man has two eyes becomes important. The two eyes produce slightly different images of the scene being viewed. The differences between the two images produced by the two eyes can be interpreted by the brain to provide a three-dimensional view of the world. It is at the optic chiasma that the signals

Fig. 6.4. A schematic diagram of the binocular visual nerve pathways. (From Padgham and Saunders (1975). Reproduced by kind permission of Bell and Hyman Ltd.)

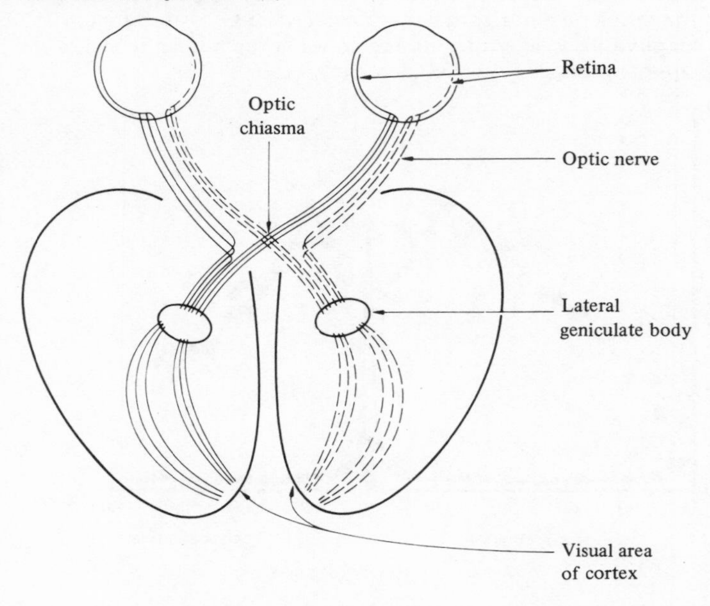

from the two retinal images in the two eyes link. The two optic nerves split so that the left half of the retina in each eye provides signals to the left hemisphere of the brain and similarly for the right halves. After the optic chiasma the ordered signals proceed through the lateral geniculate body to the visual cortex in their respective hemispheres, where the signals originating at the retina are interpreted. At the photoreceptor level of the retina and maybe even at the ganglion cell level it is possible to talk of the stimulus to the eye producing a sensation, but once the visual cortex is involved the stimulus produces a percept. The distinction between sensation and perception involves both complexity and timing. Sensation is simple and immediate; perception is more complex and involves the interpretation of the sensory information on the basis of past experience.

It should be apparent by now that the visual system is exceedingly complex. It is a system about which our knowledge changes from considerable understanding to profound ignorance as we move from the optics of the eye to the visual cortex. Although much progress has been made during the last two decades on understanding the visual system, much still remains to be discovered.

6.2 Light and colour

Light is the stimulus which allows the visual system to operate. It is only a very small part of the electromagnetic spectrum (Fig. 6.5), which in total ranges from gamma-rays to radio waves. Fig. 6.6 shows the measured relative spectral sensitivity of the rods and cones in the human retina. It can be seen that the visual system only operates over the wavelength range from about 380 nm to 760 nm and is not equally sensitive to all wavelengths within this range. It is also

Fig. 6.5. A schematic diagram of the electromagnetic spectrum. The divisions between the various regions are indicative only. (From Henderson and Marsden (1972). Reproduced by kind permission of Thorn Lighting Ltd.)

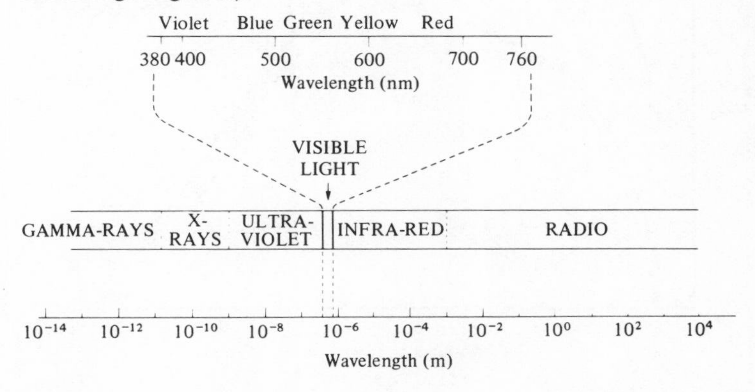

clear that the rod sensitivity is generally much higher than that of the cones and that the wavelengths for peak sensitivity are different. For the cones, which can roughly be said to operate in daytime, the peak wavelength is about 555 nm, but for the rods, which function in night conditions, the peak is about 507 nm. There is thus a shift in sensitivity to wavelength, called the Purkinje shift, which occurs as day fades to night.

Photometry

Once the spectral responses of the rods and cones have been identified it is possible to define the quantities used to characterise the light stimulus to the visual system. For electromagnetic radiation the amount of energy received at a surface in unit time is defined as the radiant flux and is measured in watts. Radiant flux is defined independently of the properties of the surface on which it impinges. Thus radiant flux tells us nothing about the effect of the incident energy on the eye. To do this the spectral sensitivity of the photoreceptor has to be introduced. This is achieved by multiplying the incident radiant flux in each wavelength interval by the spectral sensitivity of the visual system over that wavelength interval and integrating over the range of visible wavelengths. The

Fig. 6.6. Log relative spectral sensitivity for rods (filled circles), foveal cones (filled triangles) and peripheral cones (open triangles) plotted against wavelength. The peak sensitivity of the foveal cones is taken as unity. (After G. Wald (1945) Human vision and the spectrum. *Science*, **101**, 653–8. © 1945 American Association for the Advancement of Science.)

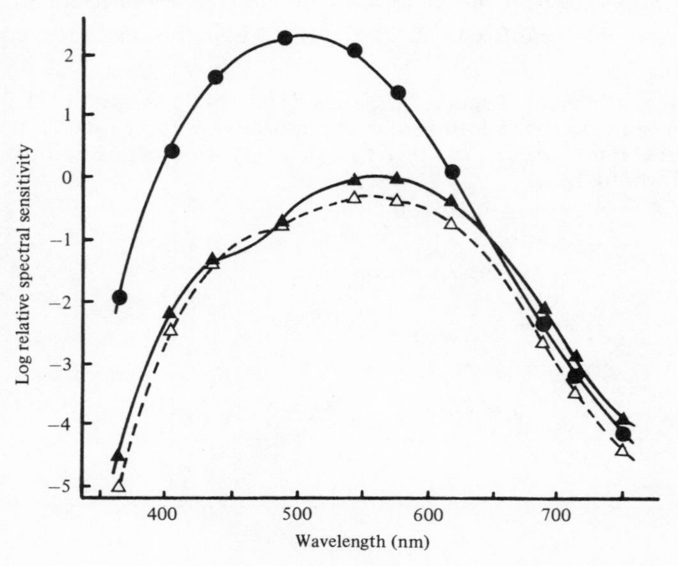

result is called the luminous flux. In fact luminous flux (ϕ) is defined by the expression

$$\phi = K_m \int_{380}^{760} \phi\lambda \; V_\lambda \; d\lambda$$

where $\phi_\lambda \, d\lambda$ = the radiant flux in a small wavelength interval $d\lambda$,
V_λ = the relative luminous efficiency function (CIE, 1970),
K_m = a constant called the maximum spectral luminous efficacy which relates the units of luminous flux to the units of radiant flux. The unit of luminous flux is the lumen.

V_λ and K_m are the two variables which introduce the spectral sensitivity of the visual system. In practice there are two sets of values of V_λ and K_m commonly used, one for conditions when the cones are operating and called photopic values (V_λ, K_m), the other used when the rods are dominant and called the scotopic values (V'_λ, K'_m). In both cases the relative luminous efficiencies (V_λ, V'_λ) follow from the spectral sensitivity curves of the photoreceptors, the sensitivity at each wavelength being expressed as a proportion of the peak sensitivity. The constants (K_m, K'_m) represent the luminous efficacies at the peak sensitivity wavelengths, i.e. when V_λ and V'_λ are 1. Currently K_m is 683 lumens/watt and K'_m 1700 scotopic lumens/watt. It should be noted that the terms V_λ and V'_λ discussed above are internationally agreed versions of man's relative spectral sensitivity, i.e. they represent standard observers. Any individual, even one with normal vision, will probably deviate slightly in spectral sensitivity from the relative luminous efficiency functions for cone and rod vision (CIE, 1978*a*). The V_λ and V'_λ curves are a necessary compromise on a standard observer, necessary because it is desirable that some agreed physical quantities be established to measure light, but still a compromise. The compromise was reached in 1924 for photopic vision and in 1951 for scotopic vision by the Commission Internationale de l'Eclairage (CIE), the body responsible for lighting standardisation (CIE, 1970).

Luminous flux is most commonly used to describe the total light output of a light source. Once luminous flux has been defined the other three main photometric quantities follow readily. The luminous flux emitted per unit solid angle from a point is called the luminous intensity and is measured in candelas (cd). The link between candelas and lumens is that one lumen is the luminous flux emitted through one steradian from a uniform point light source having a luminous intensity of one candela. As might be expected luminous intensity is most commonly used to describe the light output of a light source in a specified direction. Both luminous flux and luminous intensity have area measures associated with them. The luminous flux incident per unit area of a surface is the illuminance. The luminous intensity per unit projected area of a surface in a given direction is the luminance. Both illuminance and luminance are used in the description of the lighting of rooms, illuminance to specify the amount of light falling on a surface and luminance to quantify the light emitted by or reflected from a surface in a specified direction. Full definitions of the four fundamental photometric quantities - luminous flux, luminous intensity, illuminance and luminance - and the relationships between them are given in Table 6.1. In the past photometry has been bedevilled by a surfeit of terms and units, many of which give no indication of the dimensions of the quantities they represent. Table 6.2 is a list of some alternative terms which may be found and the conversion factors necessary to change to the approved Système Internationale (SI) unit.

One other point must be mentioned, namely the range of luminances over which photopic and scotopic vision operate. As a rule of thumb, above a luminance of about 3 cd/m^2 vision is photopic, below 10^{-3} cd/m^2 it is scotopic. Between these two limits an intermediate state, called mesopic, exists. The spectral sensitivity in this region is still being investigated (CIE, 1978*a*).

Colorimetry

So far this discussion of the visual system and photometry has considered only what is subjectively seen as brightness and has ignored the phenomenon of colour. Man has highly developed colour vision during daylight but at night no colours are perceived; the world is seen in shades of grey. This observation implies that the rods do not produce any information on the wavelength of the absorbed light quanta. However, the cones, which are operating in daylight when the rods are saturated, must provide colour information. As long ago as 1802 Thomas Young suggested that there should be three types of colour receptor in the retina. This conclusion arose from the finding that almost all colours of light could be matched by the appropriate combination of red, green and blue light. In recent years three types of cone photopigments have been identified. This finding implies that any particular incident spectral distribution of light will produce three signals from the three types of cones, which in theory at least should explain colour vision. However, electrophysiological studies have shown that three separate signals do not occur at the higher levels of the visual system. Rather the signals from the three cone types are summed to form one brightness signal, which corresponds to the photopic spectral sensitivity of the visual system, and subtracted to give two colour difference signals, red/green and blue/yellow. It is interesting to note that this model of colour vision uses both the main theories of colour vision, the trichromatic and the opponent colour, over which much ink has been spilt for more than a hundred years.

Table 6.1. *The photometric quantities*

Quantity	Definition	Units
Luminous flux	That quantity of radiant flux which expresses its capacity to produce visual sensation	lumen (lm)
Luminous intensity	The luminous flux emitted in a very narrow cone containing the given direction divided by the solid angle of the cone, i.e. luminous flux/unit solid angle	candela (cd)
Illuminance	The luminous flux per unit area at a point on a surface	lumen/metre^2 (lm/m^2)
Luminance	The luminous flux emitted in a given direction divided by the product of the projected area of the source element perpendicular to the direction and the solid angle containing that direction, i.e. luminous flux per unit solid angle per unit area	candela/metre^2 (cd/m^2)
Reflectance	The ratio of the luminous flux reflected from a surface to the luminous flux incident on it	
For a diffuse surface	Luminance = (illuminance x reflectance)/π	Luminance in candela/metre^2; illuminance in lumen/metre^2
Luminance factor	The ratio of the luminance of a reflecting surface, viewed in a given direction, to that of a perfect white uniform diffusing surface identically illuminated	
For a non-diffuse surface for a specific viewing direction and lighting geometry	Luminance = (illuminance x luminance factor)/π	Luminance in candela/metre^2; illuminance in lumen/metre^2

It is now necessary to describe how colours may be measured. In 1931 the Commission Internationale de l'Eclairage produced a system of colour measurement based on another standard observer, this time one with three spectral sensitivity curves (CIE, 1971). These three spectral sensitivities are called the CIE 1931 colour matching functions ($\bar{x}, \bar{y}, \bar{z}$) and are shown in Fig. 6.7. With these three functions and the spectral distribution of the light incident on the eye either directly or by reflection from a surface, any colour can be specified by a two-stage process. The first stage is to determine what are called the tristimulus values (X, Y and Z) of the colour by summing the product of the spectral radiant flux of the light and the appropriate colour matching function over the relevant wavelength range, e.g.

$$X = h \sum_{380}^{760} \phi_\lambda \bar{x}_\lambda \Delta\lambda$$

where $\phi_\lambda \Delta\lambda$ = the radiant flux in a wavelength interval $\Delta\lambda$,
$\bar{x}_\lambda$ = the spectral tristimulus value of the x colour matching function,
and h = a constant.

Table 6.2. *Some common photometric units and the conversion factors necessary to change to the approved SI units*

Unit	Dimensions	Multiplying factor to convert to SI unit
	(a) *Illuminance*: SI unit, lumen/metre^2	
Lux	lumen/metre^2	1.00
Metre candle	lumen/metre^2	1.00
Phot	lumen/centimetre^2	10 000.00
Foot Candle	lumen/foot^2	10.76
	(*b*) *Luminance*: SI unit, candela/metre^2	
Nit	candela/metre^2	1.00
Stilb	candela/centimetre^2	10 000.00
–	candela/inch^2	1 550.00
–	candela/foot^2	10.76
Apostilb[a]	lumen/metre^2	0.32
Blondel[a]	lumen/metre^2	0.32
Lambert[a]	lumen/centimetre^2	3 183.00
Foot-lambert[a]	lumen/foot^2	3.43

[a] These four items are based on an alternative definition of luminance. This definition is that if the surface can be considered as perfectly matt its luminance in any direction is the product of the illuminance on the surface and its reflectance. Thus the luminance is described in lumens per unit area. This definition is deprecated in the SI system.

The second stage is to express each of the tristimulus values X, Y and Z as a proportion of the sum of them all, e.g. $x = X/(X + Y + Z)$. These proportions x, y and z are called the chromaticity coordinates of the colour. As x, y and z are all proportions of the same total, once x and y are known, z is determined. Therefore, only the x,y chromaticity coordinates of a colour are normally given in its specification. Since all colours can be represented by two values it is possible to present them on a plane. Fig. 6.8 shows such a plane, the CIE 1931 (x,y) chromaticity diagram. Each colour is represented by a single point on the plane. The curved limit line is called the spectrum locus since it represents all the monochromatic colours. Points falling within the marked areas will be identified as having the stated hue. The equal-energy point at the centre of the diagram is normally taken as white. The closer a point is to the spectrum locus, and the further it is from the equal-energy point, the more saturated is the colour. The ease with which two colours can be distinguished will depend on their separation on the diagram, two points which are coincident being indistinguishable. Unfortunately the CIE 1931 chromaticity diagram has been found not to be a perceptually uniform plane, i.e. the same separation between two points will correspond to a different degree of colour difference on various parts of the plane. In 1960 the CIE reduced the degree of non-uniformity by altering the shape of the plane to form the CIE 1960 Uniform Chromaticity Scale (UCS) diagram. The process was repeated in 1978 in order to further improve the perceptual uniformity. The result was the CIE 1976 Uniform

Fig. 6.7. The CIE 1931 colour matching functions, $\bar{x}, \bar{y}, \bar{z}$ (CIE, 1971).

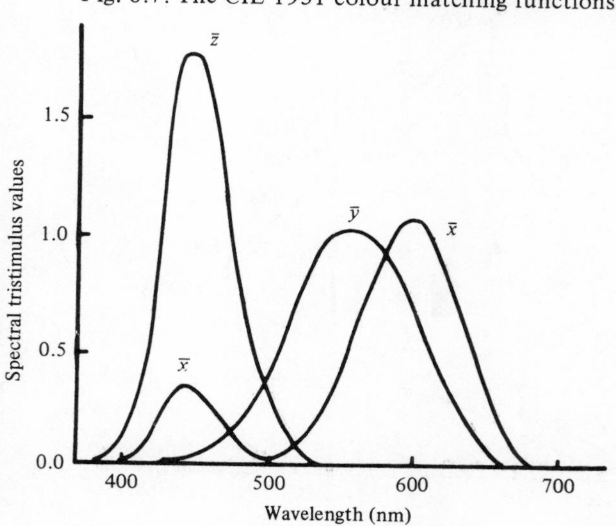

Chromaticity Scale diagram. The alteration (CIE, 1978*b*) involves a simple linear transformation of the chromaticity coordinates using the equations

$$u' = \frac{4x}{(-2x + 12y + 3)}$$

$$v' = \frac{9y}{(-2x + 12y + 3)}$$

It should be noted that a colour difference determined on the CIE 1931 chromaticity diagram or on the CIE 1976 UCS diagram only involves the colour attributes of the stimuli. If the colours to be distinguished also have brightness differences then it is better to use the CIE colour spaces and the colour difference formulae (CIE, 1978*b*).

It should also be noted that the whole CIE colorimetry system is based on a

Fig. 6.8. The CIE 1931 chromaticity diagram (CIE, 1971). The boundary curve is the spectrum locus with the wavelengths (nm) marked. The filled circle is the equal-energy point. The enclosed areas indicate the chromaticity coordinates of light signal colours which will be identified as the specified colours.

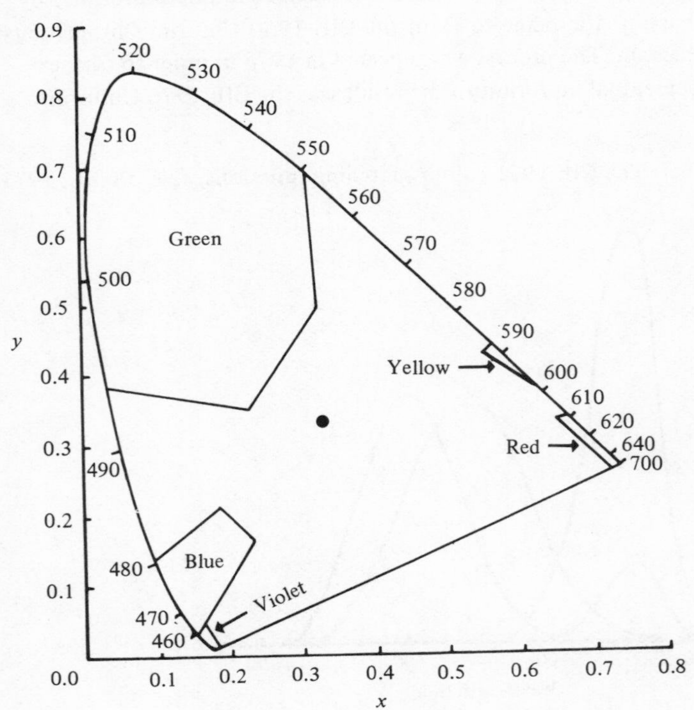

standard observer, either the 1931 version or the 1964 large field version (CIE, 1971). In the real world people do not have cones with the same spectral sensitivities as these standard observers. Nonetheless if the colour matching functions did not bear some resemblance to people's responses then colour measurement on the CIE colorimetry system would produce large departures from reality which would make the system useless. In practice the CIE colorimetry system has been found to be very useful, particularly for quantifying colours so that they can be reproduced, for defining the colours of light signals and for representing the colour properties of light sources (Wyszecki and Stiles, 1967).

One other aspect of colour measurement needs mentioning. It is not always convenient to discuss colours as numbers or positions on a diagram. For many practical purposes it is desirable to be able to have an actual colour sample to examine and to act as a standard. This need has led to the production of a range of colour ordering systems, usually in the form of atlases. The most widely used of these is the Munsell colour system. This presents some 1200 colours classified according to three independent attributes - hue, value and chroma - in a three-dimensional space. Fig. 6.9 indicates the form of the Munsell colour system. Munsell hue is the attribute of the colour which relates to the nearest spectral

Fig. 6.9. The form of the Munsell colour system. The hue letters are B, blue; BG, blue/green; G, green; GY, green/yellow; Y, yellow; YR, yellow/red; R, red; RP, red/purple; P, purple; PB, purple/blue. (From Padgham and Saunders (1975). Reproduced by kind permission of Bell and Hyman Ltd.)

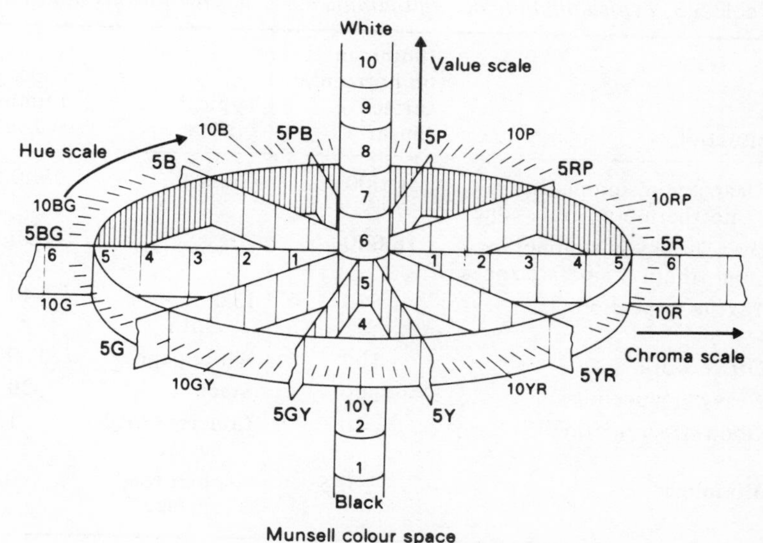

colour. There are five hues – blue, green, yellow, red and purple – and five intermediate hues, arranged in a circle. The Munsell value refers to the apparent lightness of the colour, i.e. the extent to which it contains white. Value is drawn on the vertical axis. As a rough estimate, the reflectance of a Munsell colour can be obtained from its value (ν) by the formula $\rho = \nu(\nu - 1)/100$. The final attribute, Munsell chroma, refers to the saturation or purity of the colour. Chroma is scaled as a radial distance in the direction of the appropriate hue. Colour atlases can be used to identify colours for standardisation purposes. In fact a whole range of special colour samples have been produced to standardise the description of colours found in the paint, plastic and tile industries as well as for naturally occurring skin, hair, eye, soil and plant materials.

6.3 The performance capabilities of the visual system

Adaptation

Having described the visual system and the quantities used to measure the stimuli that affect it, the next step is to summarise its performance capabilities. The first aspect of interest is adaptation. The visual system can work over an enormous range, from a luminance of 10^{-6} cd/m^2, corresponding to complete darkness, to 10^5 cd/m^2, which can occur under bright sunlight. (Table 6.3 lists the illuminances and luminances that occur for a number of practical conditions.) However, the visual system cannot cover this immense range simultaneously. Anyone who has walked from daylight into a darkened cinema will

Table 6.3. *Typical illuminance and luminance values for various situations*

Situation	Illuminance on horizontal surface (lm/m^2)	Typical surface	Luminance (cd/m^2)
Clear sky in summer in northern temperate zones	150 000	Grass	2900
Overcast sky in summer in northern temperate zones	16 000	Grass	300
Textile inspection	1 500	Light grey cloth	140
Office work	500	White paper	120
Heavy engineering	300	Steel	20
Good street lighting	10	Concrete road surface	1.0
Moonlight	0.5	Asphalt road surface	0.01

be aware of this. To cover the complete range the sensitivity has to be changed, i.e. adaptation to the prevailing conditions has to occur. This is achieved partly by adjustment in the pupil size but mainly by a change in the state of the photopigment in the photoreceptors. The time-course of this adaptation depends considerably on the direction and magnitude of the stimulus change. When going from bright sunlight into a dark interior several minutes will elapse before anything can be seen and complete adaptation may take a long time. On going from a dark room into sunlight the adaptation takes a much shorter time.

Fig. 6.10 shows the time-course of adaptation from a high photopic retinal illumination to darkness and from a scotopic retinal illumination to darkness. The course of adaptation is expressed in terms of the retinal illumination required for the observer to see a cross on a small flashing field. Retinal illumination, as its name suggests, is a measure of the illuminance on the retina. It is calculated by multiplying the luminance of the observed field (in cd/m^2) by the pupil area (in mm^2) and is expressed in units called trolands. From Fig. 6.10 it can be seen that for adaptation from photopic to scotopic conditions a two-stage process occurs. Initially the cones adapt to an approximately steady state and then the rods act to increase the sensitivity further. Complete adaptation from a high luminance to complete darkness can take up to an hour. When the adaptation required is completely within either the photopic or the scotopic range then only a single-stage process is involved and this takes less time. The time course of adaptation is important when considering such practical matters as the illuminance to be provided at the entrances of road tunnels.

Fig. 6.10. The time-course of adaptation to complete darkness after 2 minutes' exposure to a 400 000 troland adapting field or a 263 troland adapting field. (After S. Hecht, C. Haig and C.M. Chase (1937). The influence of light adaptation on subsequent dark adaptation of the eye. *Journal of General Physiology*, **20**, 831–50.)

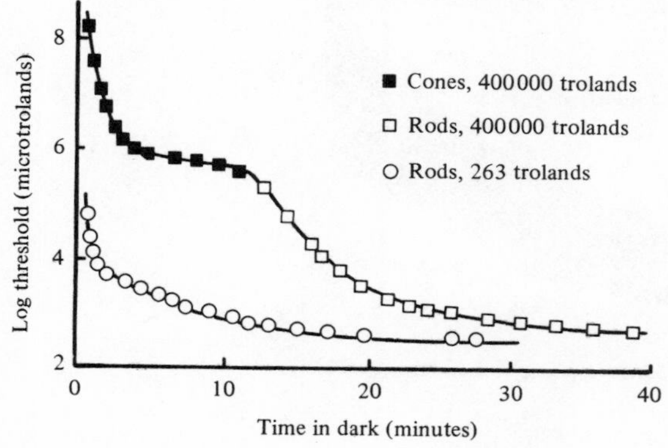

Thresholds

Once the visual system is fully adapted then the limits to the luminance range over which luminance differences can be distinguished are determined. Fig. 6.11 gives an indication of that range and how it varies with adaptation luminance. Luminances greatly below the adaptation luminance are perceived as black, whilst those greatly above are seen as glaringly white. Within these limits differences in luminance can be discriminated. This enables objects to be detected and identified from their details. It is this aspect of detection that will now be considered. The measurement of performance most usually quoted is the threshold level of the stimulus, i.e. the level of the stimulus which is detected on 50% of the occasions it is presented. This is a totally impractical level of performance for everyday use but it may be taken as an indication of the limits of performance. The detection of an object in a field of view depends upon many physical features: its contrast, size, shape, colour and edge sharpness, and the background against which it is seen. Other factors within the individual – his state of adaptation, his motivation, visual defects, intelligence, etc. – also affect the ability to detect. Of the physical factors the most fundamental is contrast. If there are no colour differences between target and background and no

Fig. 6.11. A schematic illustration of the range of object luminances within which discrimination is possible for different adaptation luminances. The boundaries of the glare and black shadow regions are approximate. (After R.G. Hopkinson and J.B. Collins (1970) *The Ergonomics of Lighting.* MacDonald Technical and Scientific, London. Reproduced by kind permission of the authors and publishers.)

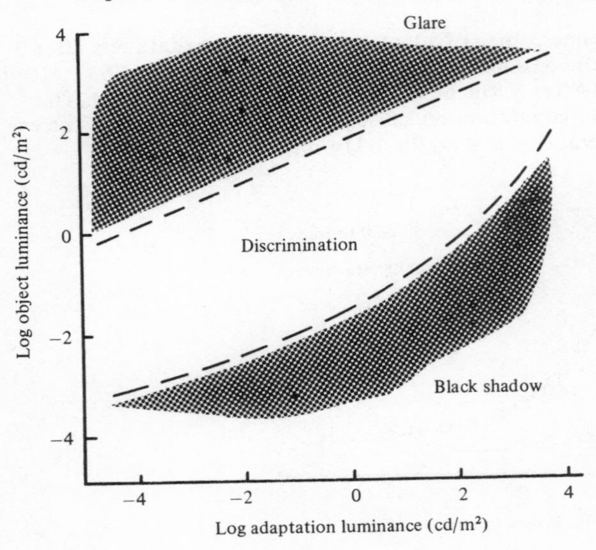

distinction between the luminance of the task and its background then it cannot be seen, no matter what values the other factors have. For a task on a uniform background, contrast (C) is conventionally defined as

$$C = \frac{|L_t - L_b|}{L_b}$$

where L_t = luminance of the target,
L_b = luminance of the background.
It is usually assumed that the visual system is adapted to the luminance of the background. It must be emphasised that contrast is a physical measure. There is no *exact* relationship between the contrast of a task and the ease with which it may be seen, although they are obviously linked.

The effects of the other physical factors can be expressed in terms of the threshold contrast. It is convenient to start with size, quantified by the angle the object subtends at the eye. As an object increases in size it becomes easier to see, i.e. the threshold contrast of the target decreases. For small stimuli, say discs less than 4 min arc diameter, the threshold contrast is inversely proportional to the area of the target. Above 4 min arc diameter the relationship gradually changes so that the threshold contrast is inversely proportional to the diameter of the disc, but above about 2° diameter there is no change in threshold contrast with increases in size (Blackwell, 1946). These results have largely been obtained on circular targets and at photopic luminances, but McNelis & Guth (1969) have shown that threshold contrast is largely independent of shape, at least at high luminances, for a range of target forms. Lamar *et al.* (1947) support this view until their rectangular targets exceed an aspect ratio of 7 : 1. For larger aspect ratios the threshold contrast is steadily increased.

It is sometimes useful to consider size as a measure of threshold performance in its own right. For a high-contrast target on a uniform field, a line only 0.5 sec arc wide can be detected (Hecht and Mintz, 1939), although detection of presence is not the most common aspect of visual work. Usually discrimination of detail is required, and for this the most common test takes the form of a grating on a uniform field: a gap between the bars of the grating of 0.5 min arc can be detected (Schlaer, 1937). Even more precise judgements can be made for the alignment of two lines on a uniform field. In this case, called vernier acuity, a misalignment of 1 to 2 sec arc can be detected (Berry, 1948).

Another relevant factor in determining threshold performance is any blurring that occurs in the retinal image, regardless of cause. The evidence on the effect of edge sharpness is confused but it appears that blurred edges considerably increase threshold contrast for very small objects, e.g. less than 40 sec arc diameter, but have little effect for large objects, e.g. greater than 20 min arc diameter (Ogle, 1961). Another way to look at the effect of edges and target structure is to consider the spatial modulation transfer function of the visual system. This rather grand name hides a simple but useful piece of data, namely the response of the visual system to different spatial frequencies. Its usefulness lies in the fact that the changes in luminance across a surface can be represented by a waveform and that that waveform can in turn be represented by a Fourier series of different frequency terms (see p. 222). Then the detectability of the features of a given luminance pattern and hence its appearance can be predicted by combining the Fourier series representing the pattern and the spatial modulation transfer function of the visual system. Fig. 6.12 shows two spatial modulation transfer functions from Campbell and Robson (1968) in terms of the reciprocal of threshold contrast (contrast sensitivity) plotted against spatial frequency, for adaptation luminances of 500 cd/m^2 and 0.05 cd/m^2. For the high luminance the sensitivity is a maximum about a spatial frequency of 3 cycles/degree, with a decline at higher or lower spatial frequencies. A similar pattern occurs at the lower luminance but with the peak sensitivity at a lower spatial frequency. This reduced sensitivity at low and high spatial frequencies explains why very gradual changes in luminance, i.e. low spatial frequencies, such as occur across a wall,

Fig. 6.12. Spatial modulation transfer functions; the reciprocal of threshold contrast (contrast sensitivity) plotted against spatial frequency. Open symbols for 500 cd/m^2, filled symbols for 0.05 cd/m^2. (After Campbell and Robson, 1968.)

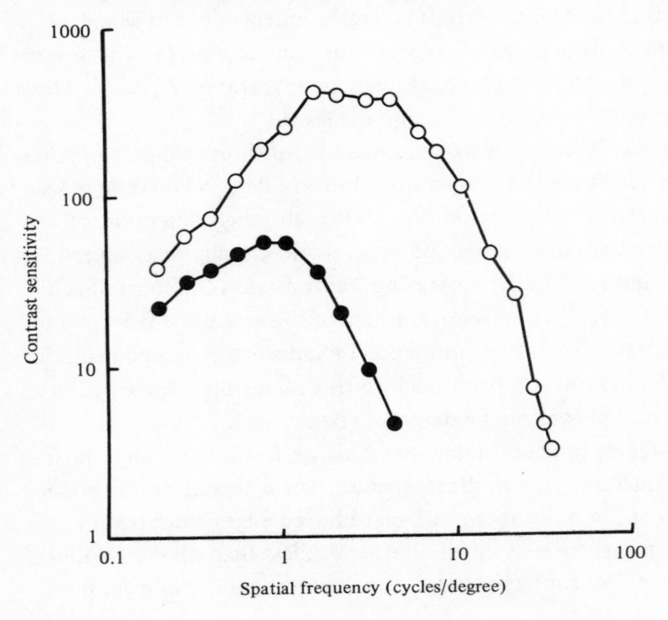

and very small details, which will have high spatial frequencies, are hard to detect.

Fig. 6.12 clearly shows that increasing the adaptation luminance increases the contrast sensitivity absolutely and extends the frequency range, particularly at high spatial frequencies. The beneficial effect of higher adaptation luminances for detection is also apparent in Fig. 6.13 for a range of discs of different sizes presented for 1 second (Blackwell, 1959). Increasing the adaptation luminance decreases the threshold contrast, but the decrease is less at higher luminances. It can also be seen that decreasing the size of the target increases the threshold contrast. Presentation time also has an important effect on threshold contrast. For times less than 0.1 second threshold contrast is inversely proportional to presentation time and for times greater than 1 second threshold contrast is virtually constant. The intermediate presentation times cover a transition region which is still being investigated. Thus, short presentation times, such as occur in some inspection tasks, increase the visual difficulty of the task.

Other factors which influence the performance possible are the size and luminance of the field of view other than that immediately adjacent to the target. Foxell and Stevens (1955) showed that visual acuity steadily improved as a uni-

Fig. 6.13. Threshold contrast for discs of different diameters presented for 1 second at different background luminances. (After Blackwell, 1959.)

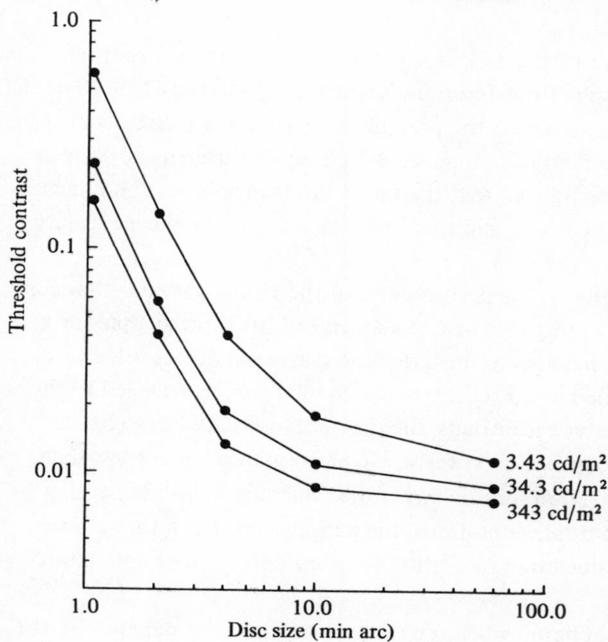

form background field increased in size up to about 15°. For such uniform background fields, the maximum visual acuity occurs when the background field luminance is slightly less than that of the immediate task surround. These are some of the reasons for the practice of lighting the work area uniformly and the task more brightly than the surroundings.

Another situation of practical interest is a very non-uniform background field. An example of this is meeting an oncoming car whilst driving at night. This situation produces what is called disability glare, the light from the source being scattered in the eye so that it forms a veil over the target, which decreases its contrast. In fact disability glare is usually quantified not by threshold contrast directly but by equivalent veiling luminance, i.e. the luminance of a uniform field which when superimposed on the target increases the luminance difference threshold by as much as the actual glaring light source does. The formula most widely used (Fry, 1954) to predict equivalent veiling luminance is

$$L = \sum_n \frac{9.2\, E_n}{\theta_n\,(1.5 + \theta_n)}$$

where L = the equivalent veiling luminance (cd/m^2),

E_n = the illuminance at the observer's eye from the nth glare source (lux),

and

θ_n = the angular deviation of the nth glare source from the line of sight (degrees).

An importance feature of the task is the part of the retina it occupies. Fig. 6.14 shows some results from Johnson, Keltner and Balestrery (1978) of the target luminance required to detect the presence of a 10 min arc disc occurring for 500 milliseconds on a field of luminance 3.2 cd/m^2 at different deviations from the visual axis. It can be seen that the target luminance is least and hence sensitivity is greatest at the fovea, and that sensitivity decreases as eccentricity increases.

A rather different aspect of the performance of the visual system is that concerned with the detection of movement. Salaman (1929) showed that for a high-contrast target moving on a uniform field people could detect velocities as low as 1 min arc per second at 10° eccentricity. Similar studies have led to the conclusion that with greater eccentricity the threshold velocity has to be increased for detection to occur. This decrease in performance is much less than the decrease in static detection sensitivity for the same target with increasing eccentricity. Thus, in photopic conditions, the periphery of the retina is much better, relatively, at the detection of motion than the detection of detail, but the fovea is best at both.

As for the detection of detail when the target is moving, this depends on the velocity. When the eye can 'lock on' to the target and follow it, e.g. a slow-flying aeroplane, performance is little different from that in static conditions. This 'locking on' starts to break down when the velocity exceeds about 50 degrees per second. At higher velocities performance deteriorates rapidly.

Another aspect of the performance of the visual system is the detection of temporal variations in luminance, or flicker as it is more commonly called. The threshold criterion of interest here is the ability to detect the presence of the oscillation. The threshold frequency for this type of performance is usually called the critical fusion frequency (CFF). CFF is directly proportional to mean luminance in the photopic and scotopic regions, higher luminances giving higher CFF, but it approaches independence in the mesopic region. Another factor influencing CFF is the area of the flickering field: the larger the field area the higher the CFF. For the most sensitive conditions of a high luminance, large area, 100% modulation flicker covering the fovea the CFF is of the order of 75 Hz (Collins and Hopkinson, 1957). Oscillations at frequencies higher than this are unlikely to be detected although there are considerable individual differences in people's flicker sensitivity.

CFF varies considerably with the percentage modulation of the flicker. This leads to an alternative way of examining the detection of flicker, i.e. to measure the percentage modulation required at each frequency in order for the flicker to be seen. Taken together such results can form a temporal modulation transfer function. Fig. 6.15 shows a set of results of this type for a large-area sinusoidal

Fig. 6.14. Log target luminances required for detection of a 10 min arc circular target on a field of luminance 3.2 cd/m^2 presented for 500 milliseconds at various deviations from the visual axis. The gap around 15° in the nasal retina corresponds to the blind spot. (After Johnson *et al.*, 1978.)

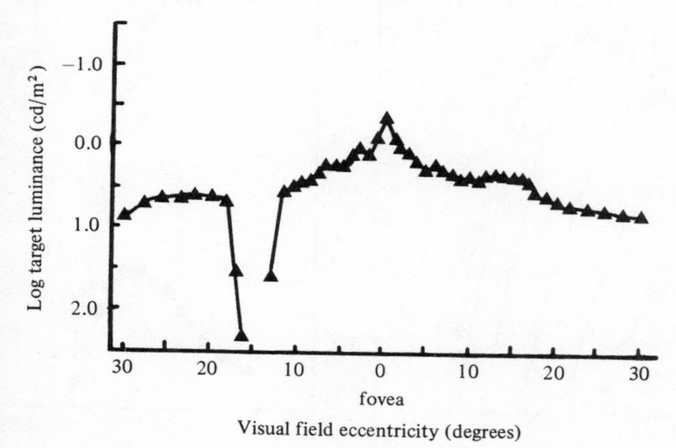

oscillation seen under different adaptation conditions (Kelly, 1961). A number of points should be noted from this diagram. First, it is apparent that as the retinal illumination increases the relative sensitivity to flicker increases. Secondly the peak sensitivity occurs around 20 Hz for high retinal illumination but moves to about 5 Hz at very low levels of retinal illumination. Thirdly, the common envelope for low frequencies shown in Fig. 6.15(*a*) means that the low-frequency response is determined by the percentage amplitude modulation. However, the common envelope for high frequencies shown in Fig. 6.15(*b*) means that the high-frequency response is determined by the absolute amplitude modulation. Such results have implications for the visibility of the flicker produced by light sources. Specifically it means that lamps on an a.c. supply and which produce both 50 and 100 Hz oscillation in light output should have a much lower absolute modulation amplitude for the 50-Hz component than for the 100-Hz component.

The final area of interest is colour. All the results that have been described above have been obtained on achromatic targets. However, colour adds another

Fig. 6.15. Flicker detection thresholds for a large field at different retinal illuminations and frequencies. The detection thresholds are expressed as percentage modulation amplitude (*a*) and absolute modulation amplitude (*b*). The retinal illuminations are (in trolands): ●, 9300; ○, 850; △, 77; ▽, 7.1; □, 0.65; ■, 0.06. (After Kelly, 1961.)

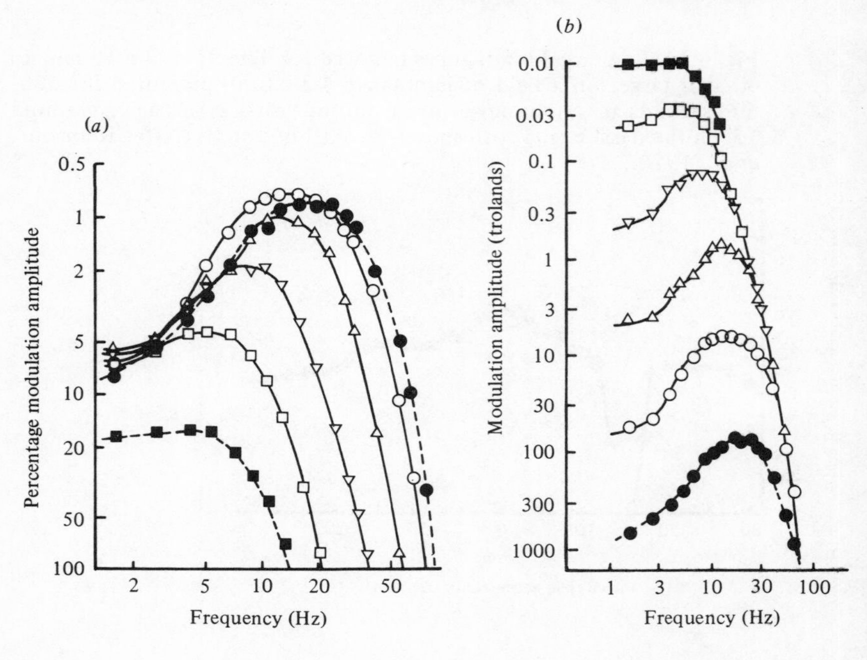

dimension to the gamut of differences that can exist between target and background. When there exists a clear luminance difference between the target and the background regardless of whether the colour of the target is the same as or different from the background, the effect of colour on detection is small (Eastmann, 1968). However, when the target and background have the same luminance the ability to detect the target will depend on any colour differences. Considering the number of colour combinations possible, little work has been done on this topic, but from our knowledge of how colour vision works it does not seem unreasonable to suggest that hue combinations that maximise the difference signals will be easiest to detect.

Measurements of the performance limits of the visual system have been made for many decades, so the above represents only a glimpse of the available data. Such topics as depth perception, form perception and the response to light flashes have not even been mentioned. Nonetheless it is hoped that the information given is sufficient to indicate the important features of a task in relation to the ultimate capabilities of the visual system. Further information can be obtained from the many excellent books on visual perception (Davson, 1962; Weale, 1963; Cornsweet, 1970; Padgham and Saunders, 1975; Carterette and Friedman, 1975).

6.4 Light and work

Basic relationships

We are now concerned with the everyday performance of everyday tasks under different lighting conditions. Such questions as 'how does the lighting in a factory or an office affect people's ability to do their work?', 'how important is the illuminance on the task?', 'when does the colour of the light become a factor to be considered?', 'is glare relevant?', indicate the topics of interest here. The first point to note is that the threshold values discussed previously are conventionally measured at the 50% level of performance. A factory inspector who detected only 1 defect in 2 would not be doing very well. Most visual work is done at much higher performance standards than 50%, and in addition usually involves some form of time limitation, either externally imposed or self-imposed. This difference has resulted in a different methodology being used to examine the above questions. Basically the approach is to take some representative task and to examine the performance achieved under different lighting conditions, usually in terms of the speed and accuracy of response. There are variations in this regarding the verisimilitude of the task, the context of the measurement and the variables considered, but virtually all the work which directly examines the effect of lighting conditions on work uses this approach. The variables used tend to be those that can be manipulated in practice, i.e. illuminance, light distribution and light colour properties.

Very few studies have been made of threshold situations taken to much higher levels of performance. This might suggest that the threshold performance data are irrelevant to practice, but this view would be mistaken. First, a few everyday tasks, such as driving at night or in fog, can involve threshold vision at the time decisions have to be taken. Secondly, because thresholds have been studied much longer and much more intensively than higher levels of performance there is a vast bank of data on the effect of just about every conceivable variable. When no information at a realistic level of performance is available these data can be used to identify what are likely to be relevant factors and their direction of effect.

Typical of the experimental approach used to examine directly the effect of lighting on work has been a study by Smith and Rea (1980). They had eight young and eight older observers checking a list of 20 numbers for agreement with a comparison list. This was done at eight different luminances for lists printed in black or grey ink on white paper. Fig. 6.16 shows the results obtained, which illustrate three important points. First, that increasing the task luminance produces an improvement in the performance, the change being in the form of a law of diminishing returns with the performance eventually saturating. Secondly, that the luminance at which saturation occurs is different for different tasks. Thirdly, that by simply increasing the luminance it is not possible to ensure that

Fig. 6.16. Mean performance scores for young (20–24 years) and older (60–69 years) observers checking lists of numbers printed in black (filled symbols) or grey (open symbols) ink on white paper at different task luminances. The performance score for each subject is calculated from the equation: Score = [(20 – number of errors)/(time taken in seconds + 5)] x 100. (After Smith and Rea, 1980.)

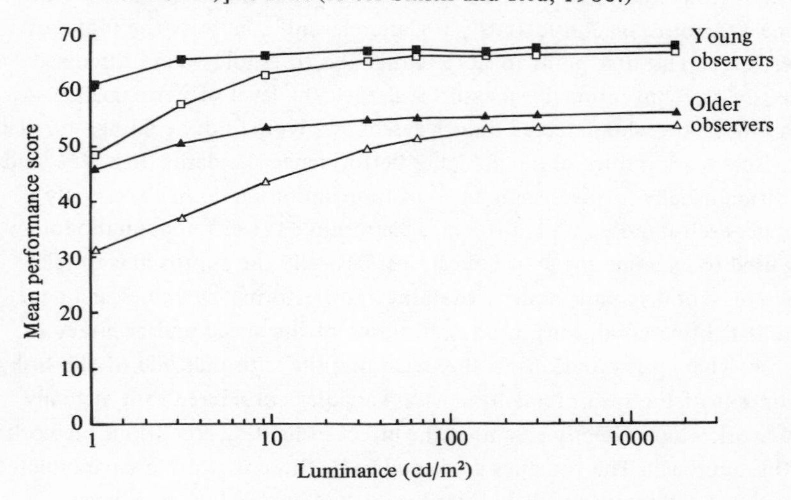

a visually difficult task is performed as well as a visually easy task. These three points have been verified by many other experimenters using different tasks (Weston, 1945; Simonson and Brozek, 1948; Muck and Bodmann, 1961; Boyce, 1973).

It should be remembered that the relationship between lighting and the performance of the task involves three components: the lighting, the task and the people. The effects of luminance considered above have been found for many different tasks. But what of the difference between people? The most widely researched personal variable has been the effect of age. Muck and Bodmann (1961) asked groups of young and older observers to do a search-task involving locating a specified number from 100 numbers scattered over a table. Fig. 6.17 shows the detection speed of the young and older observers under different luminances for the same task. As would be expected from the increased absorption of light in the eye that occurs with age, the older age-group show a greater improvement with increasing luminance than do the young age-group. Thus older people generally benefit more from higher luminances, a fact also apparent in Fig. 6.16. For both these experiments there appears to be a residual difference between the different age-groups. This may be related to the general reduction in performance that occurs in older people. In another test of the effect of age (Boyce, 1973) it was found that for very simple visual tasks older people could be brought to the same level of performance as young people by increasing the illuminance.

In the same experiment (Boyce, 1973) it was found that some visual tasks showed no significant effect of illuminance and sometimes no significant differ-

Fig. 6.17. Mean detection speed for a search task requiring young (20–30 years) and older (50–65 years) observers to locate a specified integer number from 100 integer numbers, plotted against task luminance. (After Muck and Bodmann, 1961.)

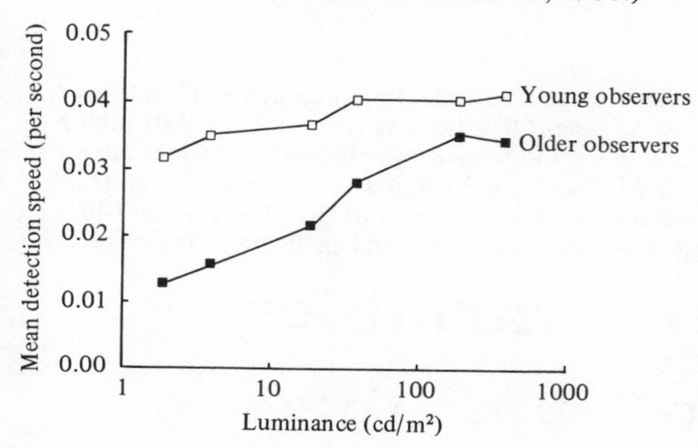

ence between the age-groups. This lack of significance demonstrates that the importance of lighting will be directly related to the importance of the visual component in the overall task. For copy-typing the visual component is large so the lighting is important, but when the visual component of the task is low or the performance of the task is limited by some other aspect then changing the lighting may well have only a limited effect, if any effect at all.

The studies of the effect of luminance/illuminance considered above have all been done under uniform lighting on matt, two-dimensional tasks where a single value can reasonably characterise the lighting conditions. However, there are many tasks which are not matt and/or in which texture or form needs to be revealed. For these tasks the spatial distribution of light incident on the task can be more important than the amount of light. Fig. 6.18 shows how the visibility of glossy print can vary for two different incident light distributions, the reduction in contrast occurring because of high-luminance reflections on the print. This dramatic reduction in task visibility goes under the name of veiling reflections. There is experimental evidence that the occurrence of veiling reflections can cause a marked reduction in task performance (Reitmaier, 1979). The two ways of dealing with veiling reflections are either to reduce the specularity of the task or to rearrange the incident light distribution so that no high-luminance specular reflections are produced towards the observer.

Veiling reflections can be regarded as the negative side of light distribution. But when tasks involving texture or form are involved then light distribution has a positive side. Fig. 6.19 shows how diffuse lighting and strongly directional lighting can give different impressions of form and texture. To the best of my knowledge no experimental work has been done to evaluate these effects as they influence task performance, but obviously when such form or texture detection is an important aspect of the task then light distribution is relevant. The exact distribution required will depend on the form or texture to be revealed.

Fig. 6.18. Differences in specular reflections and hence in visibility for two tasks lit by two different lighting arrangements. (*a*) Matt ink on matt paper lit from above and behind the observer. (*b*) Gloss ink on semi-matt paper lit from above and behind the observer. (*c*) Matt ink on matt paper lit from above and in front of the observer. (*d*) Gloss ink on semi-matt paper lit from above and in front of the observer.

(*a*) psycho (*b*) variables

(*c*) psycho (*d*) variables

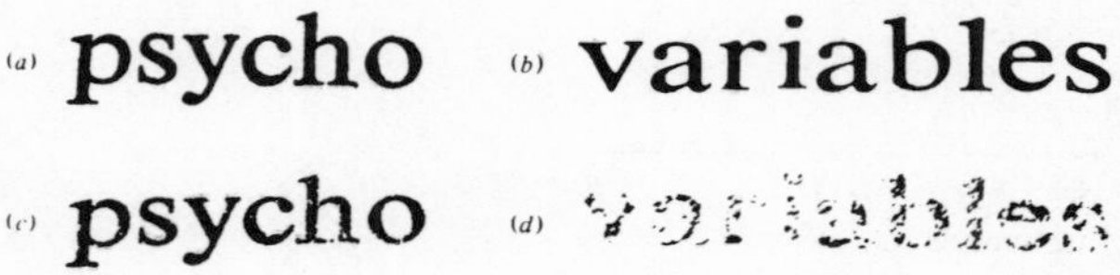

Both the above aspects of light distribution have concerned the impact of light directly on the task, but the threshold results discussed earlier suggest that the distribution of luminance around the task area can also be relevant. Unfor-

Fig. 6.19. Variation in form and texture created by different patterns of incident light: (*a*) diffuse lighting, (*b*) strongly directional lighting.

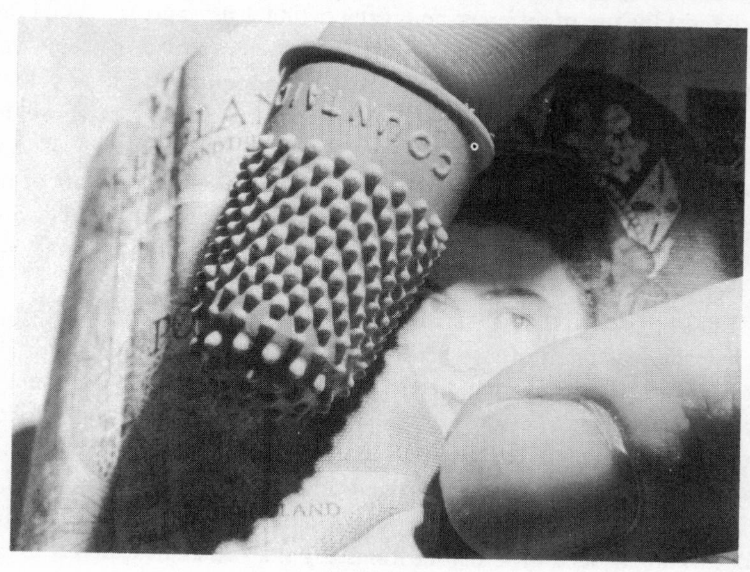

tunately very few studies of visual performance under non-uniform background conditions, e.g. glare conditions, have been done. One that has (Stone and Groves, 1968) showed no effects on task performance even for levels of discomfort glare which have been subjectively considered as 'just intolerable'. Thus it would appear that non-uniform luminances in interior lighting have little effect on task performance. However, this strange result reveals a problem with laboratory performance experiments in general: that is that with sufficient motivation people are capable of performing tasks at a high level in a wide range of what they report as uncomfortable conditions. The problem is that in an obviously experimental situation the subjects often have the necessary motivation but not in everyday life. The crucial point for evaluating the likelihood of specific lighting conditions affecting task performance is whether the conditions directly alter the difficulty of the task as seen by the worker. If they do then performance can be affected. For example, it is known that a non-uniform background in the form of very high luminances close to the line of sight can affect the performance of tasks by producing disability glare since this reduces the contrast of the task at the retina. The most commonly occurring example of disability glare is in driving, where oncoming headlights effectively reduce the probability of detecting a target. Christie and Fisher (1966) have shown that this effect is stronger in older people, probably as a consequence of the increased scattering of light that occurs in aged eyes. Disability glare very rarely occurs in interiors but some form of glare discomfort is more common.

The results so far discussed have been concerned with achromatic tasks. For such tasks the use of different nominally 'white' light sources has no effect on the performance (Smith and Rea, 1980). A different situation occurs when the target is one colour and is seen against a background of another colour, i.e. colour contrast exists. When the luminance contrast is high, say above 0.4, the colour contrast does not improve detection although it will be an aid in any classification work. However, when the luminance contrast is low or non-existent, colour contrast is the main clue to presence, and lighting which enhances colour contrast by increasing the saturation of the colours will improve detection.

But simple detection is not the only form of colour task. For lighting the important question is often whether the colour has any significance in itself. For example, in electrical wiring the colours identify the conductors. There is little point in being able to distinguish the wiring if the colours are wrongly identified. For wiring, the colour is effectively compared with a remembered criterion. This also occurs in assessing the freshness of some foods. Comparison with an existing standard occurs in many manufacturing industries where a number of components are made at different locations and then assembled. Before assembly, it is usual to inspect the components to see if any colour difference is discernible between them and if they are of the required colour. For both these types of task accurate rendering of colour is important. The accuracy with which a light source reproduces colours relative to some standard is quantified by the CIE Colour Rendering Index (CIE, 1974). Boyce and Simons (1977) showed that light sources with high colour rendering indices ensured a smaller number of errors in a task involving the sorting of 85 closely related but different hues into the correct sequence than did light sources with low colour rendering indices. There can be little doubt that good colour rendering is a desirable property for a light source to be used on a task involving fine assessment of colour. Nonetheless it is important to realise that accurate colour rendering is not always required. Accurate colour rendering is an appropriate requirement when the colour as such has some meaning. If it does not but is simply a means of separating the task from the background or from other objects then some distortion of the colours may be helpful. It is not possible to give general advice here because the direction and extent of the distortion will depend on the materials involved.

It should be noted that about 8% of males and 0.4% of females in the population have deficient colour vision. When this occurs performance for some colour combinations will be reduced. Other forms of visual defect also occur, e.g. poor acuity, and will cause problems for specific jobs (Johnston *et al.*, 1976). For example, a crane driver with poor depth perception would be something of a liability. When a job involves some particular visual skill then some form of vision screening is usually worthwhile.

The problem of the strong motivation usually induced in subjects taking part in experiments has already been mentioned. This needs to be remembered when considering the possibility of indirect effects of lighting on task performance. An indirect effect is one produced by a lighting variable that does not cause a change in the visual difficulty of the task. All the variables considered above, i.e. task illuminance, light distribution, disability glare and light colour, are known to affect directly the difficulty of tasks, either by changing the nature of the task itself or by altering the operating state of the visual system. Other features of the lighting may act indirectly by distracting attention from the task, by causing discomfort or by changing the subject's level of arousal. Experience tells us that, outside an experiment, workers who are not completely involved with their work will be distracted and made uncomfortable by extraneous variations in the environment, such as a flickering or high-luminance light source. When this occurs it is likely that performance will suffer. As far as is known no reliable experiment has been done on this possible effect of lighting. Until one is done the widely held belief in this sort of indirect effect of lighting on performance cannot be dismissed.

To summarise, the effect of illuminance on the performance of two-dimensional tasks has been thoroughly investigated. Generally, increasing illuminance leads to better performance on a law of diminishing returns, i.e. equal increments of illuminance produce smaller and smaller improvements in performance until performance saturates. The level at which this saturation occurs depends on the inherent visual difficulty of the task: i.e. its size, contrast, duration of exposure and position; the background against which it is seen; the age of the observer; and the importance of the visual component in the complete task. When tasks are specular reflectors and/or involve the detection of form or texture then incident light distribution may well be more important than illuminance.

Other lighting features such as disability glare and light source colour properties are known to affect the performance of tasks. However, disability glare rarely occurs anywhere other than in road lighting and light source colour properties are only relevant for tasks in which the colour provides information or an important contrast with the background. In addition to these factors there is also the possibility that flicker and high luminances in the background, e.g. from a light source, can have an indirect effect on task performance by distracting attention from the task, by causing discomfort or by increasing arousal. However, such indirect effects are believed, rather than known, to occur.

6.5 The visibility approach

Our knowledge of the effect of lighting conditions on task performance, as exemplified by the preceding section, is very much an *ad hoc* collection based on a need for practical information. However, over the last two decades a comprehensive model of the way in which lighting affects vision and visual performance has been developed (CIE, 1972, 1981). This model, called the visibility approach, is the subject of this section. The whole orientation of the visibility approach is to quantify how easy a task is to see rather than to describe the lighting conditions provided. This is as it should be because lighting is essentially a means to an end, namely to enable people to see well.

The essence of the visibility approach is contained in the instrument used to make the necessary measurements, the visibility meter. When using this instrument the observer views the task of interest through the visibility meter and sees superimposed on the task a field of uniform luminance called the veiling field. By increasing the luminance of this veiling field and simultaneously decreasing the luminance of the task field the observer can reduce the contrast of the task and eventually make it disappear. The reciprocal of the proportion of the total luminance represented by the task field luminance when the task disappears is taken as a measure of the task visibility, called the relative visibility. This is intuitively reasonable. For a task which is clearly visible a very strong veiling field will be necessary to make it disappear; the proportion of the total luminance represented by task field luminance when the task does disappear will be small so the reciprocal of this proportion, i.e. the relative visibility, will be large.

The essential property of the visibility meter is that the luminances of the task and veiling fields are combined in such a way that although their relative contribution to the total luminance can vary, the total luminance, to which the observer is adapted, is constant. In other words, the observer's state of adaptation is unchanged by making the measurements. The total luminance is usually set equal to the task field luminance, in the absence of a veiling field.

Now relative visibility is specific to each observer and to each task. To overcome this specificity the observer views a reference task, consisting of a 4 min arc circular disc on a uniform field, at the same total luminance as the task of interest and establishes the contrast of the reference task that is reduced to threshold by the same veiling luminance as reduced the task of interest to threshold. This contrast is called the equivalent contrast ($\overline{C}$) of the task of interest. This is a very useful measure as it not only provides in a single figure an estimate of the intrinsic difficulty of all the static aspects of the task, i.e. its size, contrast, edge sharpness, layout, texture, etc., as integrated by the observer but also provides a common scale on which the intrinsic difficulties of different tasks can be compared.

Once the equivalent contrast ($\overline{C}$) of the task of interest has been measured, the effect of luminance on task visibility can be quantified by taking the ratio of the equivalent contrast of the task of interest to the threshold contrast (C_1) of the reference task at any particular luminance. This ratio ($\overline{C}/C_1$), which is a measure of how far the reference task equivalent to the task of interest is above its threshold at any particular luminance, is called the visibility level. This too is a useful measure as it quantifies the extent to which the luminance makes the specific task visible. Luminance and visibility level are not directly linked. Two tasks with different equivalent contrasts will have different visibility levels even when they have the same luminance.

The generalising power of visibility level has been demonstrated using some results of Weston (1945). When they first appeared these results were simply plotted as performance scores against illuminance, with a different curve for each task of different size and/or contrast (see Fig. 6.22). After measuring the equivalent contrast of each task the visibility level at each illuminance was determined. When the performance scores were plotted against visibility level the results were found to fit reasonably onto a single line (Blackwell, 1980).

Following this compression of Weston's results and those of other people it was thought that a unique relationship between visibility level and visual performance existed (CIE, 1972). Further research showed this not to be so. Differences were found in the performance of visual tasks that required the observer to search and scan the visual field to various extents, even when each task had the same visibility level. Over recent years the visibility approach has been modified to take this aspect of visual work into account (CIE, 1981). This work has served to emphasise that there are a large number of different relationships between the performance of real tasks and visibility level. This is because the importance and nature of the visual component of the task will change from task to task so the effect of visibility level will also vary. Thus even though visibility level is a more complete description of the effect of lighting on a task than the conventionally used task illuminance, it is not and cannot be the complete answer to predicting task performance. Other factors such as the degree of search and scan involved and the magnitude of the visual component are of equal if not greater importance to task performance (CIE, 1981).

However, some elements of visibility measurement can be used to assess the effectiveness of different light distributions in making tasks visible. The most important factor investigated in this way has been veiling reflections produced in glossy tasks. Fig. 6.18 demonstrated the effect of light distribution on written material with different degrees of specularity. It was seen that when a target consists of glossy material, and the light is incident on it from the mirror angle, specular reflections occur. These specular reflections change the contrast of the task and hence its visibility. The way the importance of such reflections is quantified is to measure the relative visibility of the target under the lighting of interest and under reference lighting conditions, the latter being completely uniform, diffuse, unpolarised illumination. The ratio of the two relative visibilities is called the contrast rendering factor (CRF). Since reference lighting conditions produce few veiling reflections, even in glossy tasks, a CRF of much less than 1 indicates that the task is specular and that the relative positions of observer, target and lighting are such that strong veiling reflections are being produced. Practical guidance is available on how such veiling reflections can be eliminated (Boyce and Slater, 1981).

Obviously there remains a lot to be done before all the possibilities of the visibility approach have been explored. Nonetheless it already has considerable advantages over the previous *ad hoc* approach. From psychophysical observations measurements of the visibility of any task can be obtained. This information can be used to classify the intrinsic visual difficulty of tasks, to quantify the extent to which the lighting makes tasks visible or to rank different lighting installations in terms of the extent to which they produce veiling reflections. It has also been developed into the most complete model of how lighting conditions affect visual work (CIE, 1981). If this model could be shown to be useful for practical work then it would constitute an immense step forward in our understanding of the interactions of people and lighting.

6.6 Light and health

This section is concerned not with task performance but with the effect of the lighting provided for work on the health of the worker. It is not concerned with the specific diseases for which some form of light is a treatment, e.g. hyperbilirubinaemia (Jewess, 1978). Rather it is concerned with the effects of light which produce irreversible or reversible but uncomfortable changes in the visual system. This latter category is included because feelings of discomfort which occur frequently can be considered an aspect of health.

Eyestrain

The most common experience of light affecting health is when people complain of eyestrain. Eyestrain is a word which encompasses many phenomena. In its everyday sense it is used to cover all forms of ocular discomfort and the associated symptoms. The symptoms usually complained of are:

1. local irritation in the form of a conjunctival sensation, e.g. itchy, hot, aching eyes,
2. breakdown of vision, e.g. blurring of vision and double vision;
3. referred effects, e.g. headaches, indigestion, giddiness.

Most people have experienced one or more of these symptoms at some time, usually after a period of work at a difficult visual task. Equally, most people have found that after rest the symptoms disappear: eyestrain is a temporary phenomenon. It is believed to be produced centrally by trying to interpret poor-quality retinal images and/or peripherally by excessive use of the ocular motor system in an effort to overcome deficiencies in the visual system or in the task or in the lighting of the task. The ocular motor system is made up of the muscle systems that control the fixation, accommodation, convergence and pupil size of the eyes. If any element of the ocular motor system is operating at its limits, e.g. close to the near-point of accommodation, then like any other muscle system it will become fatigued and discomfort will occur.

The probability of difficulties occurring in the ocular motor system, e.g. a limited range of accommodation, increases with age, but such visual defects as short and long sight, astigmatism and poor linkage between accommodation, convergence and pupil response affect young people as well as old. Usually these deficiencies make the fixation and focussing of a task and the evaluation of the retinal image more difficult. This in turn leads to a build-up of fatigue and eyestrain. But even if the observer's vision is normal it is still possible for the lighting conditions to cause eyestrain. For example, when a difficult visual task is given a low illuminance, eyestrain is likely to occur. The reason is simply that the task will be difficult to see, which will usually lead to fatigue occurring. Also the usual response in these circumstances is to get closer to the task so that the retinal image is bigger. Unfortunately this also means that the required accommodation of the lens is greater, as is the convergence of the two eyes. This imposes a strain on the ocular motor system which may lead to other symptoms of eyestrain appearing. In addition it should not be forgotten that in order to get the eye closer to the task the normal posture may be changed, which in turn may produce back or neck strain.

High illuminances may ensure that a visual task which is difficult under a low illuminance can easily be seen at a distance that does not require extreme accommodation, but they may also require extreme operation of the pupillary response. In addition, they may require more accurate control of the convergence mechanism to avoid any disparity between the retinal images in the two eyes becoming apparent (Bedwell, 1972). Thus high illuminances may also produce eyestrain symptoms but in a different way.

Another feature of the lighting in a room is the spatial and temporal variation of luminances. The disability caused by having high luminances close to the line of sight has already been discussed in terms of the effect of disability glare on the performance of tasks. Such disability glare may cause eyestrain and will certainly cause annoyance because of the inability to see the task clearly and the element of distraction it creates. The distraction produced by a bright light source near the task is also linked to another form of glare, called discomfort glare. This form of glare does not produce any measurable disabling effects, only discomfort, although it may have an indirect effect on task performance. Discomfort glare is simply a recognition of the fact that people sometimes complain about the brightness of the light sources, seen either directly or by reflection. There is no physiological explanation for discomfort glare, although the fluctuations in pupil size have been suggested (Fry and King, 1975). At present it is considered to be mainly a psychological response and as such its occurrence will often depend on the context of the place and the purpose of the people in being there.

Visible temporal fluctuation, or flicker as it is more commonly known, is a method of attracting attention. Indeed it is specifically used to do this in visual data displays. Flicker in a light source which attracts attention away from objects of interest will be a source of frustration and annoyance, emotions which are only too likely to lead to headaches, fatigue and irritability. It is probably for this reason that flicker in discharge lamps, including fluorescent lamps, is widely regarded as undesirable. Flicker can have another extremely undesirable effect: it can cause systematic variation in the electrical activity of the brain. When this stimulation is strong it can trigger attacks in epileptics. The best rule to follow is to have no visible flicker at all in the visual world unless essential, and then only for attention-gathering purposes.

Tissue damage

The above has been concerned with the link between lighting conditions and eyestrain. It is now necessary to consider the possibility of tissue damage caused by lighting. The first condition to examine is that of illuminance. As far as is known low illuminances do not cause any anatomical or physical damage to the eye (NIOSH, 1975), although eyestrain may be produced for some people doing some tasks.

Unlike low illuminances, very high illuminances have been found to damage the structure of the eye, at least the eyes of pigeons, rats and monkeys. Unfortunately, there is some confusion as to the actual luminous flux reaching the retinas of these creatures, the latest understanding being that the luminous flux is several orders of magnitude greater than that experienced by people in buildings (NIOSH, 1975). In terms of luminous flux, people working out of doors are exposed to much higher levels than those indoors and there is no indication that such people suffer any consequent damage to the eye. Until this is at least suggested it can be concluded that the range of illuminances used in interiors does not cause damage to the eye.

Different light sources used in interiors have different spectral distributions, the incandescent sources such as daylight or a filament lamp being continuous, the discharge sources discontinuous. There appears to be no injurious effect on the eye because of this variation in the spectral distribution in the visible region. The wavelengths immediately adjacent to the visible region are more likely to cause trouble. Infra-red radiation is absorbed in the eye largely by the cornea and ocular media. Prolonged exposure to high levels of such radiation and the consequent heating effect can cause irreversible changes in the lens which may lead to opacities developing. This is one of the reasons why protective glasses are worn for industrial tasks, such as foundry work, where the eye is exposed to high levels of infra-red radiation. The situation is similar for ultra-violet radiation but in this case the quantities required are smaller. Excessive absorption of ultra-violet radiation by the cornea can cause kerato-conjunctivitis, which is a most uncomfortable although temporary condition. Protective glasses should always be worn when working in the presence of ultra-violet radiation. Threshold limiting values for exposure to light, infra-red and ultra-violet radiation have been produced by the American Conference of Governmental Industrial Hygienists (1981).

To summarise the effect of lighting on health, both high and low illuminances can produce eyestrain by forcing the visual and ocular motor systems to operate close to their limits. Low illuminances do not damage the eye itself but very high illuminances may. However, this only occurs at much higher illuminances than are normally used. Headaches and other symptoms produced by frustration and annoyance can occur when disability glare, discomfort glare and flicker from lights are present. Finally, infra-red and ultra-violet radiation can cause damage to the eyes (Sliney, 1972).

6.7 Research routes

So far attention has been given to 'what we know' in general. The purpose now is to consider how further knowledge can be obtained for specific applications. There are a number of approaches by which the relationship between light and work can be investigated, each with its own advantages and disadvantages. These approaches will be summarised and the most suitable area of application for each suggested.

In order to obtain interesting answers in research one has to ask the right questions. In order to ask the right questions one has to understand the purpose of the investigation. There are two broad classes of question which relate to lighting: 'what lighting conditions are best for the performance of this task?', and 'what is wrong with the lighting in here?' The first question is essentially concerned with defining lighting criteria, the second with rectifying poor lighting.

Field studies

For defining lighting criteria the obvious approach is to take a real task in a real situation and change the lighting. Ever since the introduction of artificial lighting there has been interest in the relationship between lighting conditions and work performance in the field. Mainly this interest has been concentrated on the effect of the illuminance since this is closely related to the number of lamps used and the amount of fuel sold. The literature is peppered with reports of increases in illuminance and consequent increases in output in factories. Unfortunately many of these reports are simply assertions with few details of the conditions given. Even amongst those which are more than assertions few stand up to close examination. Usually there are other changes made along with the illuminance; for example, the layout of the work and in one case even the nature of the work. This is not to deny that increasing illuminances can change work output. From laboratory research it is clear that performance can improve when the task becomes more visible. There is no argument that this must occur over some illuminance range: the disagreement is with the suggestion that it occurs over the illuminance range being examined. Ever since the famous Hawthorne experiments in the 1920s (Urwick and Brech, 1965) the necessity for tight experimental control in field experiments has been appreciated and the overriding importance of human relationships realised. Better lighting can change how easy the task is to see but there are many other factors that intervene between being able to see the work and producing maximum output.

Exact experimental control is very difficult to achieve in any sort of commercial situation. A factory management is hardly likely to allow changes to be made to the lighting which will decrease output, and neither will workers if their pay is linked to output. Groups of workers will not look favourably on being given different lighting conditions some of which are apparently inferior to others. Hartnett and Murrell (1973) provide a warning of some of the problems likely to be encountered in field research and suggest some solutions. Occasionally, the difficulties of field research have been overcome. Gallagher (1976) has reported a study of the effect of different road lighting schemes on drivers' detection of obstacles on the carriageway. He arranged for a section of public highway to be lit in different ways. Then, without any indication to the drivers that an experiment was taking place, he positioned marker cones on the road at various positions and concealed observers beside the road. The measurements taken were simply the distance at which the driver took avoiding action and his speed at that time. These values were converted to a time-to-target value. Fig. 6.20 shows these results plotted against the effective visibility level of the target cones at the different distances under the different lighting installations. There is a marked increase in the time-to-target at which avoiding action was taken with increasing effective visibility level. Such an experiment obviously raises ethical

Fig. 6.20. Mean time-to-target plotted against effective visibility level for drivers taking action to avoid an obstacle. Effective visibility level is visibility level corrected for any effects of veiling reflections and disability glare. (After Gallagher, 1976.)

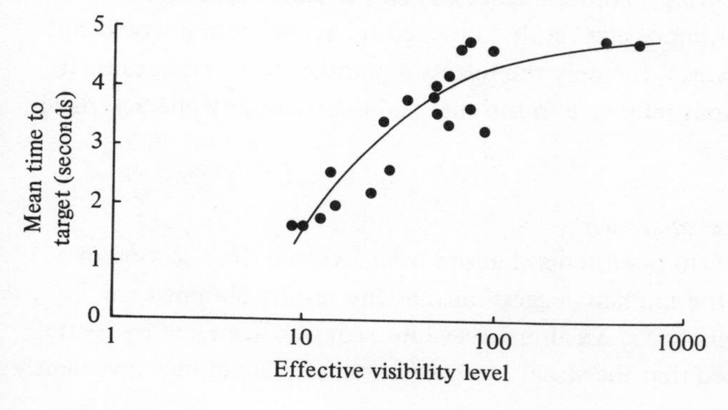

issues but there can be little doubt about the validity of the results. By careful experimental design and covert observation the experimenters were able to establish the effect of changes in lighting conditions on an important aspect of drivers' behaviour. Such information is useful in defining roadway lighting standards and in evaluating the utility of the visibility approach. This experiment shows what can be done in the field but it will only rarely be possible, or desirable, to go to such lengths to obtain the necessary information, especially as there are alternatives.

Simulated tasks

Simulated tasks involve a retreat into the laboratory, where experimental control can be strict, and the use of a representative task rather than any actual real-life task. In a sense this latter can be both a loss and a gain. The loss is in the failure to reproduce a specific task, the gain is in the possibility of generalising the result to tasks of a similar type. The other problem is that of an excess of motivation. There can be little possibility of concealing the fact that an observer is taking part in an experiment once it takes place in a laboratory. This in itself usually means the observer is well motivated to succeed, which may diminish the influence of some experimental variables.

Typical of this approach is a study by Stenzel and Sommer (1969) on people sorting screws into categories and crocheting a stole. Two groups of subjects worked for 2-hour periods at each task, each period having one of four illuminances. The conventional experimental controls of order, etc., were used. The results for both tasks show the expected increase in output and decrease in errors with increasing illuminance, although the errors increased at the highest illuminance (1700 lux) for the sorting task. Strictly these results only apply to the specific task but it is commonly asserted that they can be applied to other similar tasks. Whether this is so is a matter of speculation. Presumably, if the materials to be sorted had a similar range of cues by which the object could be classified during sorting, and these features had the same visibilities as the screws, the generalisation of the results obtained on screw-sorting would not be unreasonable. However, to apply the results to another task just because it involves sorting, without reference to the cues and their visibility characteristics, could be misleading.

The analytical approach

The types of studies discussed above have involved the selection of a particular task with the implicit suggestion that the results obtained can be applied to other similar tasks. An alternative approach was suggested by Beutell (1934) who considered that the visual component of tasks could be conveniently broken down into a critical size of detail and a critical contrast. If the effect of lighting conditions on a standard task covering a range of critical sizes and contrasts could be established then the effect of those lighting conditions on the performance of the visual part of any task with specified size and contrast could by synthesised from these data. Such an approach is another step on the road to increasing generalisation and reducing specificity. This analytical approach has been widely followed, the original set of experiments being those of Weston (1945).

Weston developed a standard task, called a Landolt ring chart, one form of which is shown in Fig. 6.21. Weston's chart consisted of a regular array of **C** targets with the gap in the **C** oriented in one of the eight major directions of the compass. The size of the gap was the critical size of detail and the contrast of the **C** against the background was the critical contrast. What the subject had to do was to read through the chart as quickly and accurately as possible, marking off all the **C**'s with gaps in a specified direction. Weston measured the time taken to do a chart, combined this with the number of mistakes made, corrected for the manual part of the task, and called the total the visual performance score. A summary of the results is shown in Fig. 6.22, plotted against illuminance. The

Fig. 6.21. A form of Landolt ring chart.

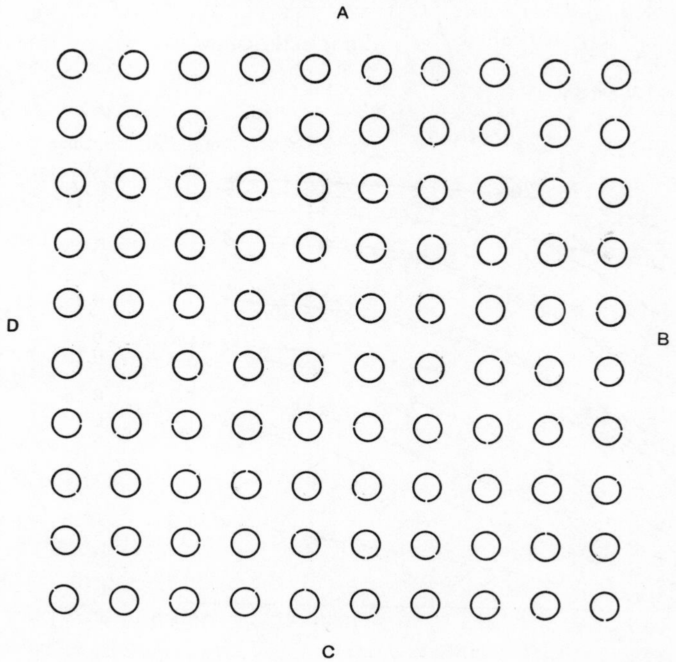

curves show very much what would be expected from the previous discussion of visual performance. The great advantage of this analytical approach is that it enables a rapid survey to be taken of the effects of any particular lighting variable over a wide range of visual difficulties. Its limitation is that it does not apply, except by assumption, to any specific real task. The advantages and disadvantages of all these research approaches are summarised in Table 6.4.

Recommendations

The analytical approach has had a very influential past. The type of results shown in Fig. 6.22 once formed the explicit basis of illuminance recommendations given in the Code for Interior Lighting produced by the British Illuminating Engineering Society (now the Lighting Division of the Chartered Institution of Building Services). This body, the professional organisation representing lighting engineers in the UK, published its first code in 1936, and has issued a revised version at irregular intervals ever since, the latest edition

Fig. 6.22. Mean visual performance scores for Landolt ring charts with rings of different critical size and contrast plotted against illuminance. The visual performance score for each individual is given by the expression

$$\left[\left(\frac{\text{Total time taken}}{\text{Number of rings correctly cancelled}}\right) \times \left(\frac{\text{Total number of rings to be cancelled}}{\text{Number of rings correctly cancelled}}\right) - \begin{matrix}\text{Normal}\\ \text{time}\\ \text{per}\\ \text{ring}\end{matrix}\right]^{-1}$$

(After Weston, 1945.)

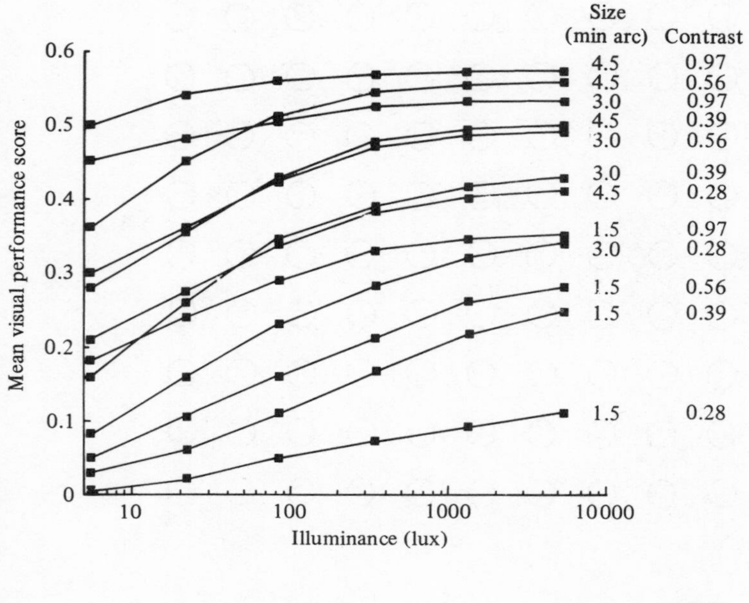

Table 6.4. *Advantages and disadvantages of different research routes to measuring the effect of lighting on work*

Research route	Advantages	Disadvantages	Practical problems
Field measurements	Realism; a real task done by a representative sample of people, sometimes without knowing they are taking part in an experiment	Results only applicable to the specific task; difficult to generalise	Exact experimental control often difficult to obtain. Ethical considerations if unwarned subjects and/or covert observation used
Simulated tasks in laboratory	Close experimental control possible. If task is chosen carefully then the results may be applied to other tasks	Possible increase in motivation produced by the context of an experiment	The selection of a task which is capable of generalisation and a representative sample of people to do it
Analytical task in laboratory	Close experimental control possible. Enables an estimate of the influence of any lighting variable to be obtained quickly for a wide range of visual difficulty	Possible increase in motivation produced by the context of an experiment. Results not applicable to any specific task	The selection of a representative sample of people to do the task

being published in 1977 (CIBS, 1977). This document is accepted in many countries as giving authoritative advice on the lighting standards for a large number of applications. The recommendations are summarised in a schedule which, for each application, specifies the illuminance on the task, the position of measurement of the illuminance, the limiting glare index necessary to control discomfort glare, the range of light sources that can reasonably be used and any statutory requirements. It also gives some notes on the specific application. Table 6.5 shows a summary of the scale of illuminance used, the types of activities to which the illuminances apply and a flow chart which modifies the recommended illuminances in certain circumstances.

It is important to understand the nature of such advice. It is not determined by some magic formula since none exists. It is not even based on experimental studies of all the applications for which recommendations are made. The real world cannot wait for the level of certainty required by researchers. Rather, the recommendations are based on a synthesis of the experimental evidence available, the experience of those working at different applications, and the cost of lighting equipment together with energy to operate it. The published lighting criteria represent consensus opinions of what is a reasonable balance between the benefits in terms of task performance and observer satisfaction and the cost of buying the light. This consensus is reflected in the status of the recommendations, which are claimed to be merely representative of good practice. Obviously a consensus opinion can change from time to time as either the nature of work or economic factors alter the balance between cost and benefit. In spite of their lack of scientific rigour the recommendations can be used with confidence. They have to face a much wider test than is applied to many research results. People actually light buildings following these recommendations. If the recommendations were to produce any deleterious effects it would not be long before this would become known. Jay (1973) discusses the aspect of practical proof in an interesting paper.

Other countries have similar sets of recommendations. These recommendations differ in form and value but they all have one thing in common: they are all based on a consensus of opinion rather than any definite understanding of the effect of lighting on people. Each area of application has a different relationship between task performance and the lighting conditions and this relationship is usually unknown in detail. This, together with the ubiquitous economic and technical considerations, means that lighting recommendations are inevitably a compromise between the cost and the benefits. Different countries and different organisations may have different ways of reaching that compromise but for all of them it is still a compromise. Table 6.6 lists some of the more widely available sets of lighting recommendations.

Table 6.5. *The illuminances recommended for different categories of work and the modifications suggested for particular circumstances*

Task group and typical task or interior	Standard service illuminance (lux)	Are reflectances or contrasts unusually low?	Will errors have serious consequences?	Is task of short duration?	Is area windowless?	Final service illuminance (lux)
Storage areas and plant rooms with no continuous work	150	—	—	—	—	150
Casual work	200	—	—	—	no — 200 / yes	200
Rough work: rough machining and assembly	300	no — 300 / yes	no — 300 / yes	300	no — 300 / yes	300
Routine work: offices, control rooms, medium machining and assembly	500	no — 500 / yes	no — 500 / yes	no — 500 / yes	500	500
Demanding work: deep-plan, drawing or business machine offices; inspection of medium machining	750	no — 750 / yes	no — 750 / yes	no — 750 / yes	750	750
Fine work: colour discrimination, textile processing, fine machining and assembly	1000	no — 1000 / yes	no — 1000 / yes	no — 1000 / yes	1000	1000
Very fine work: hand engraving, inspection of fine machining or assembly	1500	no — 1500 / yes	no — 1500 / yes	no — 1500 / yes	1500	1500
Minute work: inspection of very fine assembly	3000	3000	3000	no — 3000 / yes	3000	3000

Using local lighting, if necessary supplemented by use of optical aids, e.g. binocular loupes, magnifiers, profile projectors etc.

Source: From CIBS (1977) Reproduced by kind permission of the Chartered Institution of Building Services.

Troubleshooting

Another question to which research addresses itself is an answer to the complaint that there is something wrong with the lighting. This occurs in a rather restricted situation. The lighting already exists. What is required is some idea of where it is deficient and how it can be improved. Ideally the attack on such a problem occurs in four stages (see Table 6.7). The first stage is for the investigator to define the nature of the complaints about the lighting more precisely. Are they about the lighting generally or about one specific feature? How many people complain? Do the complaints occur for one location? Are they related to any specific task or are they unrelated to any task but simply reflect a general feeling of discomfort? The investigator should collect the opinions of the people who have experience of working under the lighting and, if a task

Table 6.6. *Sources of information on lighting criteria and design*

Document	Publisher
IES Code for Interior Lighting	Illuminating Engineering Society[a]
Technical Report No. 4: Daytime lighting in buildings	Illuminating Engineering Society[a]
British Standard CP3: Daylighting and Artificial Lighting	British Standards Institution
A Guide on Interior Lighting	Commission Internationale de l'Eclairage
Innenraumbeleuchtung mit kunstlichen Licht (DIN 5035)	Deutches Institut für Normung
The IES Handbook 1981	The Illuminating Engineering Society of North America
Interior Lighting Design	Lighting Industries Federation and Electricity Council
Technical Report No. 15: Multiple Criterion Design – A Design Method for Interior Electric Lighting Installations	Illuminating Engineering Society[a]
British Standard CP 1004 Parts 1–9: Road Lighting; and British Standard 5489	British Standards Institution
Recommendations for the Lighting of Roads for Motorised Traffic	Commission Internationale de l'Eclairage
Outdoor Lighting Handbook	Lumsden, W.K., Aldworth, R.C. and Tate, R.C.C. (Published by Gower Press)
IES Lighting Guide, the Outdoor Environment	Illuminating Engineering Society[a]

[a] Now the Lighting Division of the Chartered Institution of Building Services.

Table 6.7. *The process of investigating complaints about lighting*

Purpose	Factors to be considered	Methods
1. Define complaint precisely	Nature of complaint – general or specific; number of people affected – many or few; location of people affected – one area or different; work of people affected – one task, different tasks, or no task	Conversations with people; questionnaires; interviews
2. Identify problem by examining lighting, task and people in combination or separately as appropriate for the complaint	Lighting unsatisfactory; how does it compare with standards, is it well maintained, is it suitable for its purpose, does it cause discomfort?	Photometric survey and comparison with standards; for illuminance, illuminance uniformity, glare, and surface illuminance ratios. Also consider colour properties of light source, flicker, veiling reflections, reflected glare and directional effects
	Visual difficulty; features which need to be seen; time for which they are available; background against which they are seen	Examination of task for size, contrast, form, texture, colour and movement of details that need to be seen, shadowing by worker and complexity of background
	Possible visual defects in individuals	Observation of individual at work; visual screening for long/short sight, colour vision, depth perception, visual acuity, visual field size; ophthalmic examination

(*Continued on next page*)

Table 6.7 (*contd*)

Purpose	Factors to be considered	Methods
3. Prepare and test solution	Effectiveness; practicability; cost; implications for other services	Possibilities are to change lighting, to make tasks easier, to move individuals to other work or location; for new lighting a temporary installation can be useful
4. Monitor solution	Long-term satisfaction of people; effective operation of equipment	Conversations; interviews with people; questionnaires; photometric survey for lighting

is involved and if practicable, should do the task himself. The opinions can be collected in several different ways; conversations, questionnaires, interviews, etc. Whichever way they are obtained they will need to be interpreted and this can be aided by the experimenter's experience of the task.

The second stage is to identify the problem. The most probable cause of the complaint can often be deduced from the precise definition of the complaint. If many people doing different tasks complain about discomfort in the interior then the general room lighting is probably at fault. If the complaints are identified within a specific area then the lighting of that area should be investigated. If the complaints are related to a specific task then it is likely that the lighting is not suitable for that task. If only a few people complain about a task but many others in the same area and doing the same task do not, then it is appropriate to consider the possibility of the individual's visual defects. In this case much can be learnt from watching the people at work, particularly the posture they adopt. Visual screening for visual acuity, long/short sight, depth perception, colour vision and size of visual field should also be considered, as should a detailed ophthalmic examination.

When a specific task has been identified as causing problems it is as well to consider the properties of the task and lighting together. The task needs to be assessed in terms of its size, contrast, and movement properties, any texture or colour discrimination required, the background against which it is seen, and any shadowing produced by the worker. A thorough photometric survey of the lighting provided is also desirable. It is not enough simply to measure the illuminance on the task. Although this is valuable, other factors such as the luminances in the visual field and the luminance ratios between the task, the immediate surround and the general background are important. The researcher should also examine the light source colour properties and any flicker or veiling reflections that occur. Of particular importance for three-dimensional tasks is the directionality of the lighting (Faulkner and Murphy, 1973). Having made a survey of the lighting conditions provided, it is then worthwhile comparing them with any relevant recommendations. Often very wide differences will be found which point to the root of the problem. If such discrepancies are apparent then they can be compared with the people's opinions. If there is agreement over the source of the problem then the next stage is to change the lighting conditions to remove it. But if there is not agreement then the people's opinions should be taken as the prime data. After all, the lighting criteria may be wrong for this task or a relevant feature may not have been considered. It should be noted that a similar photometric survey and comparison with relevant standards procedure is of value when the complaint concerns the general room lighting rather than a specific task. Often discrepancies between standards and actuality will be revealed which will indicate the nature of the problem.

Assuming that the problem has been identified, the investigator now has to suggest and test a practical and effective solution. This may be a matter of changing the lighting, possibly just to repair the damage from poor lighting maintenance, or it may be a matter of altering the tasks to make them visually easier or it may involve moving people who do not have the necessary visual ability to do difficult visual tasks. If a new way of lighting a task is being suggested it is often useful to try a temporary 'lash-up' installation. However, an improvement in the lighting conditions is not an end in itself but only a means to an end. The real test of the solution to the problem is for the complaints to disappear. The only people who can withdraw the complaints are the people who made them. This final stage therefore calls for more conversations and interviews so that the level of long-term satisfaction can be assessed.

A good example of this type of investigation is given by Carlsson, Knave and Wibom (1977). They examined lighting for one particular task: manual teeming in casting bays and foundries. Their report starts with a description of the job, which is to pour molten metal from a small ladle into small moulds for precision casting. This task was considered to be very difficult, particularly getting the stream of molten metal into the small opening in the mould. The feature of the task that makes it difficult is the low contrast between the mould opening and the mould, typically about 0.05. This, together with the glare from the high luminance of the molten metal (25 000 cd/m^2) and the necessity of wearing of dark goggles, makes the point to be hit by the stream of molten metal very difficult to identify. Thus there is often a danger and a waste from spilt molten metal. Carlsson *et al*. (1977) measured the lighting conditions in four foundries in which manual teeming was done. They were found to vary widely, one to be better and the rest worse than the appropriate lighting recommendations. This occurred not because the installations were designed to be worse but because severe fouling of the luminaires had occurred. Having identified the lighting and the task contrast as the basic problem, one solution suggested was to increase the luminance around the task and thereby increase contrast sensitivity. An alternative would have been to increase the contrast of the pouring hole itself but this does not seem to have been practically possible. To test the proposed solution, a trial installation was set up at one foundry using suitable floodlights that provided a series of three illuminances all much higher than that previously available. The effects of these illuminances were measured in terms of the contrasts produced at the pouring hole and the worker's visual acuity as determined using a test chart with figures of contrast 0.10, seen with and without goggles. The results obtained indicate that there is a steady increase in contrast and in attainable visual acuity with the increased illuminances produced by the floodlights. Standardised interviews were conducted with some of the workers on the teeming line after using the experimental installation. There was a strong indication that the new lighting arrangements were much better than the old. The report concludes with recommendations for the type of lamp and luminaire to be used as well as the lighting conditions. This is worthwhile because it offers a complete solution to the practical problem rather than the partial solution of simply saying what conditions are required.

Photometric surveys

Having stated that a thorough photometric survey is often necessary when investigating complaints associated with lighting, it is appropriate to review briefly the instruments available and their functions. It would be very useful if a visibility meter were widely available but unfortunately it is not, although a simple comparator has been suggested (O'Donnell, Critchley and Chapman, 1976). In fact the most widely available and most misused instrument is the illuminance meter, or light meter as it is generally known. The illuminance meter is used to measure the illuminance on a plane. It is cheap, simple and inaccurate. British Standard BS 667, which relates to these illuminance meters, calls for two classes, one with an error of not more than ± 15% and the other not more than ± 10%. To maintain even this level of accuracy, illuminance meters should be calibrated at least once a year. For general use there are two other features of an illuminance meter which are important. The response must be corrected to take account of light falling on it from oblique angles and the sensor should have a spectral response corresponding to the appropriate relative luminous efficiency function V_λ. These two features are known as cosine correction and colour correction respectively. Cosine correction is usually built into the construction of the meter. Colour correction can be achieved either by filters or by a series of multiplying factors for different light sources. In almost all instruments the colour correction is for photopic conditions. It is worthwhile knowing that in addition to conventional planar illuminance measurements, the illuminance meter can be used to determine scalar illuminance and illumination vector, both of which measures are involved in quantifying the directionality of lighting (CIBS, 1977). Scalar illuminance, which is the mean illuminance over the surface of a small sphere at a point, can be approximated by the average of the illuminances on the four sides of a regular tetrahedron. The illumination vector is a vector with a magnitude defined by the maximum difference of illuminance between two sides of a small plane and a direction defined by the normal to the plane when the maximum illuminance difference occurs. The illumination vector can be calculated by the vectorial combination of the differences in illuminance for the three pairs of opposing faces of a cube.

Colour correction also applies to the other instrument commonly used in lighting surveys, the luminance meter. This device, which looks rather like a cine-camera, measures the average luminance of a small area, typically 1° or less, on any surface it is pointed at. This instrument together with a standard reflectance plate can also be used to measure the reflectance or luminance factor of surfaces. Fig. 6.23 shows several illuminance and luminance meters. Reviews of the range of equipment commercially available are given elsewhere (Jewess, 1976; IES, 1981). Armed with these two instruments, a degree of common-sense and the ability to use their eyes, anyone can make a reasonable photometric survey. However, four points should always be remembered. First, most lighting installations take some time to reach a stable condition after switch-on. Twenty minutes is typical for fluorescent systems and it may be longer for other discharge sources. Secondly, light output is very closely linked to supply voltage. Thirdly, daylight and sunlight can produce large variations in measurements very rapidly. Finally – a practical point backed by long experience – many lighting installations are poorly maintained. This means that occasional lamps will have failed, some will have been replaced, whilst others may be operating inefficiently. Photographs and written records of the lighting taken at the time of the survey are often a useful aid to interpreting the results obtained.

Fig. 6.23. A range of photometric instruments. Those above the horizontal line are luminance meters, those below are illuminance meters.

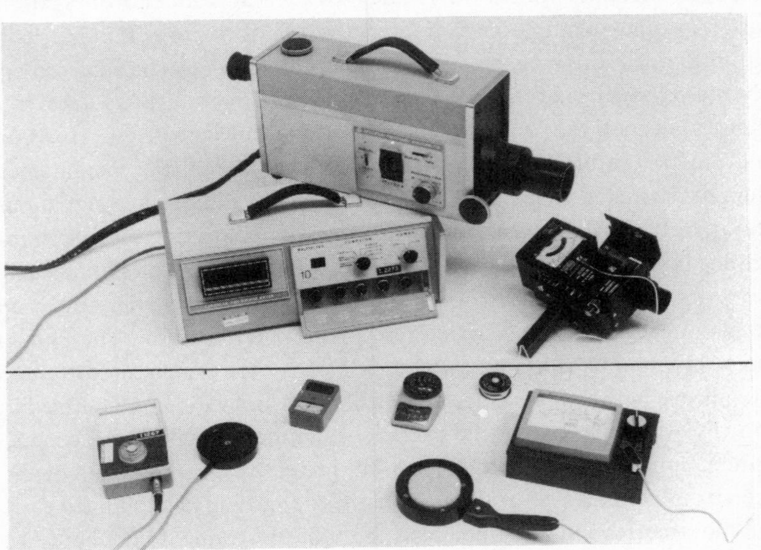

Physiological indices

One final aspect of the methods whereby the interaction between people and lighting has been studied must be mentioned. Throughout the discussion the methods considered have all used the output produced by the worker as a response measure. There has been no consideration of the physiological cost or, in everyday terms, the effort, required to perform the task. At first this may seem a surprising omission but in fact it merely reflects the level of information. The failure is not for want of trying. There have been many studies which have tried to measure the effort involved but the problem is that the lighting conditions change the mental component of the work not the physical. Whilst there are several well-established measures of physical effort, the search for a reliable measure of the mental effort involved in visual work has been long and ultimately fruitless.

Starting with Luckiesh and Moss, who used heart rate, blink rate and finger pressure (Luckiesh, 1944), researchers have attempted to find a measure that could be used to quantify the effort or stress and ultimately the fatigue experienced under different lighting conditions. Direct measurements of visual functions such as the accommodation time (Collins and Pruen 1962), and range (Baumgardt and Le Grand, 1956), the critical flicker fusion frequency (Grandjean and Perret, 1961) and the control of eye movements (Brozek and Simonson, 1952) have all been used. More recently there have been attempts to use more sophisticated variables such as the level of keto-steroids as measures of stress (George, 1972). Unfortunately none of these attempts has been completely successful. Many have failed to produce any changes and those that have, have not been reproducible. This is probably because many of the responses being measured can be produced by a multitude of stimuli, many of which are beyond the control of the experimenter. There is also the point that the pattern of response produced by stress is specific to each individual (Schnore, 1959). This implies considerable inter-individual differences in response, a factor which obscures any effect on a group of people. Even where repeatable differences have been obtained, the approach has foundered on the rock of interpretation. What does a 10% change in accommodation time mean: is it good or bad, or within the range of normal variation?

The sad fact is that the search for a physiological measure of mental effort has been a chimaera, but there is an alternative available. In the real world people will usually complain and even take avoiding action over tasks and lighting conditions that require excessive effort. Further, they will take these actions under conditions less extreme than those required to produce any measurable physiological response. Considerable experience is available on the collection of subjective and behavioural information for the comparable fields of acoustics and thermal conditions. There seems to be no reason why these methods should not be applied to the examination of lighting conditions. To do so would certainly be more productive than to continue working through the list of potential physiological measures.

6.8 Applications

Having examined the methods whereby information about the effects of lighting can be obtained, it is necessary to consider where such information may be required. This is a very wide field. For almost all aspects of modern living that are likely to be of interest to the ergonomist the provision of some form of lighting would be beneficial. Given that some form of lighting is desirable, the next question is what form the lighting should take. There is plenty of advice available on the lighting conditions desirable for different activities and on methods of designing lighting installations to provide these conditions (Table 6.6). By far the most commonly used design method for electric lighting is the lumen method. It is usually followed by a calculation of the IES glare index, a measure which predicts the extent of discomfort glare that will be experienced under a specific lighting installation. Other countries have different glare prediction methods but virtually all use the lumen method. The lumen method basically tells one how many luminaires of a specific type are needed in order to provide the illuminance uniformly over a horizontal working plane. It is short and simple to use but it is also limited. It can only be used for repetitive arrays of luminaires, which may not always be sufficient. The design of daylighting for interiors is much more varied, depending on the form of the building, the number of floors, the roof height and the thermal and noise implications of the windows (Hopkinson, Petherbridge and Kay, 1966; Lynes, 1968).

Regardless of whether the lighting is provided by electric light or daylight an important feature of most of the available lighting recommendations and the design methods is that they concentrate on uniform lighting over a large area, e.g. a complete engineering workshop, where a wide range of tasks is being carried out. The ergonomist is unlikely to be involved at this general level. He is more likely to be concerned with specific tasks. It can hardly be otherwise. The essence of ergonomics is matching the task to the man. Unless the task is specified there cannot be a match. A number of studies have been made of the lighting problems that occur in particular application areas and how they may be overcome. Table 6.8 lists some of these areas. Others are described in Lynes (1981). In evaluating such studies it is important to be aware of their country of origin because methods of doing a job may be different in different countries, as may the recommended lighting standards. Nonetheless where the working method is the same as that being investigated, the actual lighting techniques suggested should be valid even if some modification in quantity is required.

An examination of such studies will show that one of the major problems in industry is getting the light in the right place. This problem usually arises when the general lighting is designed without any thought to the obstruction and shading of tasks produced by large machinery and overhead conveyors. This type of problem can usually be solved by simply identifying what has to be seen and where it is. With some thought, suitable lighting can then be arranged using a rough classification of the task difficulty (see Table 6.5). Ideally it is necessary to establish what the lighting should be by thorough field or laboratory studies. This is rarely possible but some *ad hoc* tests with any proposed lighting equipment often are, and these can be very informative.

Table 6.8. *Sources of recommendations for lighting in specific areas*

Area of application	Reference
Art galleries and museums	Individual Illuminating Engineering Society Guides available from Chartered Institution of Building Services, Delta House, 222 Balham High Road, London SW12 9BS
Building and civil engineering sites	
Hospitals and health care buildings	
Lecture theatres	
Libraries	
Shipbuilding and ship-repair yards	
Sports areas	
Areas containing visual display units	
Glass making	*Light and Lighting*, April 1973
Printing	*Light and Lighting*, January 1973
Bakeries	In *Illuminating Engineering Society Lighting Handbook*, 5th edn, available from Illuminating Engineering Society of North America, 345 East 47th Street, New York, NY 10017, USA
Bottling and canning	
Car assembly	
Clothing manufacture	
Confectionery production	
Dry-cleaning	
Farms	
Flour mills	
Foundries	
Fruit and vegetable packaging	
Metal machining	
Milk bottling	
Petro-chemicals	
Printing and graphic arts	
Sheet metal shops	
Shoe manufacture	
Steel mills	
Textiles	
Tyre manufacture	
Logging and sawmills	In *Journal of Illuminating Engineering Society* (1977) **6**, 67

Another problem which often occurs is veiling reflections masking what is to be seen. It is a frequent event in industry where polished metal scales are often found and Perspex sheets are used to protect control panels, and is becoming common in offices with the arrival of visual display units. Fig. 6.24 shows a vernier scale rendered unusable by veiling reflections. Looking at the scale from another position would make it visible but it is hardly good practice to make the man fit the machine. The problem can be overcome either by using matt materials, which is not always practicable, or by rearranging the lighting in relation to the task and the direction from which it is viewed. In general such veiling reflections will be reduced if small, fixed, high-luminance luminaires are avoided and there is a high degree of inter-reflection in the interior.

Another feature that often causes difficulty is shadowing, either because it means an important part of the work is in relative darkness or because the shadows are distracting or confusing. Drawing boards are an example where both problems can occur (see Fig. 6.25).

There also exists a whole series of tasks, usually classified as visual inspection, the visual features of which may not be easy to identify. Typical inspection tasks involve looking not for a single defect but for several different ones. For example, in inspecting glassware the potential defects are cracks, slips, globules of glass stuck to the sides and spikes of glass on the base. All these are seen by an experienced inspector in a few seconds. Visual inspection is an area that the ergonomist may well be called upon to investigate. The essence of the problem is

Fig. 6.24. The effect of veiling reflections on a vernier scale.

to identify the cues used as a basis for inspection and then to make these cues as easy to see as possible. This may involve separating the lighting of the inspection area from the general room lighting by constructing an inspection booth (Fig. 6.26). There are many different techniques for easing the difficulties of visual inspection. Bellchambers and Phillipson (1962) discuss a range of solutions but

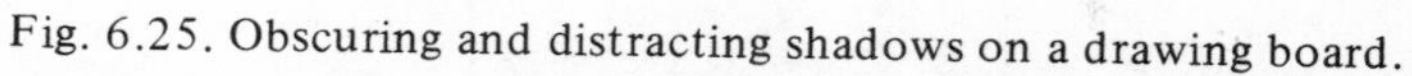

Fig. 6.25. Obscuring and distracting shadows on a drawing board.

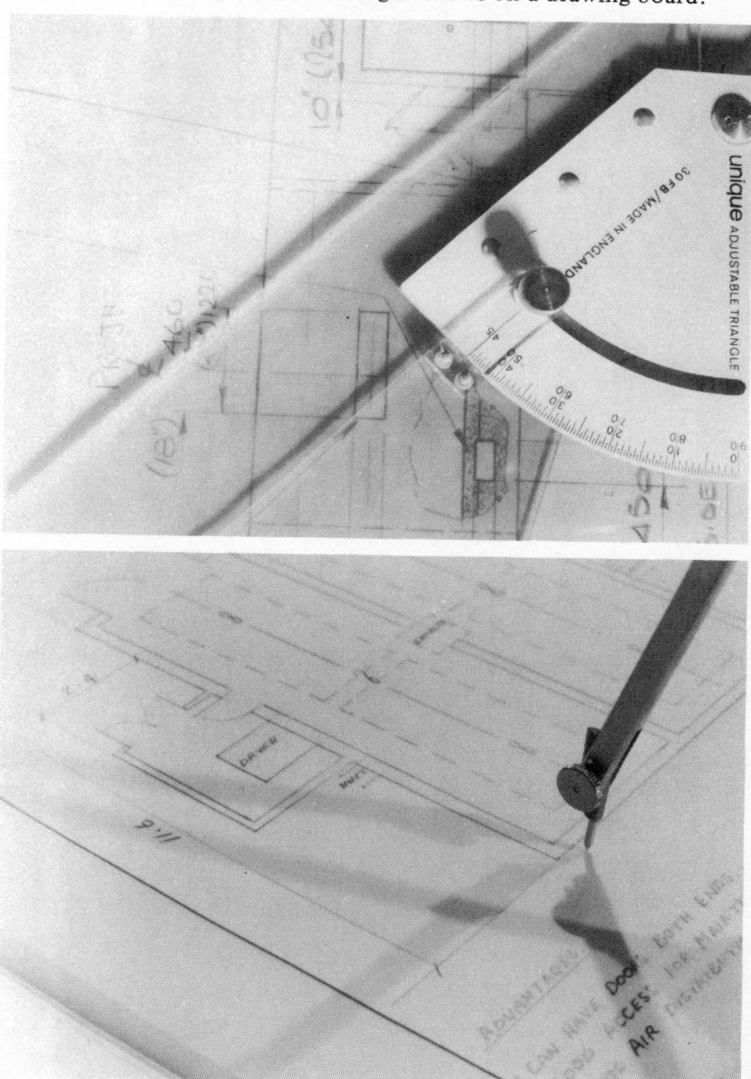

Fig. 6.26. Some of the techniques available for visual inspection. (From Henderson and Marsden (1972). Reproduced by kind permission of Thorn Lighting Ltd.)

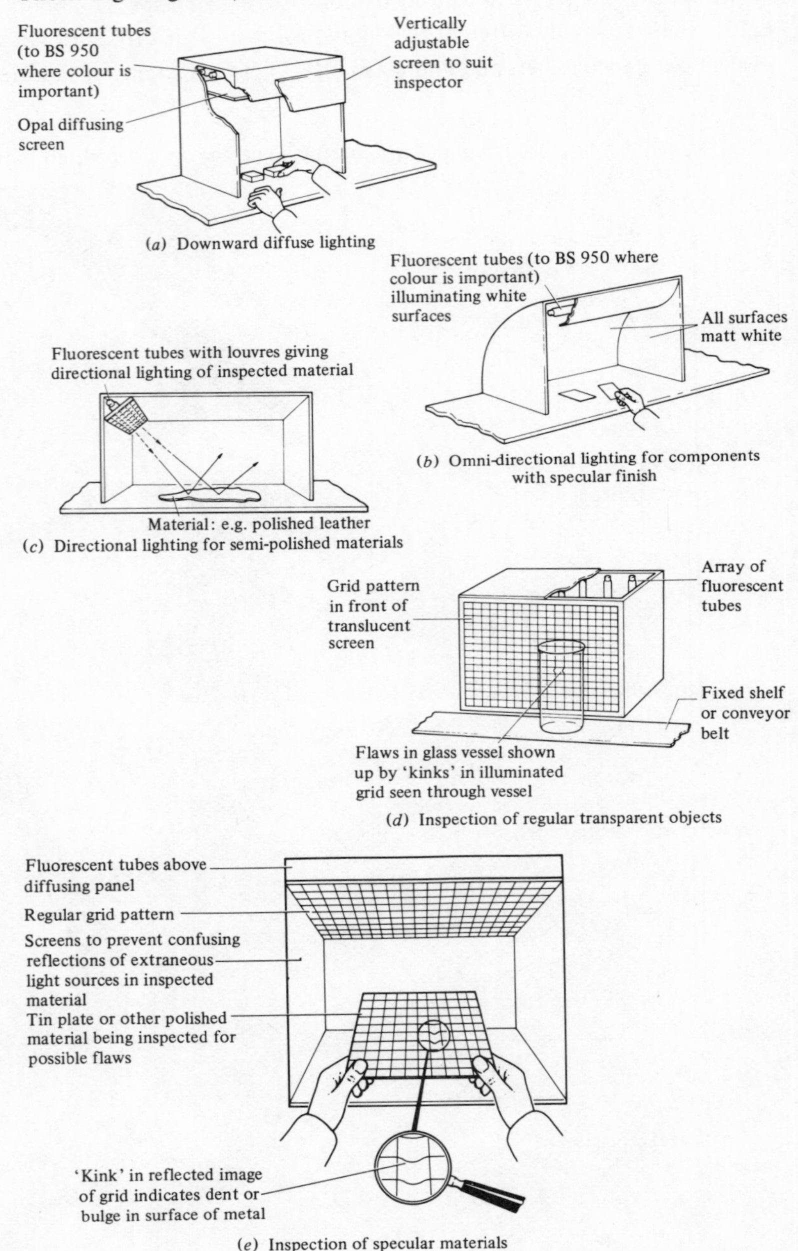

there is no guarantee that even their wide experience will embrace them all. Fig. 6.26 illustrates some of the methods and Table 6.9 classifies suitable approaches to different sorts of task. A special case is inspection involving accurate colour matching or discrimination. For this situation light sources as specified in British Standard BS 950 part 1 should be used. It is important to ensure that light from other sources does not reach the inspection area, which suggests an inspection booth, and that any surrounding surfaces have a neutral matt finish.

Although this is relevant for colour appraisal work, each visual inspection method has to be considered in its own right. There is no certainty that recommended general lighting conditions are relevant for inspection lighting since they may not be of much use in identifying the defects being sought. Laboratory studies based on a field analysis of the tasks are a reasonable way of investigating the problem.

All the above application areas have been concerned with interior lighting. However, exterior lighting should not be forgotten. Roadway lighting for safety, and security lighting of premises would seem to offer opportunities for ergonomic investigation (Janoff, Freedman and Koth, 1977). Special situations such as emergency lighting are also possible fields of investigation. Certainly little is known about emergency lighting (Simmons, 1975). Whatever the ergonomist is investigating he has a choice of method to use. Whichever method is chosen, a willingness to listen and observe will be very useful aids to understanding.

6.9 Epilogue

This chapter has attempted to present what is known about the interaction between lighting and people at work, to suggest some methods by which this relationship can be investigated, and to indicate those areas of application where the ergonomist is most likely to be involved. The tables provide and the following bibliography gives references to other sources of information on a wide range of topics which space does not allow to be discussed in detail. By following up whichever of these is relevant it should be possible to gather more detailed and authoritative advice. One last point is important. Lighting is about more than making things easy to see. It is about elegance and sometimes beauty. There is an aesthetic side to lighting as well as a functional side. If the latter can be provided without destroying the former then everyone will be the richer.

Table 6.9. *Classification of visual tasks and lighting techniques*

Classification of visual task: general characteristics	Example: Description	Example: Lighting requirements	Lighting technique: Luminaire type (see p. 360)	Lighting technique: Locate luminance (see p. 360)
		Part I: Flat surfaces		
Opaque materials				
1. Diffuse detail and background				
(*a*) Unbroken surface	Newspaper proof-reading	High visibility with comfort	2 or 3	To prevent direct glare and shadows (A)
(*b*) Broken surface	Scratch on unglazed tile	To emphasise surface break	1	To direct light obliquely to surface (C)
2. Specular detail and background				
(*a*) Unbroken surface	Dent, warps, uneven surface	Emphasise unevenness	5	So that image of source and pattern is reflected to eye (D)
(*b*) Broken surface	Scratch, scribe, engraving, punch marks	Create contrast of cut against specular surface	3 or 4 or 5 when not practical to orient task	So detail appears bright against a dark background So that image of source is reflected to eye and break appears dark (D)
(*c*) Specular coating over specular background	Inspection of finish plating over under-plating	To show up uncovered spots	4 with colour of source selected to create maximum colour contrast between two coatings	For reflection of source image toward the eye (D)
3. Combined specular and diffuse surfaces				
(*a*) Specular detail on diffuse, light background	Shiny ink or pencil marks on dull paper	To produce maximum contrast without veiling reflections	3 or 4	So direction of reflected light does not coincide with angle of view (A)
(*b*) Specular detail on diffuse dark background	Punch or scribe marks on dull metal	To create bright reflection from detail	2 or 3	So direction of reflected light from detail coincides with angle of view (B)
(*c*) Diffuse detail on specular light background	Gradations on a steel scale	To create a uniform, low-brightness reflection from specular background	3 or 4	So reflected image of source coincides with angle of view (B or D)

Classification of visual task: general characteristics	Example		Lighting technique	
	Description	Lighting requirements	Luminaire type (see p. 360)	Locate luminaire (see p. 360)
Part I: Flat surfaces (contd)				
(*d*) Diffuse detail on specular dark background	Wax marks on car body	To produce high brightness of detail against dark background	2 or 3	So direction of reflected light does not coincide with angle of view (A)
Translucent materials				
1. With diffuse surface	Frosted or etched glass or plastic, lightweight fabrics, hosiery	Maximum visibility of surface detail	—	Treat as opaque, diffuse surface
		Maximum visibility of detail within material	2, 3 or 4	Transilluminate behind material (E)
2. With specular surface	Scratch on opal glass or plastic	Maximum visibility of surface detail	—	Treat as opaque, specular surface
		Maximum visibility of detail within material	2, 3 or 4	Transilluminate behind material (E)
Transparent materials				
Clear material with specular surface	Plate glass	To produce visibility of details within material (such as bubbles) and details on surface (such as scratches)	1 and 5	Transparent material should move in front of 5 then in front of black background with 1 directed obliquely. 1 should be directed to prevent reflected glare
Transparent over opaque materials				
1. Transparent material over diffuse background	Instrument panel	Maximum visibility of scale and pointer without veiling reflections	1	So reflection of source does not coincide with angle of view (A)

(*Continued on next page*)

Table 6.9 (*continued*)

Classification of visual task: general characteristics	Example		Lighting technique	
	Description	Lighting requirements	Luminaire type (see p. 360)	Locate luminaire (see p. 360)
Part II: Three-dimensional objects				
	Varnished desk-top	Maximum visibility of detail on or in transparent coating or on diffuse background; emphasis of uneven surface	5	So that image of source and pattern is reflected to the eye (D)
2. Transparent material over a specular background	Glass mirror	Maximum visibility of detail on or in transparent material	1	So reflection of source does not coincide with angle of view; mirror should reflect a black background (A)
		Maximum visibility of detail on specular background	5	So that image of source and pattern is reflected to the eye (D)
Opaque materials				
1. Diffuse detail and background	Dirt on a casting or blow-holes in a casting	To emphasise detail	2 or 3	To prevent direct glare and shadows (A)
			1	In relation to task to emphasise detail by means of highlight and shadow (B or C)
			2 or 3 as a u.v. source when object has a fluorescent coating	To direct u.v. radiation to all points to be checked
2. Specular detail and background				
(*a*) Detail on the surface	Dent on silverware	To emphasise surface unevenness	5	To reflect image of source to eye (D)
	Inspection of finish plating over under-plating	To show up areas not properly plated	4 plus proper colour	To reflect image of source to eye (D)
(*b*) Detail in the surface	Scratch on a watch case	To emphasise surface break	4	To reflect image of source to eye (D)

Classification of visual task: general characteristics	Example		Lighting technique	
	Description	Lighting requirements	Luminaire type (see p. 360)	Locate luminaire (see p. 360)
Part II: Three-dimensional objects (contd)				
3. Combined specular and diffuse				
(*a*) Specular detail on diffuse background	Scribe mark on casting	To make line glitter against dull background	2 or 3	In relation to task for best visibility; adjustable equipment often helpful Overhead to reflect image of source to eye (B or D)
(*b*) Diffuse detail on specular background	Micrometer scale	To create luminous background against which scale markings can be seen in high contrast	3 or 4	With axis normal to axis of micrometer
Translucent materials				
1. Diffuse surface	Lamp-shade	To show imperfections in material	2	Behind or within for trans-illumination (E)
2. Specular surface	Glass enclosing globe	To emphasise surface irregularities	5	Overhead to reflect image of source to eye (D)
		To check homogeneity	2	Behind or within for trans-illumination
Transparent materials				
Clear material with specular surface	Bottles, glassware: empty or filled with clear liquid	To emphasise surface irregularities	1	To be directed obliquely to objects
		To emphasise cracks, chips and foreign particles	4 or 5	Behind for transillumination; motion of objects is helpful (E)

After IES (1972). Reproduced by kind permission of the Illuminating Engineering Society of North America.

Key to location of luminaires (see Table 6.9)

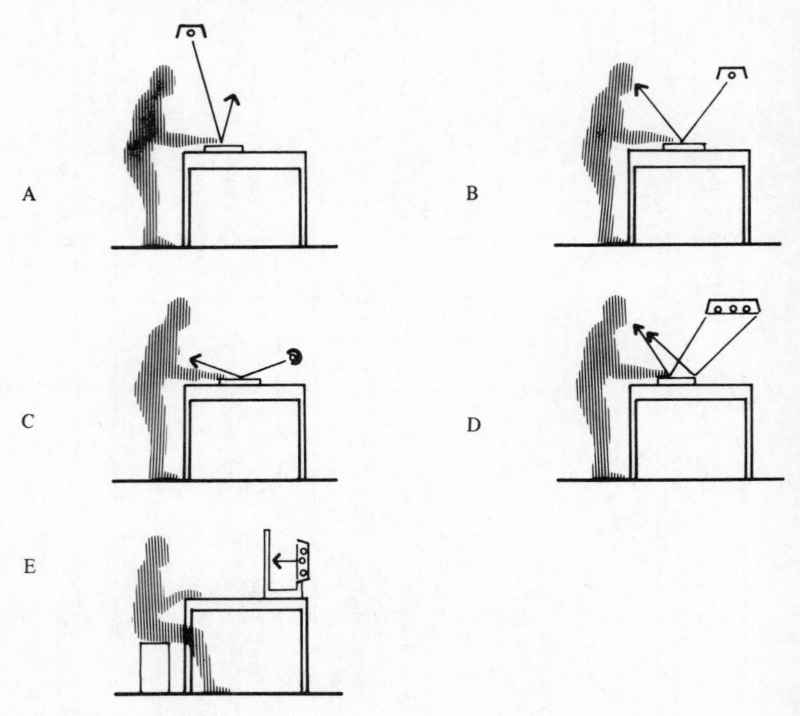

Key to luminaire types (see Table 6.9)

1. Narrow, concentrated beam e.g. spotlights, fluorescent luminaires with narrow light distribution
2. Widespread beam from high-luminance, small-area luminaires, e.g. incandescent or high-pressure discharge lamps in diffusing luminaires
3. Widespread beam from moderate luminance, large-area luminaires, e.g. fluorescent luminaires with wide light distribution
4. Uniform moderate luminance luminaire (variation in luminance <2 : 1) e.g. fluorescent lamps behind a diffusing panel
5. Uniform moderate luminance luminaire (variation in luminance <2 : 1) with pattern superimposed (see Fig. 5.26*d*)

References

American Conference of Governmental Industrial Hygienists (1981) *Threshold Values for Physical Agents*. Cincinatti: ACGIH.

Baumgardt, E. and Le Grand, Y. (1956) Sur la fatigue visuelle. *Cahiers du Centre Scientifique et Technique du Batiment*, **27**, 230–6.

Bedwell, C.H. (1972) The eye, vision and visual discomfort. *Lighting Research and Technology*, **4**, 151–8.

Bellchambers, H.E. and Phillipson, S.M. (1962) Lighting for inspection. *Transactions of the Illuminating Engineering Society (London)*, **27**, 71–87.

Berry, R.N. (1948) Quantitative relations among vernier, real depth and stereoscopic depth acuities. *Journal of Experimental Psychology*, **38**, 707–21.

Beutell, A.W. (1934) An analytical basis for a lighting code. *Illuminating Engineering (London)*, **27**, 5–16.

Blackwell, H.R. (1946) Contrast thresholds of the human eye. *Journal of the Optical Society of America*, **36**, 624–43.

Blackwell, H.R. (1959) Specification of interior illumination levels. *Illuminating Engineering*, **54**, 317–53.

Blackwell, H.R. (1980) *Analysis of 20 Sets of Visual Performance Data in Terms of a Quantitative Method for the Prediction of Visual Performance Potential as a Function of Reference Luminance.* Illuminating Engineering Research Institute Report. New York: IERI.

Bodmann, H.W. (1962) Illumination levels and visual performance. *International Lighting Review*, **13**, 41-7.

Boyce, P.R. (1973) Age, illuminance, visual performance and preference. *Lighting Research and Technology*, **5**, 125–39.

Boyce, P.R. and Simons, R.H. (1977) Hue discrimination and light sources. *Lighting Research and Technology*, **9**, 125–41.

Boyce, P.R. and Slater, A.I. (1981) The application of CRF to office lighting design. *Lighting Research and Technology*, **13**, 65–79.

Boynton, R.M. and Boss, D.E. (1971) The effect of background luminance and contrast upon visual search performance. *Illuminating Engineering*, **66**, 173.

Brozek, J. and Simonson, E.L. (1952) Visual performance and fatigue under conditions of varied illumination. *American Journal of Ophthalmology*, **35**, 33–46.

Campbell, F.W. and Robson, J.G. (1968) Application of fourier analysis to the visibility of gratings. *Journal of Physiology (London)*, **197**, 551–66.

Carlsson, L., Knave, B. and Wibom, R. (1977) Industrial lighting. VI. Workplace lighting in casting bays and foundries: manual teeming. *Ljuskultur*, **49**, 4–11.

Carterette, E.C. and Friedman, H.P. (1975) *The Handbook of Perception: Seeing.* London: Academic Press.

Christie, A.W. and Fisher, A.J. (1966) The effect of glare from street lighting lanterns on the vision of drivers of different ages. *Transactions of the Illuminating Engineering Society (London)*, **31**, 93–107.

CIBS (1977) *IES Code for Interior Lighting.* London: Chartered Institution of Building Services.

Collins, J.B. and Hopkinson, R.G. (1957) Intermittent light stimulation and flicker sensation. *Ergonomics*, **1**, 61–76.

Collins, J.B. and Pruen, B. (1962) Perception time and visual fatigue. *Ergonomics*, **5**, 533–8.

CIE (1970) *Principles of Light Measurement.* CIE Publication 18. Paris: Commission Internationale de l'Eclairage.

CIE (1971) *Colourimetry*. CIE Publication 15. Paris: Commission Internationale de l'Eclairage.

CIE (1972) *A Unified Framework of Methods for Evaluating the Visual Performance Aspects of Lighting.* CIE Publication 19. Paris: Commission Internationale de l'Eclairage.

CIE (1974) *Method of Measuring and Specifying Colour Rendering Properties of Light Sources*. CIE Publication 13.2. Paris: Commission Internationale de l'Eclairage.

CIE (1978*a*) *Light as a True Visual Quantity.* CIE Publication 41. Paris: Commission Internationale de l'Eclairage.

CIE (1978*b*) *Recommendations on Uniform Colour Spaces, Colour Difference Equations and Psychometric Colour Terms.* CIE Publication 15, Supplement 2. Paris: Commission Internationale de l'Eclairage.

CIE (1981) *An Analytical Model for Describing the Influence of Lighting Parameters upon Visual Performance.* CIE Publication 19/2. Paris: Commission Internationale de l'Eclairage.

Cornsweet, T.N. (1970) *Visual Perception.* London: Academic Press.

Davson, H. (1962) *The Eye.* London: Academic Press.

Eastmann, A.A. (1968) Colour contrast versus luminance contrast. *Illuminating Engineering,* **63**, 613–20.

Faulkner, T.W. and Murphy, T.J. (1973) Lighting for difficult visual tasks. *Human Factors*, **15**, 149–62.

Foxell, C.A.P. and Stevens, W.R. (1955) Measurements of visual acuity. *British Journal of Ophthalmology*, **39**, 513–33.

Fry, G.A. (1954) A re-evaluation of the scattering theory of glare. *Illuminating Engineering*, **49**, 98–102.

Fry, G.A. and King, V.M. (1975) The pupillary response and discomfort glare. *Journal of the Illuminating Engineering Society*, **4**, 307–24.

Gallagher, V.P. (1976) A visibility metric for the safe lighting of city streets. *Journal of the Illuminating Engineering Society*, **5**, 85–91.

George, W.Z. (1972) An investigation into a possible physiological effect of high levels of illumination. Postgraduate thesis, Loughborough University.

Grandjean, E. and Perret, E. (1961) Effects of pupil aperture and of the time of exposure on the fatigue induced variations of flicker fusion frequency. *Ergonomics*, **4**, 17–23.

Hartnett, O.M. and Murrell, K.F.H. (1973) Some problems of field research. *Applied Ergonomics,* **4**, 219–21.

Hecht, S. and Mintz, E.U. (1939) The visibility of single lines at various illuminations and the retinal basis of visual resolution. *Journal of General Physiology,* **22**, 593–612.

Henderson, S.T. and Marsden, A.M. (1972) *Lamps and Lighting.* London: Edward Arnold.

Hopkinson, R.G., Petherbridge, P. and Kay, J.D. (1966) *Daylighting.* London: Heinemann.

Hultgren, G.V. and Knave, B. (1974) Discomfort glare and disturbances from light reflections in an office landscape with CRT display terminals. *Applied Ergonomics,* **5**, 2.

IES (1981) *IES Lighting Handbook,* 6th edn. New York: Illuminating Engineering Society of North America.

Janoff, M.S., Freedman, M. and Koth, B. (1977) Driver and pedestrian behaviour: the effect of specialised crosswalk illumination. *Journal of the Illuminating Engineering Society,* **6**, 202–8.

Jay, P.A. (1973) The theory of practice in lighting engineering. *Light and Lighting,* **66**, 303–6.

Jewess, B.W. (1976) Portable illuminance meters reviewed. *Light and Lighting,* **69**, 28–31.

Jewess, B.W. (1978) Medical uses of radiation from lamps. In *Proceedings of the IES National Lighting Conference 1978.* London: IES.

Johnson, C.A., Keltner, J.L. and Balestrery, F. (1978) Effects of target size and eccentricity on visual detection and resolution. *Vision Research,* **18**, 1217–22.

Johnston, A.W., Cole, B.L., Jacobs, R.J. and Gibson, A.J. (1976) Visibility of traffic control devices; catering for the real observer. *Ergonomics,* **19**, 591–609.

Kehliborou, T. and Aronasou, B. (1973) On the choice of illumination for increasing the contrast of colours in the grading of tobacco leaves. *Colour, 73.* (London: Hilger).

Kelly, D.A. (1961) Visual responses to time dependent stimuli. I. Amplitude sensitivity measurements. *Journal of the Optical Society of America,* **51**, 422–9.

Lamar, E.S., Hecht, S., Schlaer, S. and Hendly, C.D. (1947) Size, shape and contrast in the detection of targets by daylight vision. I. Data and analytical description. *Journal of the Optical Society of America,* **37**, 531–45.

Luckiesh, M. (1944) *Light, Vision and Seeing.* New York: Van Nostrand.

Lynes, J.A. (1968) *Principles of Natural Lighting.* Amsterdam: Elsevier.

Lynes, S.L. (1981) *Handbook of Industrial Lighting.* London: Butterworth.

Lythgoe, R.J. (1932) *The Measurement of Visual Acuity.* MRC Special Report 173. London: HMSO.

McAdam, D.L. (1942) Visual sensitivities to colour differences in daylight. *Journal of the Optical Society of America,* **32**, 247–74.

McNelis, J.A. and Guth, S.K. (1969) Visual performance: further data on complex test objects. *Illuminating Engineering,* **64**, 99–102.

Milova, A. (1971) The influence of light of different spectral composition on visual performance. In *CIE 17th Session Proceedings, Barcelona.*

Muck, E. and Bodmann, H.W. (1961) Der Bedeutung des Beleuchtungsniveaus bei praktischer Sehtätigkeit. *Lichttechnik,* **13**, 502–7.

NIOSH (1975) *The Occupational Safety and Health Effects Associated with Reduced Levels of Illumination.* New Publication (NIOSH) 75–142. Cincinatti: National Institute of Occupational Safety and Health.

O'Donnell, R.M., Critchley, S.H. and Chapman, R. (1976) A sector disc visibility comparator. *Lighting Research and Technology,* **8**, 113–14.

Ogle, K.N. (1961) Foveal contrast thresholds with blurring of retinal image and increasing size of test stimulus. *Journal of the Optical Society of America,* **51**, 862–9.

Padgham, C.A. and Saunders, J.E. (1975) *The Perception of Light and Colour.* London: Bell and Hyman.

Reitmaier, J. (1979) Some effects of veiling reflections on papers. *Lighting Research and Technology,* **11**, 204–9.

Rowlands, E., Waters, I., Loe, D.L. and Hopkinson, R.G. (1973) *Visual Performance in Illumination of Different Spectral Quality.* UCERG Report. London: University College.

Salaman, M. (1929) *Some Experiments on Peripheral Vision.* MRC Special Report 136. London: HMSO.

Schlaer, S. (1937) The relation between visual acuity and illumination. *Journal of General Physiology,* **21**, 165–88.

Schnore, M.M. (1959) Individual patterns of physiological activity as a function of task differences and degree of arousal. *Journal of Experimental Psychology,* **58**, 117–28.

Simmons, R.C. (1975) Illuminance, diversity and disability glare in emergency lighting. *Lighting Research and Technology,* **7**, 125–32.

Simonson, E. and Brozek, J. (1948) Effect of illumination level on performance and fatigue. *Journal of the Optical Society of America,* **38**, 384–97.

Sliney, D.H. (1972) Non-ionising radiation. In *Industrial Environmental Health,* ed. L.V. Cralley, pp. 171–241. London: Academic Press.

Smith, S.W. and Rea, M. (1980) Relationship between office task performance and rating of feelings and task evaluation under different light sources and levels. In *CIE Publication 50, Proceedings of 19th Session.* pp. 207-11. Paris: Commission Internationale de l'Eclairage.

Stenzel, A.E. and Sommer, J. (1969) The effect of illumination on tasks which are largely independent of vision. *Lichttechnik,* **21**, 143–6.

Stone, P.T. and Groves, S.D.P. (1968) Discomfort glare and visual performance. *Transactions of the Illuminating Engineering Society (London),* **33**, 9–15.

Stratton, G.M. (1900) A new determination of the minimum visible and its bearing of localisation and binocular depth. *Psychological Review,* **7**, 429.

Stiles, W.S. (1929) The effect of glare on the brightness difference threshold. *Proceedings of the Royal Society of London, Series B,* **104**, 322.

Taylor, J.H. (1969) *Factors Underlying Visual Search Performance.* Scripps Institute of Oceanography Report 69-22.

Urwick, L. and Brech, E.F.L. (1965) *The Making of Scientific Management,* vol. 3, *The Hawthorne Investigations.* London: Pitman.

Weale, R.A. (1963) *The Ageing Eye.* London: H.K. Lewis.

Weston, H.C. (1945) *The Relation between Illumination and Visual Performance.* Industrial Health Research Board Report 87. London: HMSO.
Wyszecki, G. and Stiles, W.S. (1967) *Colour Science.* New York: Wiley.

7 HEARING AND NOISE

D. Roy Davies and Dylan M. Jones

Noise is usually defined as unwanted sound, and there seem to be certain stimulus characteristics of noise that contribute particularly to its unwantedness (Kryter, 1973). These include the masking of wanted sounds, especially speech, excessive loudness, a general quality of 'bothersomeness', and the capacity of certain noises to produce startle responses on the one hand and auditory fatigue – together with damage to the auditory system – on the other.

Three main behavioural consequences of exposure to noise may be distinguished. First, a temporary, or, with repeated exposures over a prolonged period, a permanent and irreversible hearing loss may result; secondly, noise may produce feelings of annoyance and irritation; thirdly, noise may affect the efficiency with which a variety of different tasks is performed, usually impairing but sometimes also enhancing efficiency. These three effects are not necessarily related. Noise which produces hearing loss may not impair efficiency; noise which improves efficiency may nevertheless be annoying; and noise may be found acceptable even though it produces hearing loss (Broadbent, 1970).

7.1 Sound

Sound is propagated through an elastic medium such as air, transmission being effected by molecular collision which gives rise to oscillatory movements around the 'stable' positions of the molecules. An initial disturbance from a vibrating object will force air molecules surrounding it to move. In this way successive collision causes the motion of the object to be transmitted as a sound wave. Molecules are squeezed closer together then pulled further apart than normal. This results in alternate regions of molecular compression and rarefaction. Sound is thus propagated by a pressure wave, moving outwards from the vibrating body. If the motion of molecules is observed at any given point in air, they will bunch and spread as time passes. Molecules do not advance with the wave but transmit their motion to adjacent molecules. In simple terms the regularly occurring zones of compression and rarefaction can be represented by a cyclic or sinusoidal variation in pressure around ambient air pressure (see Fig. 7.1). Such a simple waveform might arise from a properly struck tuning fork and can be described by specifying three parameters: (1) the *frequency* or the number of complete variations or cycles that occur in one second, expressed in hertz (Hz); (2) the *amplitude,* which is the amount of pressure variation about the normal; (3) the *phase*, the portion of the cycle through which the wave has progressed relative to some fixed starting point. Of these parameters frequency and amplitude are particularly relevant to the study of noise and need some further elucidation.

Frequency

Sounds which are heard in everyday life are usually composed of pressure variations at a number of different frequencies. The range of frequencies to which the ear is sensitive is known as the audio-frequency range and may encompass frequencies in the range 20 to 20 000 Hz. Frequencies below this range are called infrasounds and those above it are called ultrasounds. Pure tones, like those produced by a struck tuning fork, are rare. However, everyday sounds, though more complex than pure tones, do demonstrate periodicity. These periodic sounds share with a pure tone the ability to evoke the sensation of pitch. Pitch is the subjective property of a tone and, for sine waves at least, pitch is

Fig. 7.1. Pressure variations with distance for a sound wave of a single frequency showing areas of compression and rarefaction (above) and a conventional sinusoidal display (below).

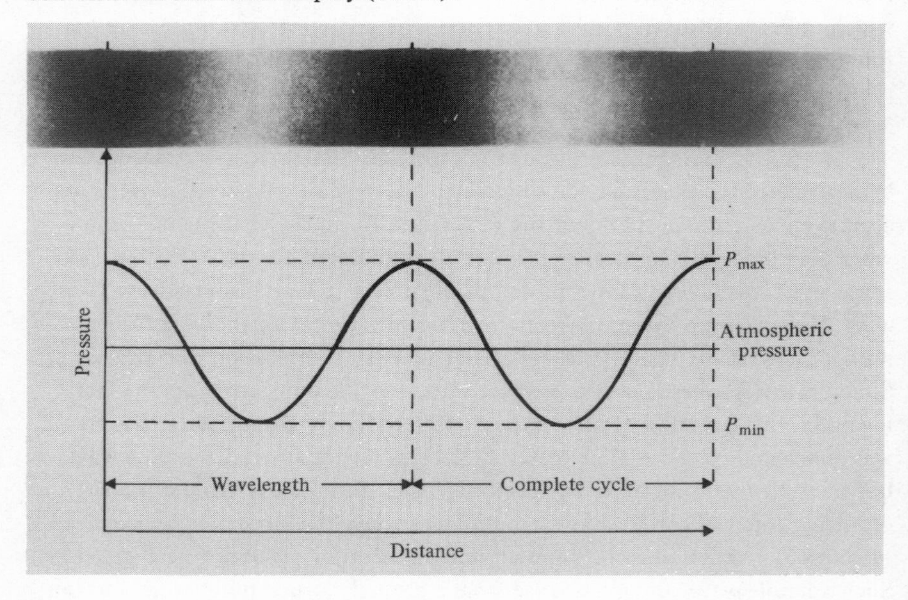

closely related to frequency: the higher the frequency, the higher the pitch. The pitch of a sound containing complex tones is taken to be the frequency of a sine wave which matches it in pitch. A periodic waveform is one that repeats itself after some interval of time. The rate of this repetition is the *fundamental frequency* of the waveform. Complex periodic waves are usually made up of energy at the fundamental frequency and at integral multiples of the fundamental, known as harmonics of that fundamental. However, a periodic waveform need not have energy at its fundamental; for example, a waveform containing sinusoids at 1000, 1020 and 1040 Hz is periodic, the fundamental being the greatest common denominator of these frequencies, i.e. 20 Hz.

For a full physical description of a complex tone, additional information about its constituent parts is needed. Within a complex sound there are stable and regularly repeated patterns. A measure of the relative contributions of each of these repeated patterns to the overall sound may be accomplished by the method of Fourier analysis. This enables the complex tone to be resolved into its constituent pure tones: the fundamental, and a number of higher tones known as harmonics, which are some multiple of the fundamental.

Amplitude

The ear can respond to an enormous range of sound intensities. The ratio of the power of the loudest sound we can hear to the power of the weakest sound we can detect is approximately 100 000 000 000 000 : 1. In order to encompass this range within a scale of manageable numbers a *logarithmic* scale expressing the *ratio* of the two intensities is used. One intensity is chosen as a reference level, I_0, which is expressed relative to the other intensity I_1. Initially, the bel was used to measure intensity, with one bel corresponding to a ratio of 10:1. However, the bel was found to be an impracticably large unit and therefore the decibel (dB) was introduced which could specify smaller ratios. Simply, the decibel represents one-tenth of the intensity ratio in bels: that is, the number of dB = $10 \log_{10} (I_1/I_0)$. The number of decibels does not represent the absolute intensity but the intensity or power *ratio*. The most commonly used reference level is an intensity of 10^{16} watts per square centimetre (W/cm^2). Thus it is usual to specify the level in relation to this reference intensity.

Instruments which are used to measure sound levels usually respond to changes in air pressure, their output being proportional to the amplitude of the sound. It is convenient to adapt the decibel notation to express pressure ratios in addition to intensity ratios. The reference sound pressure is expressed as 20 μPa (micropascals); a sound specified with this reference level is referred to as Sound Pressure Level (SPL). Intensity is proportional to the square of the pres-

Table 7.1. *Sample noise levels: typical operator/listener distances are employed unless otherwise specified*

	Overall level (dB(A))	Industrial and military	Community or outdoor	Home or indoor
130	*Uncomfortably loud*	Armoured personnel carrier, 123 dB(A)		
		Oxygen torch, 121 dB(A)		
120		Scraper-loader, 117 dB(A)		Rock-n-roll band, 108–114 dB(A)
		Compactor, 116 dB(A)	Chain-saw operator, 110 dB(A)	
		Riveting machine, 110 dB(A)	Snowmobile rider, 105 dB(A)	Symphony orchestra, 110 dB(A)
		Textile loom, 106 dB(A)	Jet flyover, 1000 ft, 103 dB(A)	
100	*Very loud*	Electric furnace area, 100 dB(A)	Power mower, 96 dB(A)	Inside subway car, 35 mph, 95 dB(A)
		Farm tractor, 98 dB(A)	Compressor, 20 ft, 94 dB(A)	
		Newspaper press, 97 dB(A)	Rock drill, 100 ft, 92 dB(A)	
90			Motorcycles, 25 ft, 90 dB(A)	Cockpit of light aircraft, 90 dB(A)
		Cockpit of prop aircraft, 88 dB(A)	Propeller aircraft flyover, 1000 ft, 88 dB(A)	Food blender, 88 dB(A)
		Milling machine, 85 dB(A)		
		Cotton spinning, 83 dB(A)	Diesel truck, 40 mph, 50 ft, 84 dB(A)	
		Lathe, 81 dB(A)		
80	*Moderately loud*		Passenger car, 65 mph, 25 ft, 77 dB(A)	Garbage disposal, 80 dB(A)
				Clothes washer, 78 dB(A)
				Living room music, 76 dB(A)
				Dishwasher, 75 dB(A)
70			Near freeway, auto traffic, 64 dB(A)	Television, 70 dB(A)
				Conversation, 65 dB(A)

60		Air-conditioning unit, 20 ft, 60 dB(A) Large transformer, 200 ft, 53 dB(A)
50	*Quiet*	Light traffic, 100 ft, 50 dB(A)
40	*Very quiet*	
30		
20	*Just audible threshold*	
10	*of hearing*	
0	*(1000–4000 Hz)*	

From Ward (1977).

sure and the relation between intensity (I) and amplitude or pressure (A) is as follows:

$$\text{Number of decibels} = 10 \log_{10} (I_1/I_2) = 20 \log_{10} (A_1/A_2)$$

Sample noise levels for different environments are shown in Table 7.1.

7.2 Hearing

The peripheral portion of the human auditory system, which is primarily concerned with the transformation of sound into neural impulses, can be conveniently divided into outer, middle and inner portions (see Fig. 7.2). The outer ear consists of the visible pinna which surrounds the opening of the external auditory canal and effectively channels sounds into it. The auditory canal itself is a tube about 2.5 cm in length and about 7 mm in diameter at the end of which is situated the eardrum or tympanic membrane. Together the pinna and the auditory canal play an influential role in the perception of sound. They aid auditory localisation (see below) and selectively promote certain audio-frequencies. The resonant properties of the auditory canal contribute much to the sensitivity of the ear in the frequency range 2000–6000 Hz, enhancing sound pressure levels by as much as 12 dB.

Sounds travelling down the auditory canal cause sympathetic vibration of the tympanic membrane, which forms the boundary between the outer and the middle ear. There are three ways in which vibrations may be conducted from the tympanic membrane to the receptor cells of the auditory nerve: (*a*) by air conduction, (*b*) by bone conduction and (*c*) by direct mechanical transmission. The last is by far the most important for normal hearing and is effected by three bones or ossicles – the malleus (hammer), the incus (anvil) and the stapes (stirrup) – which bridge the chamber of the middle ear and transmit vibrations from the tympanic membrane across the middle ear to the membrane of the oval window, an opening in the temporal bone which serves as the entrance to the inner ear. Impedance matching is accomplished through the mechanical advantage provided by the simple lever action of the ossicles and by the difference in effective area between the tympanic membrane and the membrane of the oval window. The pressure concentrated on the latter membrane is thus quite large, per unit of area about 25 to 30 times as great as that occurring at the eardrum (Geldard, 1972). The middle ear can also act as an attenuator, through the damping action of the muscles attached between the wall of the middle ear and the ossicles. The action of these muscles in response to intense sound is known as the acoustic reflex and provides as much as 20 dB attentuation. However, the latency of the acoustic reflex is between 60 and 120 milliseconds and little protection is thus provided against impulsive sounds such as gunshots.

Fig. 7.2. (*a*) The structure of the peripheral portion of the auditory system. Inserts (*b*) and (*c*) show progressively greater detail of a cross-section of the cochlea.

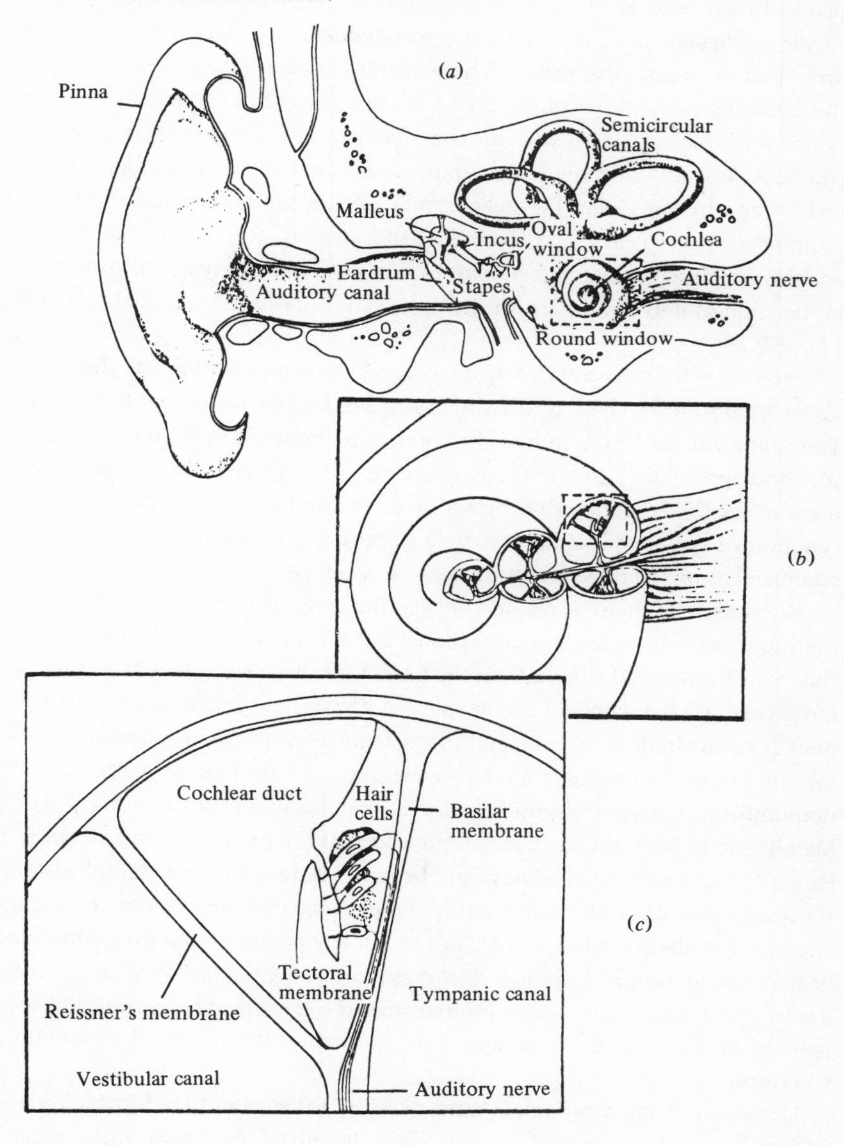

The inner ear is encased within the temporal bone and principally comprises the cochlea, a coiled structure resembling the shell of a snail and consisting of two and three-quarter turns of a gradually tapering cylinder (see Fig. 7.2*b*). The cochlea is divided along its length by two flexible membranes: Reissner's membrane and the basilar membrane. These membranes form three chambers within the cochlea – the scala vestibuli, the scala media or cochlear duct, and the scala tympani – which are filled with incompressible fluids. The two outer chambers (the scala vestibuli and the scala tympani) are connected at the apex of the coil by an opening called the helicotrema. Within the self-contained central chamber (the cochlear duct) is situated the organ of Corti, that part of the cochlea which senses pressure variations due to sound and relays information, in the form of nerve impulses, to the higher auditory centres of the brain (see Fig. 7.2*c*).

The oval window is situated at the base of the scala vestibuli and there is also a window at the base of the scala tympani, known as the round window. The round window, by virtue of the connection between the two chambers and the incompressibility of the fluids contained within them, reflects any movement in the fluids induced by the action of the stapes at the oval window. The function of the round window is thus to relieve pressure in the cochlea by complementing the movements of the oval window.

Movements of the oval window set the fluids in the cochlea into motion and the sound wave becomes a compressional wave in the fluids of the cochlea. In the case of sinusoidal stimulation, the basilar membrane displays two types of movement: (i) the whole of the membrane vibrates in sympathy with the frequency of the imposed sound, and (ii) differences in phase of the membrane's motion produce a travelling wave (von Békésy, 1960). This wave travels with non-uniform velocity from the base of the cochlea to its apex (see Fig. 7.3). Mainly due to inherent mechanical properties of the basilar membrane along its length (it has a narrow, rigid base and becomes wider and more flaccid toward the apex), sounds of different frequencies produce different points of maximum amplitude in the travelling wave. Thus frequency of sound is converted to place of maximal vibration. Although it is by no means the whole of the story, the basilar membrane behaves like a Fourier analyser, capable of separating component parts of complex pressure variations into characteristic places of maximum vibration.

Detection of this vibration is carried out by the organ of Corti, which is about 34 mm long and lies on, and runs the whole length of, the basilar membrane. In the organ of Corti are four parallel rows of hair cells (a total of about 17 000 in all) which are situated immediately below the relatively stiff tectoral membrane (Fig. 7.2*c*). When the basilar membrane is set into motion, shearing forces cause deformation in the hair cells and neural activity is thereby initiated. Below 5 kHz the firing of the nerve fibres linking the auditory receptors to the brain is phase-locked, that is, firings tend to occur at a particular point of the stimulating waveform so that a nerve may not fire on every cycle of the stimulus but at some regular rate which is a submultiple of its frequency.

Loudness

The frequency-to-place conversion outlined earlier is not a complete account of the transduction of auditory information. As loudness and pitch are considered in detail other subsidiary mechanisms will become apparent, although the process is not yet fully understood. Loudness is the subjective attribute of a sound whose primary determinant is intensity, but loudness also changes with

Fig. 7.3. Progress of a travelling wave along the cochlea. The time between successive curves is 0.05 msecond, each curve representing the wave at intervals following delivery of the acoustic signal with darkest curves being most recent. (Redrawn from Teas, Eldredge and Davis, 1962.)

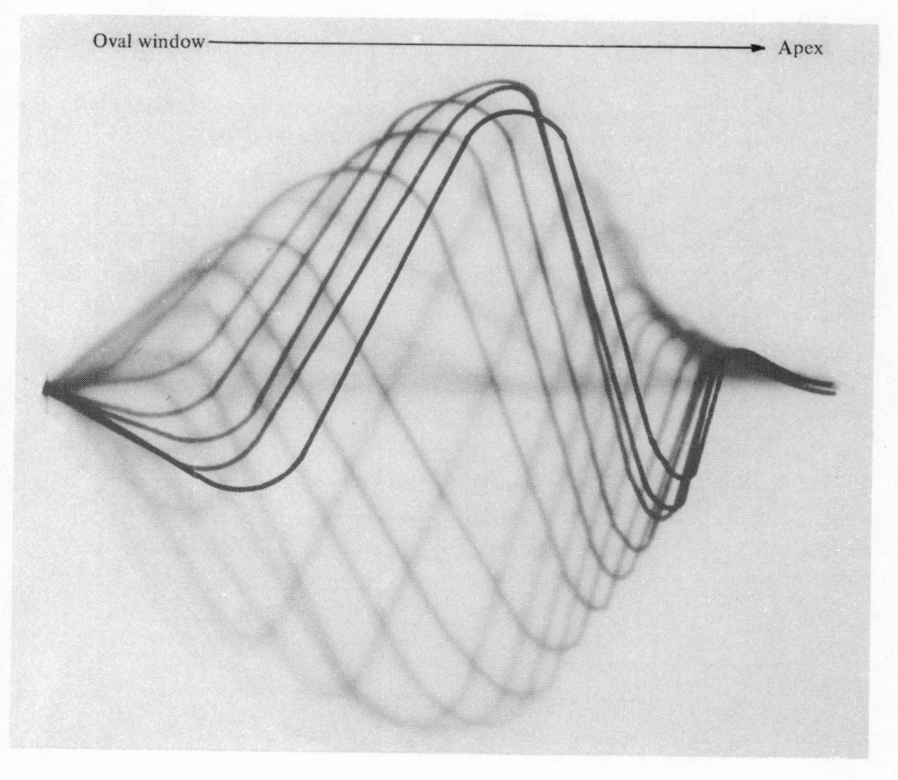

frequency. Several methods are available for the subjective scaling of loudness, all of them to some degree unsatisfactory (see Stevens, 1975, for a discussion). The scale most often employed is one derived from an estimation of loudness level, involving a matching in loudness of two tones of different frequency. A standard reference tone of 1000 Hz may be presented at 60 dB SPL; the subject is then presented with a tone of a different frequency and asked to adjust its intensity until it sounds as loud as the standard reference tone. Frequencies above and below the standard may be compared in this way, giving an overall picture of the perception of loudness as a function of frequency. Thus the loudness level of a tone (measured in phons) is that SPL at which it sounds as loud as a standard 1000-Hz tone. Psychophysical measurements of this sort yield a family of curves known as equiloudness contours; an example is given in Fig. 7.4 with typical data for loudness levels from 10 to 120 phons.

The 40-phon contour, for example, shows that at 30 Hz almost 80 dB SPL is required to produce a level of loudness equivalent to the standard. At high frequencies the discrepancy is less marked, so that at 8000 Hz nearly 50 dB SPL matches the standard in loudness. The contours become flatter with increases in intensity, a factor that can be incorporated into the design of sound level meters in order to simulate the ear's sensitivity to sound intensity at different frequencies.

Fig. 7.4. Equiloudness contours for a range of loudness levels. The dashed line represents the absolute threshold (minimum audible field). (Based on data of Robinson and Dadson, 1956.)

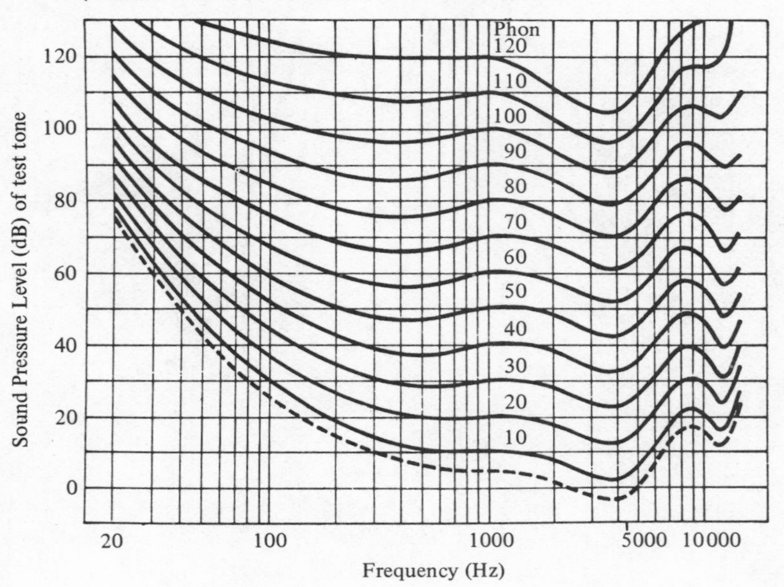

Fig. 7.4 also shows the maximum sensitivity of the ear to different frequencies. This is known as the absolute threshold for hearing, sounds on this contour being just barely detectable. The dashed line shows the sensitivity averaged over a large number of subjects. In practice there are marked variations due either to the method of measurement (of the order of 5–10 dB, arising from the use of either headphones or loudspeakers), or to individual differences (some 20 dB around the mean, especially at high frequencies). The reference SPL used in the decibel scale is a value close to the average absolute threshold. Occasionally investigators employ the actual threshold of an individual as the reference for the sound of interest, expressing the intensity of the sound as its sensation level.

Practical purposes require a less cumbersome method of estimating loudness than that required by the phon scale. Ideally such a scale should involve a simple calculation procedure, following the measurement of the sound spectrum, to yield a numerical estimate of the sound's loudness. For this we need to know, in addition to the way loudness varies with different frequencies, how loudness grows with intensity. Stevens (1957) advocated the sone scale of loudness based on the technique of magnitude estimation. The subject is presented with two tones at 1000 Hz and asked how many times louder one sound is than the other. Results using this procedure show that loudness increases with the cube root of intensity: the judgement of loudness, J, is related to the physical intensity, I, by a power law of the form $J = KI^{0.3}$. As a simple approximation, a 10-dB increase in intensity increases the loudness by a factor of 2. The unit of loudness is the sone. By definition, 1 sone is equal to the loudness of a 1000-Hz tone at 40 dB SPL; therefore the loudness of a 1000-Hz tone at 50 dB SPL is equivalent to 2 sones. The sone scale has been used by several workers in models which can be used to calculate the loudness of complex sounds (see, for example, Stevens, 1956; Kryter, 1973; Zwicker, 1958; Schultz, 1972).

Masking

Inability to hear sounds can result from the presence of other sounds; in such situations sounds are said to mask one another. Given that each sound produces a corresponding peak of activity of the basilar membrane (see above), it seems plausible that sounds which are close together in pitch will mask one another more readily than sounds further apart in pitch.

One way of investigating this phenomenon is to present a masker of fixed intensity and frequency, and on each occasion to adjust the intensity of test tones above and below the frequency of the masker until they are just audible. This procedure yields a series of masking curves each corresponding to a particular intensity of the masker (see Fig. 7.5*b*). The most notable feature of these curves is their asymmetry; the effect of the masker stretches well above its own frequency but the downward extent of masking is restricted. In addition, the amount of masking increases non-linearly on the high-frequency side, each increase in intensity producing progressively greater masking. The reason for this asymmetry is the pattern of excitation on the basilar membrane produced by the masker and the test tone. The envelope of excitation produced by a sound on

Fig. 7.5. Masking. (*a*) The envelope of vibration patterns on the basilar membrane to a 400 Hz tone. Note that the wave travels from the oval window to the apex, that is, from right to left on the illustration. (Based on data of von Békésy, 1960.) (*b*) Masking curves for a narrow band of noise centred at approximately 400 Hz. Each curve shows an idealised elevation in pure-tone threshold as a function of frequency for a particular level of masker. (Redrawn from Egan and Hake, 1950.)

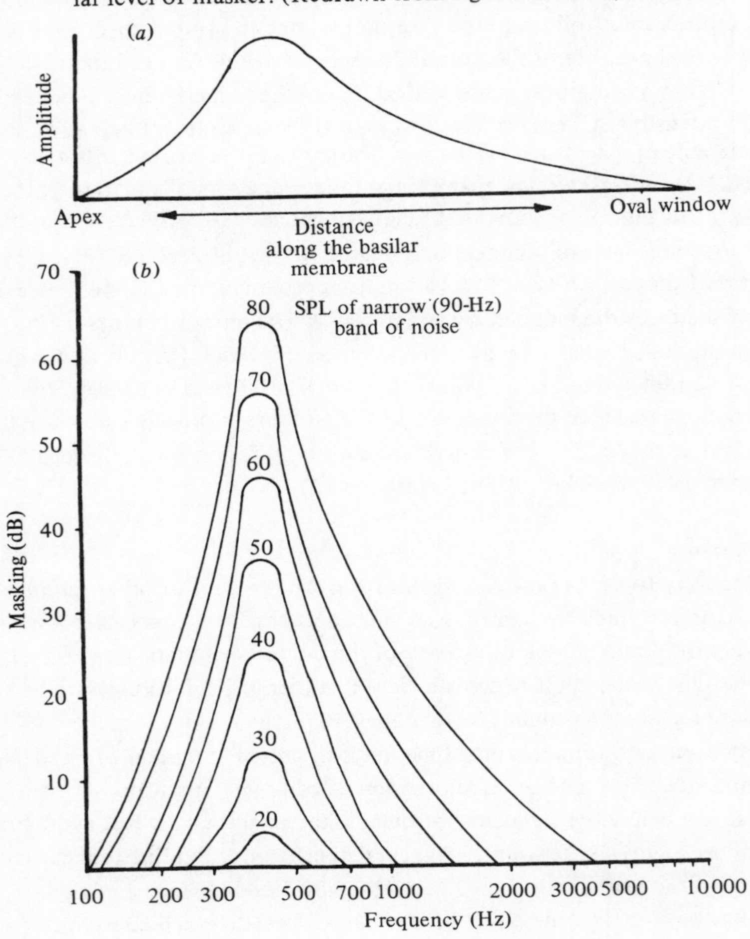

the basilar membrane is itself asymmetrical. A tone of, say, 400 Hz, will produce a peak of activity a little over half-way up the basilar membrane (see Fig. 7.5*a*), but the envelope will fall off much more slowly towards the *high*-frequency region (toward the oval window) than in the *low*-frequency region (toward the apex). This means that a tone introduced at a frequency higher than that of the masker will overlap more with the activity produced by the masker than will a tone of lower frequency.

This type of masking can be described as *simultaneous peripheral* masking. Another type of peripheral masking occurs when tone and masker are not presented simultaneously: perception of a tone may be influenced by a preceding masker and vice versa (see Elliot, 1971). Masking may also be *central*, when an elevation in the threshold of a signal presented to one ear is produced by a masker presented to the other ear. This effect is obviously not due to the coincidence of basilar membrane acitivity but is the result of the way in which information from both ears is combined centrally (see Zwislocki, 1971).

The critical band

Up to a critical point the loudness of two simultaneously presented tones of similar frequencies remains constant as their frequency separation increases (given a constant average frequency), but beyond this point loudness increases with frequency separation. Within a critical frequency region sound energies interact with each other and these frequencies are said to be in the same *critical band*. The loudness of a narrow band of sound can be seen to act in the same way. For a constant total energy level, the loudness of the sound remains constant as the breadth of the band of sound increases beyond a certain point. Beyond the critical separation the loudness increases as the band becomes wider. Critical bands are continuous rather than discrete and it is appropriate to speak of a critical band around any frequency of interest.

Within a critical band sound *energy* is combined, resulting in a marginal increase in loudness (e.g. the doubling of energy may result in a 20% increase in loudness). Outside the critical band it is *loudness* that is combined; for instance, if two sounds of the same loudness are presented this results in a doubling of loudness. The critical band is of obvious importance to any physical system of measurement (such as a sound level meter) which purports to give an indication of the loudness of a complex sound. In addition, the critical band appears to be important in the perception of the dissonance of sounds.

Pitch

Pitch is a subjective quality of a sound related to its repetition rate; for simple sinusoidal waveforms this corresponds to frequency, for complex tones to the fundamental frequency. The standard Western musical scale of pitch is logarithmic, each octave being twice the frequency of the previous octave. However, perceived pitch does not correspond to this scale of musical pitch. Raising or lowering the frequency by an octave does not double or halve the perceived pitch. The unit of pitch is the mel, and by definition a tone of a 1000 Hz (at 60 dB SPL) has a pitch of 1000 mels. As mentioned earlier, a change from high to low frequency is accompanied by a shift in the point of maximum vibration of the basilar membrane from near to the oval window toward the apex. There is good correspondence between the physical distance separating the positions of maximum vibration along the basilar membrane and the psychological distance between pitches as described by the mel scale (see Fig. 7.6). This further suggests that pitch is coded by place of excitation on the basilar membrane, but a number of phenomena are not accounted for by this explanation. Consider the case of the *missing fundamental.* If two tones of 1000 Hz and 1100 Hz are added, the overall pattern of the waveform will take the form of a regular rise and fall of the sound energy whose rate of repetition (100 Hz in this case) is equal to the differences between the frequencies of the two component sine waves. The change in amplitude of the wave envelope is heard as a *beat*. In other words, the beat frequency is the fundamental of the combined tone. There is no energy present at this fundamental frequency and thus according to place theory the basilar membrane is only stimulated at locations corresponding to the 1000 and 1100 Hz components. Nonetheless the beat is heard.

Two mechanisms can be adduced to explain the phenomenon of the missing

Fig. 7.6. Psychological distance between pitches (shown by the mel scale: continuous line) and the locus of maximum vibration on the basilar membrane (dashed line). (Based on Zwislocki, 1965.)

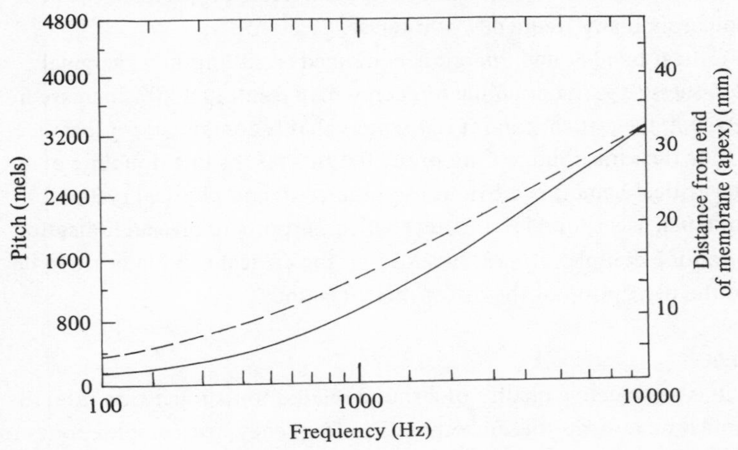

fundamental. The first, which is inconsistent with place theory, proposes that the complex sound produces waxing and waning of neural discharge which serves as a basis for the perception of pitch. A second explanation proposes that the transduction process before the basilar membrane produces non-linear distortion in the form of sum and difference tones. According to this view the basilar membrane *is* stimulated at a place corresponding to the fundamental.

A number of elegant experiments have been designed to discriminate between these alternatives. Schouten (1940), for example, found that the introduction of an additional sine wave, at a frequency close to the fundamental frequency, did not produce beats. He concluded that there was no energy present in the stimulus at the fundamental frequency. More compelling are those studies involving attempts to mask the fundamental pitch (e.g. Patterson, 1969). The rationale of these studies is as follows: if the fundamental is *not* physically present on the basilar membrane then it should be immune from disruption by a narrow band of noise at the fundamental frequency; if, on the other hand, the fundamental frequency is present on the basilar membrane it will be masked by noise. Since these studies show that noise fails to mask the fundamental, some doubt is cast on place theory as a complete explanation for the perception of pitch.

Auditory localisation

By listening with two ears (binaural listening) rather than one we can more effectively localise sounds in acoustic space and, in addition, selectively attend to different sound sources. If we consider a sound source lying on one side of the head, the acoustic shadow cast by the head means that sounds reaching the farther ear will be (i) less intense and (ii) delayed in time relative to the ear near the source. These two cues form the basis for binaural localisation, but their effectiveness is frequency dependent.

Phase differences provide useful cues to localisation for sources whose frequency is below 1500 Hz. If a sound source is directly to one side of the head the sound reaches the nearer ear almost immediately. In order to reach the farther ear the sound has to travel around the head, a journey which takes approximately 0.8 milliseconds. In this case one ear will be exposed to sound which is out of phase with the sound reaching the other ear. Ambiguity arises when the wavelength of the sound is less than the path difference between the two ears. The position of any sound which takes less than the 0.8 milliseconds or so to traverse the distance between the two ears cannot be satisfactorily resolved because information is lacking concerning the number of intervening cycles. Even when a tone is below 1500 Hz, perfect localisation using phase differences alone is relatively successful; a tone to one side of the head has the same phase shift whether it is in a position forward of the head or in the same relative position to the rear. In practice this ambiguity can be resolved by using head movements.

The use of phase for the localisation at low frequencies is complemented by the use of intensity differences between the ears at high frequencies. At low frequencies, sounds will be diffracted around the head with the result that the head will be acoustically transparent, little or no intensity difference at the two ears being produced. However, at high frequencies intensity differences are marked; for example, at 6000 Hz little diffraction occurs and there can be as much as 20 dB difference between the intensity at the two ears when the source is located directly to one side (Feddersen *et al.*, 1957). Above 3000 Hz the intensity difference forms a reliable basis for auditory localisation. Stereophonic recording uses the intensity differences between channels to generate the illusion of spatial separation. Thus, binaural localisation is achieved by a duplex system, time differences being used in the localisation of low-frequency sounds and intensity differences being used in the localisation of high-frequency sounds. In the region between these two extremes (1500–3000 Hz) binaural localisation is relatively poor.

Localisation of sound may be as good using one ear as when using two, particularly if the sound is complex and head movements are permitted. The pinna also plays a role in aiding monaural judgements of localisation. One interesting advantage given by binaural listening is that under certain circumstances it can improve the clarity of auditory perception. The binaural cues for localisation can act as the basis for selective attention; thus we may listen in to one of many conversations during a party. Another way in which binaural listening can be seen to improve clarity is through the phenomenon of *masking level difference.* In this case reception improves when noise and signal are in different phase relations at the two ears. The most striking effects of binaural masking level difference are to be found below 1500 Hz, in the region in which phase relationships are most successful in localising sounds. The power of the auditory system to localise sound can militate against the masking effect of noise. For example, if a signal in the form of a sine wave from a single source is presented simultaneously to each ear, accompanied by noise to both ears (again from a single source), then, as expected, the tone will be masked in proportion to the signal-to-noise ratio. However, a lower estimate of masking will be obtained by simply inverting the signal (making it out of phase) at one ear. A similar advantage can be conferred by presenting noise to both ears and presenting the signal to one ear only (see Gebhardt and Goldstein, 1972).

In natural environments as much as 90% of the energy of a sound will arrive at the ear via indirect paths in the form of echoes. If it were not for the ear's ability to give less weight to these echoes the problem of localisation would be made much more difficult. For a pair of similar sounds such as clicks, two sounds between 1 and 5 milliseconds apart are heard as a fused sound, localisation being largely determined by the first sound; this is known as the precedence effect. This effect may, however, be marginally reduced as the difference in location between clicks is increased or if the second click is much more intense than the first.

The measurement of sound

Objective measurement of overall sound level is usually accomplished by the use of commercially available sound level meters (see Bruel, 1976, for a general review). While the range of functions differs from model to model, a number of features are common to all. First, a microphone transforms sound pressure fluctuations into voltages which are eventually presented on an analogue or digital display. Secondly, weighting networks offer the facility of differentially attenuating the sensitivity to various frequencies. The most frequently encountered weighting networks provide the response characteristics A, B and C (shown in Fig. 7.7). Historically, these weightings were intended to simulate the frequency response characteristics of the human ear at low, medium and high levels of noise respectively. As Fig. 7.7 shows, a higher SPL is required to produce any given numerical reading on an A-weighted network than on a C-weighted network. Generally, the A and C weightings have found greatest favour, B now being rarely specified. A relatively new addition to the range of weightings is

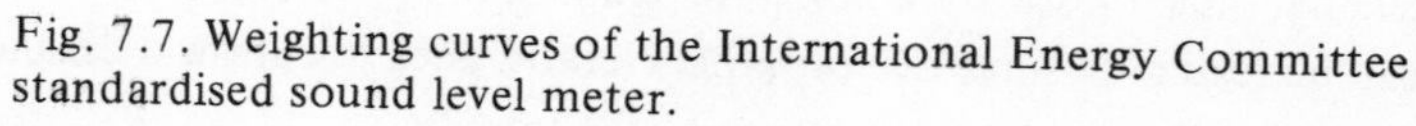

Fig. 7.7. Weighting curves of the International Energy Committee standardised sound level meter.

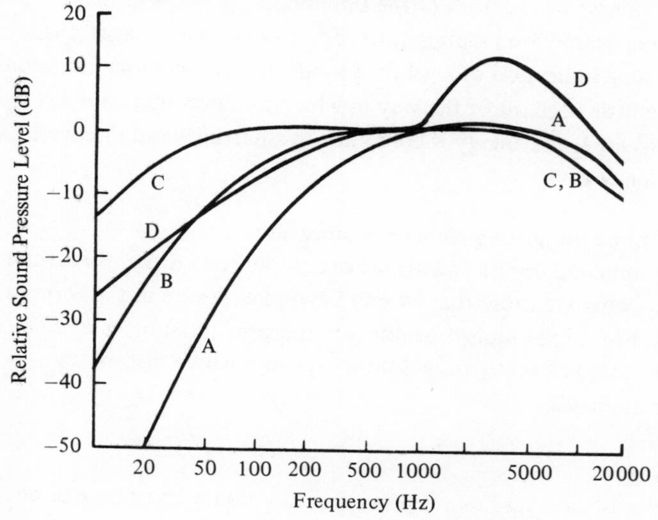

the D level, which attempts to correspond more closely to the frequency response of the human ear.

More sophisticated versions of the sound level meter often incorporate a series of filters which allow measurement of SPLs within certain specified frequency bands. The measurement of overall SPL is usually less informative than a frequency analysis derived from a number of such band-pass filters. Two sounds may give the same reading on the C-scale of a sound level meter but be made up of entirely different frequency components. It is useful, therefore, to examine the distribution of sound over the audio-frequency spectrum. In the simplest form of this type of analysis, sounds from a range of frequencies covering one octave only are recorded (that is, between two frequencies whose ratio is 2 : 1). If the SPL for the centre frequency of each octave band is plotted then it is possible to see at a glance the distribution over the audio-frequency range. For a more detailed analysis of the composition of sounds, bandwidths of a quarter or a third of an octave may be necessary.

The indication of sound pressure displayed by a sound level meter is the root mean square value averaged over a specified interval. Usually such an instrument permits a choice between 'fast' and 'slow' response characteristics, the former averaging sound pressure over a shorter period than the latter. Although these settings can satisfactorily encompass many of the fluctuations found in continuous noise, the examination of rapidly occurring noise (or impulse noise) is best achieved using an oscilloscope. Some sound level meters incorporate a facility for assessing impact noise by holding the reading at its maximum, so that this value can be read at leisure. Even so, this method does not allow a detailed examination of the precise time course of the impulse.

Generally speaking, sound level meters suffer from two disadvantages. First, they only give a reliable indication of level for sounds of relatively long duration. Secondly, they take little account of the way in which loudness of a complex sound depends on whether the energy is contained in a narrow band or covers a wide range of frequencies.

7.3 Effects of noise on hearing and communication

Hearing is impaired during and after exposure to loud noise. By far the most serious impairments are those that extend beyond exposure in the form of hearing loss, and hence noise-induced deafness is the primary subject of this section. In addition, those effects arising during exposure which disrupt communication will be discussed.

Hearing loss

Three types of hearing loss resulting from exposure to noise can be distinguished: first, a *temporary threshold shift* (TTS) involving a short-lived and reversible elevation of threshold; secondly, a *noise-induced permanent threshold shift* (NIPTS), a cumulative effect of noise exposure where thresholds remain elevated and do not return to their pre-exposure levels; and thirdly, a type of threshold shift not often encountered in occupational settings, that from *acoustic trauma* resulting from a single or very short exposure to extremely intense noise arising from gunfire or major explosions (Burns, 1973).

Temporary threshold shifts (TTS)

Although the temporal course of TTS can be resolved into several component parts (Ward, 1976), it is usual to specify the threshold shift 2 minutes after the cessation of the noise (TTS_2). The extent of TTS_2 is a function of the frequency, intensity and duration of the noise. Exposure to a restricted range of frequencies will produce modest elevation of threshold in that range but the *maximal* elevation may be in frequencies outside the range. After exposure to pure tones, TTS_2 is found at progressively higher frequencies as intensity is raised; the maximal shift will occur one-half to one octave, and occasionally two octaves, above the fatiguing tone (Van Dishoeck, 1948). Exposure to broader spectra (one or two octaves wide) also produces an upward progression of maximal TTS_2, generally one-half to one octave above the upper frequency limit of the noise (Ward, 1962).

TTS_2 also increases with the intensity of noise exposure. Figure 7.8 shows the hypothesised curves for TTS_2 resulting from exposure to various levels of noise. One effect, not clearly shown in these curves, is the tendency for noise beyond a certain intensity to have a disproportionate impact on the TTS. For example, Hood (1950) found that intensities above 90 dB produced especially large elevations in threshold. The growth of TTS_2 to a constant noise level reaches an asymptote in 8–12 hours. Provided that TTS_2 does not exceed 25 dB, recovery is usually complete in 16 hours, which means that the worker on a regime of 8 hours in 24 will have recovered in time for the next exposure.

Aside from the loss of hearing, TTS may be accompanied by tinnitus, a ringing in the ears, the pitch of which is related to the frequency of the fatiguing noise (Loeb and Smith, 1967).

Noise-induced permanent threshold shifts (NIPTS)

Evidence for NIPTS in man has largely been gleaned from retrospective studies of occupational hearing loss. These studies have attempted to establish the extent of hearing loss as a function of noise intensity and exposure duration. Typically a group of workers is subjected to audiometric examination and the hearing loss is estimated by subtracting the loss expected from other factors (such as age) from that recorded.

Generally the greater the noise intensity to which an individual is exposed the greater the NIPTS. Moreover, from the evidence of the best-documented studies, which have usually examined the effects of continuous broad-band noise, the development of occupational hearing loss follows a predictable sequence of events. Early in the period of exposure, loss is localised in the region between 3000 and 6000 Hz. As exposure continues there is a progressively greater loss in this region, which sometimes 'peaks' at about 4000 Hz. After very long exposure there is a gradual spread of effect to adjacent frequencies, the exact pattern largely depending on the spectral characteristics of the exposure. Taylor *et al.* (1965) studied hearing loss in a group of jute weavers whose time of exposure to noise ranged from 1 to 52 years. Fig. 7.9 shows that in the 4000-Hz region thresholds are markedly elevated over the first 10 years, but in the years thereafter there is a modest loss. The elevation of threshold for lower frequencies is slower at the outset, but for the cases of 2000 and 3000 Hz is just as pronounced in the long term (see also Nixon and Glorig, 1961).

Fig. 7.8. Growth of TTS_2 (at 4 kHz) after single continuous exposures to various levels of noise. The vertical arrow indicates the region of possible acoustic trauma. (Based on hypothetical curves drawn by Miller, 1974.)

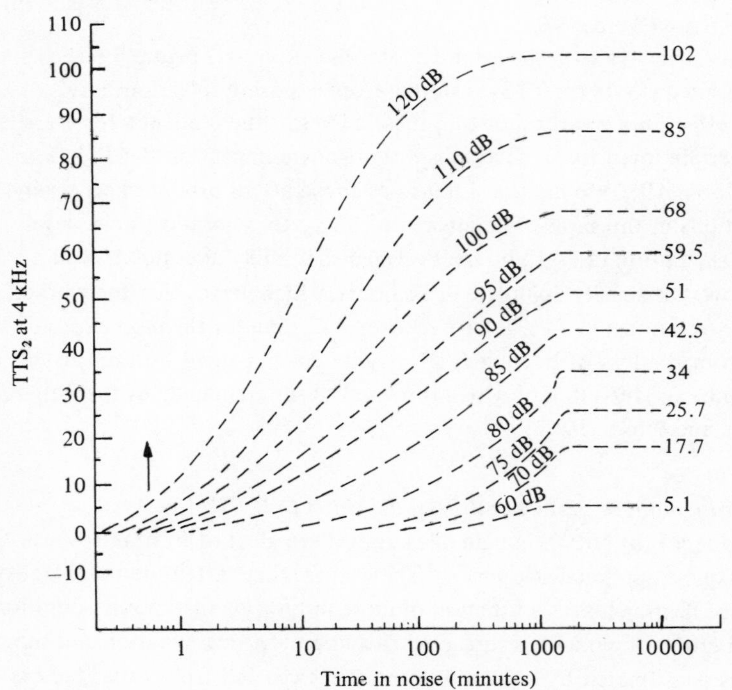

Burns (1973) noted that the locus of NIPTS seems to be less dependent on the exposure spectrum than is that of TTS. He based this conclusion on the effects of exposure to different types of noise spectra on NIPTS. Spectra with SPL increasing as a function of frequency were contrasted with those whose intensity decreased with frequency. Audiograms revealed that loss of hearing for both types of spectra showed a similar dip at around 4000 Hz.

Any strong relation between TTS and susceptibility to NIPTS would be of great prognostic value. However, no firm evidence for such a relationship exists. Kryter (1970) proposed that several factors suggest a relation between TTS and NIPTS; however, after an extensive review of empirical work he concluded that: 'The hypothesis that there might be some identifiable abnormal biological or temporary physiological weakness on the part of those developing NIPTS is without foundation at the present time' (Kryter, 1973, p. 172). This conclusion remains to be contradicted (Ward, 1976).

Limits to occupational noise exposure

The establishment of practical and useful standards of noise exposure is fraught with difficulties. While most people would agree that a low standard is ethically desirable, the realisation of such standards is largely influenced by technical and economic constraints (see Ollerhead, 1973; Bruce, 1976). Furthermore, there has to be agreement on a desirable goal for the conservation of hearing: should the standards be aimed at the prevention of *any* noise-induced hearing loss or at the prevention of any *material* handicap, for example in the hearing of everyday speech? The first of these options poses the empirical question of

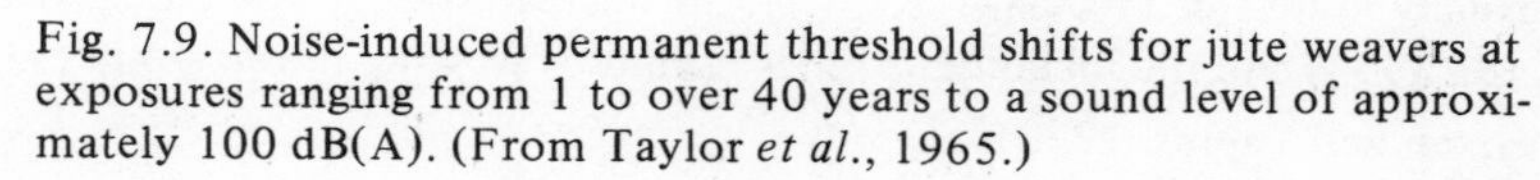

Fig. 7.9. Noise-induced permanent threshold shifts for jute weavers at exposures ranging from 1 to over 40 years to a sound level of approximately 100 dB(A). (From Taylor *et al.*, 1965.)

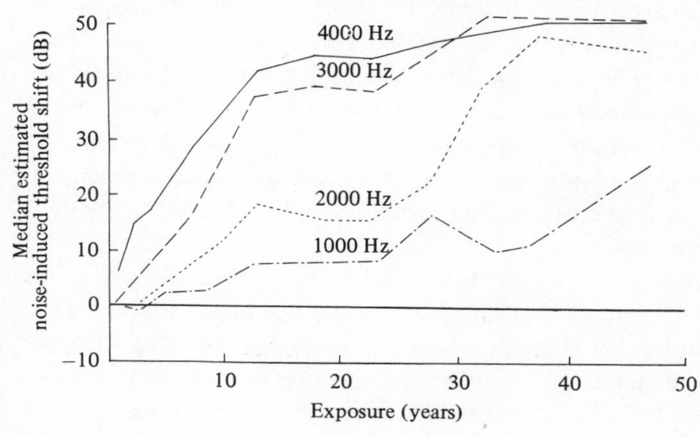

what constitutes 'normal' hearing; it requires a charting of the hearing loss brought about by age (presbyacusis) and by sounds to which the individual exposes himself willingly (socioacusis). The second option requires some consensus on what constitutes material impairment in hearing. Hearing losses resulting from NIPTS mainly will occur in the range 3–6 kHz, and there is by no means general agreement that even moderately large impairments of sensitivity at these frequencies lead to handicap in either the perception of speech or the appreciation of music (Kryter, 1973). A related problem concerns the estimation of impairment; present practice infers impairment in the perception of everyday speech from measures of sensitivity to pure tones and, to date, no direct measure of material handicap has been standardised.

As implied above, the accurate assessment of the extent of occupational noise-induced hearing loss in individuals or in groups of workers exposed to high noise levels over long periods depends upon the availability of a reference standard for hearing which takes into account the loss of auditory sensitivity resulting from the process of ageing. Any reference standard should thus incorporate a presbyacusis correction factor (the term presbyacusis (or presbycusis) referring to all age-related changes in auditory sensitivity). Audiometric surveys of non-noise-exposed individuals (that is, individuals who have no experience of head injury and no history of exposure to high-intensity noise or to gunfire) have obtained evidence of age-related declines in hearing levels (e.g. Glorig and Nixon, 1960) in populations below the age of 60 years in which the effects of ageing might be thought to be small. Furthermore, the age-related decline in auditory sensitivity appears to be much less marked in populations living in remote, virtually noise-free areas: for example, the Mabaan tribe of the Sudan (Rosen *et al*., 1964) and the inhabitants of the island of Westray in the Orkneys (Kell, Pearson and Taylor, 1970).

In the light of such evidence, it has sometimes been suggested that a socioacusis correction factor, which would reflect the effects on hearing of the noise levels commonly encountered in everyday living in industrial societies, should also be applied in the assessment of occupational noise-induced hearing loss. However, it is now recognised that the effects of age on hearing appear much earlier than was previously supposed (see Corso, 1976, 1977), and in consequence many of the effects that might be attributed to socioacusis can be subsumed under the heading of presbyacusis.

Attempts to determine the form that a presbyacusis correction factor should take have generally assumed that the hearing losses due to age and to noise exposure are additive. Yet in a comprehensive review of the available evidence, Corso (1976) noted that the additivity hypothesis suffers from certain limitations and recommended that an additive presbyacusis correction factor, based on the data assembled by Spoor (1967; see also Lebo and Reddell, 1972) should be employed for men between the ages of 35 and 60 years and for women between the ages of 40 and 65 years but discarded thereafter, since it appears that hearing impairments attributable to occupational noise exposure show little subsequent increase while age-related losses continue to accumulate. Up to the ages of 35 years for men and of 40 years for women, Corso suggested that a small, constant presbyacusis correction factor should be introduced, since the extent of age-related hearing loss, at least within the frequency range 250–3000 Hz, tends to be very slight at these ages, although occupational noise exposures may already have begun to produce irreversible structural damage to the ear.

The various national standards on noise exposure limits are given in Table 7.2 and are based on a résumé by Hay (1975). While this table gives an indication of the dimensions along which standards are assessed, the reader is advised to consult the current legislation in the country of interest.

Standards are usually couched in terms of the steady-state noise level over a specified duration (8 hours in most cases). Several studies have shown a systematic relation between loss of hearing and exposure to continuous noise throughout an 8-hour working day (Robinson, 1970, 1971; Passchier-Vermeer, 1974). There are, however, large individual differences in the rate of onset of hearing loss. Table 7.3 shows the loss expected as a result of continuous noise exposure to various A-weighted sound levels. On the basis of this table, a limit of 85 dB(A) over an 8 hour day will eventually lead to a maximum loss of 7 dB for 90% of the population (this represents the mean hearing level over four frequencies: 0.5, 1, 2 and 4 kHz). At 4 kHz the corresponding loss will be 19 dB. For a

Table 7.2. *National standards for occupational noise exposure limits*

Country	Steady noise level (dB(A))	Time exposure (hours)	Halving rate (dB(A))	Over-riding limit (dB(A))	Impulse peak SPL (dB)	Impulses (no. per day)
Australia	90	8	3	105	–	–
Belgium	90	40	5	110	140	100
Canada (Fed.)	90	8	5	115	140	–
Denmark	90	40	3	115	–	–
France	90	40	–	–	–	–
Germany (GFR)	90	8	–	–	–	–
Irish Republic	90	–	–	–	–	–
Italy	90	8	5	115	140	–
Sweden	85	40	3	115		
USA (Fed.)	90	8	5	115	140	100
UK	90	8	3	135	150	–

Based on Hay (1975).

Table 7.3. *Noise-induced permanent threshold shifts, NIPTS (dB) expected during a continuous long-term (40 years, 8 hours per day) exposure to various values of the A-weighted average sound level*

	Average 0.5, 1, 2 kHz	Average 0.5, 1, 2, 4 kHz	4kHz
		80 dB(A) for 8 hours	
Max. NIPTS 90th percentile	1	4	11
Max. NIPTS 10th percentile	0	0	2
		85 dB(A) for 8 hours	
Max. NIPTS 90th percentile	4	7	19
Max. NIPTS 10th percentile	1	2	5
		90 dB(A) for 8 hours	
Max. NIPTS 90th percentile	7	12	28
Max. NIPTS 10th percentile	2	4	11

Based on pure-tone threshold. Data presented by Eldredge (1976).

standard of 90 dB(A) the mean loss over 0.5, 1, 2 and 4 kHz will reach a maximum of 12 dB for 90% of the population. In this case the 90th percentile loss at 4 kHz is 28 dB (Eldredge, 1976).

Most standards apply the 'equal-energy' rule. That is, when daily exposures are shorter than 8 hours, because of periods of rest or quiet, higher intensities may be tolerated as long as the total energy is no greater than that for a continuous 8-hour exposure. In complementary fashion, any increase in intensity is traded off with length of exposure: each increase in level of 3 dB is commonly (though not universally) accompanied by a halving of exposure duration (halving rate: see column 4, Table 7.2). In addition to level, exposure duration and halving rate, an overriding limit which no noise level must exceed is usually specified (usually measured on the 'fast' response of the sound level meter). Such large excursions in noise level usually arise from impulse noise, characterised by sharp initial transients where an initial peak is followed by an exponentially decaying envelope of sound energy. More rarely, impulse noises produced by sound pressure changes in free field conditions (e.g. a pistol shot) show a simple N configuration. Most laws, codes and criteria take account of the additional hazard posed by impulsive noise by at least counselling caution or at most limiting the number of impulses at the overriding level (see column 4, Table 7.2).

Whether impulse noise obeys the equal-energy rule is a matter of some contention (see Ward, 1976 for a detailed discussion). On the one hand several studies have shown that TTS from impulse noise fails to follow the equal-energy rule (Fletcher and Loeb, 1967; Walker, 1970). However, it appears that this picture is somewhat complicated by the elicitation, under certain circumstances, of the acoustic reflex, which gives temporary protection. Those studies evaluating the effects of long exposures to impulse noise conclude that noise-induced hearing loss can be predicted in terms of the equal-energy concept for industrial impulse noise to peak sound levels of 150 dB.

Effects of noise on communication

Although noise may mask useful acoustic cues in the operation of machinery and impair the perception of auditory alarm signals, its primary masking effect is on speech. Speech is composed of complex acoustic patterns varying as a function of both frequency and time. These patterns are very resistant to corrupting influences due, in large part, to the multidimensional nature of speech coding. However, the sheer intensity of commonly encountered industrial noise is usually sufficient to make conversation difficult if not impossible. Communication systems (a telephone link for example) are also liable to disruption by noise arising from electrical sources. These sources of noise primarily produce their effects on intelligibility, but they also produce second-order effects on the effort required to be heard and through the restriction placed on the subtlety of conversation.

The effect of the type of material employed is often as great as changes in the ratio of signal (speech) to noise. Thus the influence of acoustic factors (such as the spectral characteristics of the noise or its intensity) is largely dependent on the nature of the message. In assessing the effect of increasing the intensity of noise, Miller, Heise and Lichten (1951) found that, as expected, the more intense the speech in relation to the noise the greater the percentage of messages correctly reported. They found, however, that if the message (a one-syllable word) was one of few possibilities, then the effect of noise was less than when the message could be one of many possibilities. Generally, satisfactory verbal communication can be achieved where the average speech level exceeds the noise by about 6 dB. Articulation scores of 50% can be reached at 0 dB signal-to-noise ratio (that is, when speech and noise levels are equal). Speech can be intelligible when the signal-to-noise ratio is negative (when speech is below the level of noise), but this is only true for connected discourse when the listener is familiar with the subject matter.

The degree of interference with speech also depends on the spectral distribution of the sound. The speech spectrum ranges from 100 to 8000 Hz, with most of the energy concentrated in the region between 100 and 6000 Hz. Noise within this region will produce maximum masking but no one particular frequency in the range is critical for the perception of speech. For example, filtering out from the speech signal frequencies above about 2000 Hz still results in intelligible speech. If, instead, frequencies below 2000 Hz are filtered out, speech is equally intelligible. Because of the multidimensional nature of the speech stimulus the listener can resort to a variety of cues, spread over the range of speech sounds, to maintain intelligibility.

The above considerations are important in establishing the intelligibility of speech in a system such as a telephone line. In a practical setting, such as in an office or on the shop floor, it is face-to-face communication which suffers in noise. The level of speech is usually increased to overcome the effect of noise but increases in vocal effort above a certain level will have side-effects on the intelligibility of the shouted voice. Keeping the signal-to-noise ratio constant, Pickett (1956) found that the intelligibility of speech was optimal in the range 50–80 dB (measured 1 m from the talker). However, despite increases in vocal effort, noise may in some cases overwhelm the spoken words (see Fig. 7.10). At levels of 95 dB(A) background noise (with the listener some 2 m away), and maximum vocal effort, communication may be judged as impossible. For communication to be judged as satisfactory, background levels need to be below about 50 dB(A) for a listener 2 m away, with the *practical* range for the unraised voice extending to 65 dB(A) (Webster, 1969).

The effects of periodic interruption of speech by noise are complex, but they can be assessed by using a range of interruption rates, keeping the noise-time fraction constant over the range. Using this technique and a noise-time fraction of 0.5 Miller and Licklider (1950) examined the effects of interruption in the range 0.1 to 1000 interruptions per second. Typically, low rates of interruption (up to 2 per second) mask whole words or syllables. When the interruption rate exceeds 30 per second the masking effects are equivalent to those found in continuous noise, due, probably, to the spread of masking over time.

The data thus far reviewed on the effects of noise on speaking and listening are based on young (usually male) articulate subjects with normal hearing. Criteria for disruption by noise are likely to be more stringent for the very young, the very old, persons with poor articulation, females (whose overall

Fig. 7.10. Effects of background A-weighted sound level on the quality of speech communication with variations in talker-to-listener distances. (From Miller, 1974.)

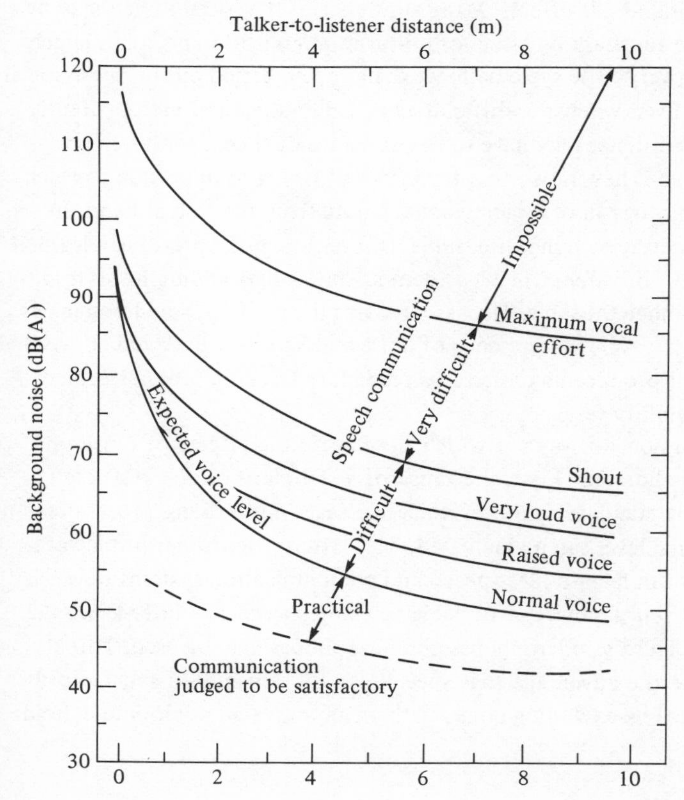

intensity of speech is some 3 dB lower than that for males) or people speaking in an unusual dialect. Similarly the criteria take little heed of the social and situational factors accompanying verbal discourse. Visual cues to speech can produce significant improvements in intelligibility. An investigation by Waltzman and Levitt (1978) found that face-to-face communication using visual cues markedly improved intelligibility: criteria based on the absence of visual cues overestimated the interfering effect of noise in face-to-face communication by as much as 20 dB (based on speech interference levels: see Webster, 1969, 1973). Most improvement was found with speaker-to-listener distances of below 1½ m for a given signal-to-noise ratio. Despite the enormous practical value of such a finding much work remains to be done on both the effects and the development of speech-reading skills. There has been no systematic study of the habits developed to counteract the interfering effect of noise on communication. Certainly, familiarity with the voice which is masked often improves intelligibility and there is the possibility that individuals can be trained to hear signals submerged in noise (see Tobias and Irons, 1973).

In addition to the direct effects of noise on communication there are a number of undesirable side-effects. Many subtleties of the social milieu may be lost, when, due to the masking effect of noise, communication becomes largely non-verbal. Again very little systematic work has been carried out on these social effects of noise. Even when the difficulties in communication are apparently circumvented, the listener may have to pay some residual cost for his effort. Rabbitt (1966, 1968) has shown that the effect of trying to understand speech in noisy conditions may have repercussions remote from the task at hand. In trying to understand speech in noise, subjects forgot material previously learned in quiet. Similarly, Broadbent (1958*a*) demonstrated that adding noise to a speech signal had much the same effect as removing the noise-masked frequencies from the signal. However, the presence of noise had an effect over and above that of masking since it produced a deficit in a secondary tracking task performed during the intelligibility test.

There are a number of ways in which the interfering effects of noise on speech can be ameliorated. First, the range of vocabulary can be restricted, which involves the standardisation of voice messages and talking procedures. Secondly, the signal level can be increased. This latter procedure is not always possible since in both face-to-face speech and communication systems power is usually limited. Even if power is available, making speech too intense would overload the ear. Thirdly, microphones and headphones may be used. Throat microphones have the advantage that speech can be transmitted satisfactorily while at the same time excluding noise. With an air-activated microphone, holding a microphone close to the mouth will improve intelligibility. Fourthly, the use of ear-defenders can considerably attenuate noise and thus lessens TTS and militates against NIPTS. However, it is obvious that if earplugs or earmuffs attenuate all frequencies equally, then the signal-to-noise *ratio* is the same as that for the unprotected ear. There are two consequences for the *listener* whose ears are defended: (*a*) at low levels of noise (levels below 70 dB, as is the case when the noise is temporarily in abeyance) speech is attenuated below the threshold of audibility; (*b*) for noise levels above 85 dB there will be no additional degradation of speech as a result of wearing ear-defenders. At levels of around 85 dB there may be a slight improvement due, presumably, to the absence of distortion at the cochlea at high noise levels. In addition, recent evidence indicates that the speech intelligibility of the *wearer* may suffer as a result of wearing ear-defenders. This may arise either because in these circumstances the level of speech drops by between 2 to 3 dB and is thus less intelligible to other listeners, or because the wearing of ear-defenders reduces the distinctiveness of speech (Howell and Martin, 1975). These difficulties may be partly overcome by amplitude-sensitive earplugs which increase attenuation of noise above a certain level (see Mosko and Fletcher, 1971). Apart from these acoustic considerations the attractiveness of ear-defenders may be lowered because of the discomfort they produce, especially in hot and humid conditions (see Karmy and Coles, 1976, for an overview of factors affecting the use of ear-defenders).

7.4 Noise and efficiency

In this section we shall discuss the effects of noise on the efficiency with which various tasks are carried out and some of the explanations which have been advanced for those effects which can be shown to be reliable. Much of the research to be reviewed is laboratory-based since, as we shall see, studies of noise and efficiency in industrial settings are relatively rare and in many cases suffer from methodological deficiencies which render their results virtually uninterpretable. Noise may well exert significant effects on task performance outside the laboratory but, with a few notable exceptions, such effects have not been satisfactorily demonstrated. For much of our present understanding of noise and efficiency, therefore, reliance must be placed on the data and theories generated by laboratory studies. We begin with an examination of some of the methodological difficulties which beset the interpretation of data from noise experiments and then provide a summary of the evidence from laboratory and industrial studies concerning the effects of noise on efficiency. Finally, some current theoretical issues are briefly discussed.

Methodological considerations

Although industrial studies possess the undoubted advantage that they are carried out on people accustomed to working in noise, one of the major difficulties they face is the 'Hawthorne effect', whereby the investigation of behaviour in an industrial setting appears itself to exert a powerful effect upon the outcome. Thus many studies of noise in industry suggest that reductions in noise levels enhance efficiency, but some investigations have also found that restoring the noise to its original level produces the same result (see Broadbent, 1957*a*). As Broadbent (1957*a*) has also pointed out, in many industrial studies noise levels are not the only variables to be manipulated during the course of an investigation; changes in other features of the work situation such as lighting conditions or temperature levels or a move to a new building often occur at the same time. It is therefore extremely difficult, if not impossible, to distinguish the changes due to the manipulation of noise levels from those due to the manipulation of these other potential influences. Finally, in many cases, the measures of efficiency used in industrial studies are somewhat suspect, although, as noted below, this criticism also applies to some laboratory studies.

The interpretation of data from laboratory studies of the effects of noise, and generalisations to the industrial situation made on the basis of such data, are also frequently limited by methodological considerations. First, as in other areas of research on human performance, subjects for noise experiments have been selected from a rather restricted sample of the population. For the most part such subjects have been students or military personnel and have generally been under the age of 30. There is some doubt, therefore, as to whether they may be considered to be representative of the industrial workforce. Secondly, for most areas of human performance research, adequate task taxonomies do not exist, although important contributions to the development of such taxonomies are being made by Fleishman and his associates (Fleishman, 1975) and these seem likely to prove useful in assessing the effects of noise on performance (Theologus, Wheaton and Fleishman, 1974). One consequence of this lack of adequate task taxonomies is that in many cases it is difficult to specify with any confidence the practical implications of laboratory findings, since it is often unclear what aspect of general efficiency is being tapped by a particular task. Furthermore, when noise is observed to exert no significant influence upon performance at a particular task, it may be that performance is genuinely unaffected by noise or that the measure of performance selected, or the task yielding the measure, is insufficiently sensitive to the chosen level of noise. Poulton (1965) and Wilkinson (1969) discuss a number of factors that should be taken into account in order to ensure that a task is maximally sensitive to environmental and other variables.

Thirdly, as reviewers of the effects of noise on performance have often noted, for many studies only the overall sound level meter reading has been reported and the band spectra or the spectrum levels of the noise used have seldom been specified (Grether, 1970; Kryter, 1973; Broadbent, 1979). In many early studies a recording of some common noise was used, for example office, factory or traffic noise, and this was then amplified to the desired intensity. In this method of sound generation the frequency composition of the noise is not directly controlled, as it is when an electronic sound generator is employed. In most recent laboratory studies of the effects of noise, investigators have used 'white' noise, in which the constituent frequencies, taken from a wide range of the audio-frequency spectrum, are of equal intensity. However, the levels of noise adopted for 'noise' and 'quiet' conditions vary widely between studies. In some experiments the sound level of the noise condition may be as low as 50 dB, while in others that of the quiet condition may be as high as 80 dB. It is also sometimes unclear whether an unweighted or an A-, a B- or a C- weighted sound level meter has been used to assess noise levels. Many discrepancies between experimental results may thus be attributable to the use of different kinds or levels of noise.

Finally, the choice of experimental design is of some importance for noise experiments, and the adoption of a repeated measures design may lead an investigator to draw inappropriate conclusions concerning the effects of noise on performance in a particular task situation (Poulton and Freeman, 1966; Poulton, 1970, 1973). The reason for this is that asymmetrical transfer effects between treatments may be present and the operation of such effects may reduce the probability of finding a reliable difference between experimental and control conditions. However, the presence of certain kinds of asymmetrical transfer effects can be detected fairly easily and, where necessary, conclusions concerning the effects of noise modified.

Laboratory studies of the effects of noise

Performance at many tasks seems to be relatively unaffected by continuous loud noise (that is, at sound levels at or above 90 dB) and may even improve slightly at more moderate noise levels. Intermittent loud noise, particularly if it is aperiodic, is somewhat more likely to produce impairments of performance at these tasks. Performance at certain kinds of task, however, shows consistent and reliable changes during loud noise, and the greater part of this discussion is concerned with a description of these changes. But first some examples are given of loosely defined task categories in which performance is generally unimpaired by loud noise. Because of limitations of space, task categories in which the results at present available permit no clear conclusion to be drawn, for example the Stroop Test (Houston and Jones, 1967; Houston, 1969; O'Malley and Poplawski, 1971; Hartley and Adams, 1974) and various tests of time judgement (Jerison and Smith, 1955; Hirsh, Bilger and Detherage, 1956; Jerison, Crannel and Pownall, 1957; Loeb and Richmond, 1956; Jerison and Arginteau, 1958) are omitted, as are studies comparing the effects of varying noise (Kirk and Hecht, 1963) or music (Davies, Lang and Shackleton, 1973) with those of steady noise. Continuous loud noise appears to exert no significant adverse effect on performance at what may be described as intellectual tasks, a category which includes tests of clerical ability (Wilbanks, Webb and Tolhurst, 1956), coding and sorting tasks (Stevens, 1972; Conrad, 1973), mental arithmetic (Samuel, 1964; Nuckols, 1969), cancellation tasks (Davies and Davies, 1975) and standard intelligence tests (Corso, 1952). However continuous noise of 100 dB (Broadbent, 1958*b*) or between 107 and 110 dB (Harris and Sommer, 1971) reliably impairs mental arithmetic performance if the task also imposes a memory load. Intermittent noise of 100 dB also adversely affects performance on this task (Woodhead 1964).

Performance at psychomotor tasks is also little affected by continuous loud noise. Hand steadiness remains unchanged in 115dB simulated aircraft noise (Stevens, 1972), while Weinstein and Mackenzie (1966) found that 100 dB noise improved performance at a simple test of manual dexterity (the Minnesota Rate of Manipulation Test). Tracking performance, whether compensatory or pursuit, is generally unimpaired either by continuous or intermittent noise at levels of between 105 and 122 dB (Plutchik, 1961; Grether *et al.*, 1971; Stevens, 1972), although if the task is extremely complex, performance may deteriorate at levels between 80 and 90 dB (Eschenbrenner, 1971). Simpson, Cox and Rothschild (1974) found that pursuit motor performance reliably deteriorated at a sound level of 80 dB(A) but that a preloading of 18 mg of glucose in 100 ml of water prevented this deterioration from occurring. However, the same preloading of glucose impaired performance in a quiet condition of 50 dB(A). Davies and Gill (1980) found that aperiodic intermittent broad-band noise at a level of 92 dB(A) reliably increased solution times in a self-paced problem solving task compared with a control condition but that a preloading of glucose at the same dosage as in the study of Simpson *et al.* abolished this effect. Other task situations in which performance is generally unimpaired by continuous or intermittent loud noise include simple and choice reaction time, unless the task is continuous (Stevens, 1972; Broadbent, 1979), single-source monitoring, where detection rate may even improve slightly (Jerison and Wallis, 1957; Broadbent and Gregory, 1965; Davies and Hockey, 1966; Poulton and Edwards, 1974), tasks requiring inspection and search, provided that the task is relatively easy and the rate of pacing moderate (Warner, 1969; Warner and Heimstra, 1971,

1972, 1973; Cohen *et al.*, 1973), and many short-term memory tasks, although here the evidence is conflicting (Schwartz, 1974; Daee and Wilding, 1977; Eysenck, 1977; Jones, 1979).

From the brief outline given above, it may be noted that there is some indication that performance tends to deteriorate in noise if the difficulty of the task is increased, particularly if the task is performed continuously without rest pauses between responses. Thus performance at complex tracking tasks, at highly paced search and inspection tasks where the target is made more difficult to locate, and at mental arithmetic tasks to which a memory load is added is likely to be impaired by loud noise. The continuous performance task which has been most extensively studied in connection with loud noise is the five-choice serial reaction task (Leonard, 1959). This task consists of five small light bulbs and five brass discs, each bulb corresponding to a disc. In some versions of the task the bulbs and the discs are set into the same wooden board, while in others the light display is separated from the response panel (Broadbent, 1976). The subject's task is to tap with a metal stylus the disc corresponding to a lighted bulb, whereupon the bulb is extinguished and another bulb is lit. The lighting of a bulb is contingent upon a disc, not necessarily the correct one, being tapped. Three scores are possible: the number of correct responses, the number of errors, and the number of long response times (1500 mseconds or longer), which are known as 'gaps'. The task duration is usually about 30 minutes. The important features of this task are that it is continuous, being performed without a break, and that it is self-paced, that is, the subject controls his own rate of work.

With this task, significant impairments of performance in loud noise are most frequently obtained in terms of increased errors, and the number of long response times may also increase (Broadbent, 1953; Wilkinson, 1963; Hartley, 1973), although this latter finding appears to depend upon which version of the task is used (Poulton, 1976). Errors also occur more frequently in noise as the task progresses. Loud noise in which the higher frequencies predominate is more likely to impair serial reaction performance than is noise containing predominantly lower frequencies (Broadbent, 1957*b*), and impairments generally are found only at sound levels at or above 95 dB(A) (Broadbent, 1978). At a level of 102 dB(C), error rates may even be reduced (Poulton and Edwards, 1974). As Broadbent (1978) observed, this level is equivalent to one of 82–85 dB(A). Thus when a sequence of rapid actions has to be performed, in response to an unpredictable sequence of inputs, efficiency deteriorates in loud noise. However, the rate of work is not affected by noise since the number of correct responses shows no reliable change in noise in any of the experiments conducted with this task. The increase in errors in noise therefore probably results from interruptions of the normal level of efficiency, and performance remains otherwise unaffected.

A similar effect of noise may be present in some tracking situations. Grimaldi (1958) found that noise levels of 90 and 100 dB did not affect the number of times the track was traversed during the testing period but did produce more errors, less precision and longer response times.

A second effect of loud noise has been observed in monitoring situations, where noise appears to affect the confidence with which reports of the detection of faint, infrequent and unpredictable signals are made. When subjects are asked to assess their level of confidence in a detection response, the proportion of doubtful responses falls in 100 dB (Broadbent and Gregory, 1963, 1965) and in 102 dB(C) (Poulton and Edwards, 1974) noise, while the proportion of confident responses rises. That is, subjects are more likely to say that a signal definitely was or was not presented. If, therefore, as a result of the instructions they have received, or perhaps because they have come to believe that the occurrence of a signal is extremely rare, subjects only report signals when they are quite sure that one has been presented, detection efficiency may show no change or even a slight improvement in noise. If, on the other hand, they are instructed to report anything which bears the remotest resemblance to a signal, or if signals occur, on the average, very frequently, then detection efficiency may well deteriorate in noise. Performance at monitoring tasks with a high signal rate has been found to be reliably worse, compared with a 'quiet' condition, at levels of 112.5 dB (Jerison and Wing, 1957), 100 dB (Broadbent and Gregory, 1965) and 80 dB low-frequency noise (Benignus, Otto and Knelson, 1975), although other experiments have found that signal frequency and noise (at levels of 95 and 70 dB) do not interact (Davies and Hockey, 1966). It is possible that monitoring tasks with a verbal component are especially vulnerable to the effects of noise (Jones, Smith, and Broadbent, 1979).

A third effect of noise appears in multicomponent situations. Broadbent (1950, 1951, 1954) used two multisource monitoring tasks – the 20 dials task and the 20 lights task – and compared performance in noise (100 dB) and quiet (70 dB). In the 20 dials task he found that noise had no effect on the total number of signals detected but did reduce the number of 'quick founds', that is, signals detected within 9 seconds of onset. Some impairment of noise was also found with the 20 lights tasks, but in both cases this impairment tended to be restricted to those areas of the display which were not in the direct line of vision, the displays being arranged to form three sides of a square with the subject facing the central portion. One possible interpretation of this result is that noise produces a change in the way in which attention is allocated to the different components of a multicomponent task, high-priority components being given more attention and low-priority components less.

Support for this interpretation comes from a series of studies by Hockey

(1970*a,b,c,* 1973). Hockey (1970*a*) required subjects to perform a combined tracking and multisource monitoring task for 40 minutes. In the instructions given to subjects the tracking task was designated as the 'high-priority' task and the monitoring task, which required the detection of the onset of lights at different spatial locations, as the 'low-priority task'. Tracking performance was unaffected by 100-dB noise and signals appearing at central sources were detected more frequently in noise. However, signals appearing in peripheral locations were detected less frequently. Hockey (1970*b*) showed that this differential detection of signals at different spatial locations resulted from the expectation that signals were more likely to appear at central locations than at peripheral ones. In a third experiment Hockey (1970*c*) found that sleep deprivation produced changes which could be interpreted as being the opposite of those obtained with noise, impairment of performance being greater on the high-priority task (tracking).

Hockey (1973), using a three-source monitoring task developed by Hamilton (1969), which required the subject to make a sampling response in order to obtain a brief glimpse of the present state of one of the three sources which might or might not contain a signal, found that 100-dB noise produced increased sampling of the source on which signals had a high probability of appearing, while the opposite effect was again found with sleep deprivation. Noise also reduced the frequency with which repeat observations of a source were made before a signal was reported (an effect similar to the reduction of doubtful responses in monitoring noted above) while sleep deprivation increased it.

Although failures to replicate Hockey's original findings (1970*a*) have been reported (Forster and Grierson, 1978; Loeb and Jones, 1978), his results, and those of others (for example, Woodhead, 1964, 1966; Glass and Singer, 1972) have been interpreted in terms of Easterbrook's (1959) cue utilisation hypothesis. It has been suggested that noise produces a structured change in the way in which attention is distributed over the different components of a task, resulting in an increase in selectivity, whereby the range of cues that a subject uses in the performance of a task becomes progressively restricted, so that attention is diverted more and more away from irrelevant or subjectively unimportant aspects of the task. An alternative explanation of the effects of noise in dual task situations, expressed in terms of the reduction of spare information processing capacity (Boggs and Simon, 1968; Finkelman and Glass, 1970), is not necessarily incompatible with the selectivity hypothesis. Noise also appears to increase selectivity in memory situations, where intentional memory for words has been shown to remain unchanged or to improve slightly in 80- and 95-dB noise, while incidental memory for the locations in which the words were presented deteriorated markedly (Hockey and Hamilton, 1970; Davies and Jones, 1975; Fowler and Wilding, 1979).

Nevertheless the effects of noise on attentional selectivity appear to be quite complex and are almost certainly related to other factors such as stimulus salience and task difficulty (Smith, 1980; Smith and Broadbent, 1980). It is unlikely, therefore, that noise invariably affects selectivity in a uniform fashion.

As an incidental feature of these results, several studies have shown that the effects of loud noise on performance may extend beyond the period of exposure. The experience of noise as the first condition in an experiment influences performance on subsequent (noise-free) trials. In this way, performance may be depressed (Broadbent, 1954; Jerison, 1959) or elevated (Frankenhaeuser and Lundberg, 1974) in the long term. More recently these after-effects have received systematic scrutiny in a number of studies which have examined the quality of performance over a brief period after the noise ceased. Hartley (1973), for example, required subjects to work at the five-choice serial reaction task for either 40 or 20 minutes in quiet, in 100-dB noise, or after 20 minutes of exposure to the noise while they read a book. Hartley found that performance was impaired by approximately the same amount either by 20 minutes of prior exposure to noise or by 20 minutes of prior task performance, and 40 minutes of exposure to noise produced twice as much impairment as 20 minutes. The effects of noise and prior performance did not interact; their effects were independent and additive. Hartley concluded that noise exerts an influence on performance which is that of 'a slowly accumulating stress, the effects of which take a similar time to dissipate' (p. 260).

In a series of experiments, Glass and Singer (1972) observed reliable after-effects of 25–30 minutes of exposure to intermittent noise at a level of 108 dB(A). These took the form of reduced tolerance to frustration in a problem-solving task and of impairments in the resolution of cognitive conflict in a version of the Stroop test and, less consistently, in the detection of errors in a proof-reading task. However, when subjects were given the impression that they could exercise control over the noise, or when the predictability of the noise was increased, for example by signalling its onset, the subsequently observed after-effects of noise exposure were attenuated. Furthermore, after-effects of exposure to noise do not depend upon task performance during exposure being degraded by the noise. In several of their experiments, Glass and Singer (1972) found that 108-dB(A) intermittent noise exerted little or no effect upon performance, and Wohlwill *et al.* (1976) observed no effect of continuous noise at a level of 80–85 dB(A) on performance at a dial monitoring task. Nevertheless, in both studies reliable after-effects of noise were obtained.

Cohen (1980), in an extensive review of research and theory concerned with the after-effects of various stressors, including noise, considered a number of theoretical explanations of such effects. These included the adaptive-cost hypothesis, first put forward by Glass and Singer (1972), Cohen's own information overload hypothesis (see Cohen, 1978), learned helplessness theory (Seligman, 1975), arousal theory, which is referred to below in relation to the effects of different stressors in combination, and explanations couched in terms of coping strategies, frustration, cognitive dissonance and self-perception. Cohen concluded that most explanations of the after-effects observed following exposure to stressors such as noise are essentially variants of the original adaptive-cost hypothesis, which emphasises the psychological, and to a lesser extent the physiological, costs involved in the adaptation to a stressor. However, different explanations describe these costs in somewhat different terms.

Studies of the after-effects of exposure to different stressors place the effects of noise in a broader perspective. It is clear that the effects of noise may extend beyond the period of exposure. But just how long these after-effects last, and whether they appear as a result of exposure to noise in an occupational setting, are two issues which remain to be explored. All the evidence from laboratory studies suggests that if these effects are persistent they may have repercussions on the life-style of the individual and on the transactions which he has with the world in general (see Jones and Chapman, 1979, for an overview).

As has already been mentioned, noise may be regarded as a stressor, but, outside the laboratory, environmental stressors rarely occur in isolation. Generally, a combination of stressors is encountered, the most frequently occurring combination probably being noise and heat. In addition 'host' factors such as sleep deprivation, fatigue or drugs may be present and may interact with externally imposed stressors to produce effects on performance. As several authors have emphasised (Poulton, 1966, 1970; Wilkinson, 1969; Grether, 1971), different stressors are likely to exert their effects through quite different mechanisms and the effects on performance of two combined stressors cannot be predicted from a knowledge of the effects of either stressor administered alone. Two stressors may combine to produce effects on performance which are additive (where the effect on performance of the two stressors in combination is the sum of their independent effects), greater than additive (synergistic) or less than additive (antagonistic or interactive). Most combinations of environmental stressors appear to be additive in their effects on performance, a few are antagonistic and scarcely any are synergistic (Grether, 1971).

The variables most frequently studied in combination with noise include heat, sleep deprivation, alcohol and vibration. In addition the effects of knowledge of results and monetary incentives have been simultaneously examined. Research on the effects of different stressors in combination has produced the 'arousal theory of stress' (Broadbent, 1963, 1971), which assumes that there is a general state of arousal or reactivity which is increased by incentives or by loud noise and reduced by loss of sleep or boredom. It further assumes that inefficiency is high when arousal level rises beyond an optimal point, which is probably related to the chronic level of the individual performing the task and to the demands of the task itself. Inefficiency is also assumed to be high when arousal level is too low; in other words, the relationship between the level of arousal and the level of efficiency is thought to follow an inverted U. In some respects this version of the arousal model is too simple and does not satisfactorily account either for the effects of all stresses or for all aspects of the experimental results obtained. Nevertheless the arousal model seems to provide an adequate account of improvements in performance with noise (Poulton, 1977*a*). Broadbent (1971) provided a critical appraisal of the arousal theory of stress and put forward a two-stage arousal model which attempts to cope with the problems faced by less complex formulations.

Industrial studies

As noted earlier, satisfactory studies of the effects of noise in industry are relatively rare. However, some well-designed investigations have succeeded in demonstrating that noise reduction can exert a beneficial effect upon efficiency. Weston and Adams (1935), for example, showed that the wearing of earplugs by weavers in a textile mill resulted in an overall increase in efficiency of about 12%; moreover, this improvement occurred for every one of the ten workers studied. In a more recent study Broadbent and Little (1960) investigated the effects of noise reduction upon the efficiency of operators engaged in film perforation. At the beginning of this study the noise level at the workplace was about 99 dB and this was reduced to 89 dB by placing absorbent material between the machines. Half of the available machines were treated in this way and the remaining half left untreated so that the efficiency of the same operators, moving from treated to untreated workbays and hence working at the two noise levels, could be compared. Broadbent and Little found that the number of errors (assessed by the number of broken rolls of film and equipment shutdowns) was greatly reduced in the treated workbays, while the rate of work changed to an equal extent in both treated and untreated workbays. Apart from providing the confirmation of some laboratory results the important finding of this study is that noise effects on efficiency appear in individuals who are accustomed to the noise and who have had plenty of experience in the work situation.

It was mentioned earlier that noise appears to affect the distribution of attention in such a way that relevant or important task components receive more attention while irrelevant or unimportant components receive less. One consequence of this may be that unexpected signals or pieces of information are inefficiently dealt with. It is possible, therefore, that accidents may be increased by noise, since they result, in part at least, from the misperception of potentially dangerous situations (Broadbent, 1970). Cohen (1973), in an investigation of industrial accident rates in two factories, did in fact find that accident rates were very substantially higher for people working in noise at levels of 95 dB(A) or above than for people working at levels of 80 dB(A) or below. Kerr (1950) examined possible correlates of accident rates in a number of different work settings and obtained a significant correlation of 0.40 with noise level, a higher correlation than for any of the 40 variables studied except a measure of job mobility. These findings indicate that noise level may be an important determinant of accidents in industry, although clearly more evidence is needed.

Theoretical issues

There are three main explanations of the effects of noise on performance: arousal, which was mentioned above, distraction (Broadbent, 1958*a*), and masking (Kryter, 1970, 1973; Poulton, 1976, 1977*b*, 1978). Arousal theory can handle both improvements and, through the operation of over-arousal and the resultant increase in selectivity, deteriorations of performance, although, as implied earlier, the theory faces certain difficulties and distraction appears a more useful way of explaining the brief periods of lowered efficiency which occur in some task situations. However, it was suggested some time ago by Kryter that adverse effects of noise on performance could be explained by the masking of task-relevant auditory cues, a point of view which has recently been taken up with some vigour by Poulton (1976, 1977*b*, 1978). Poulton has also suggested that where task-relevant auditory cues are absent, as they are, for example, in most memory experiments, adverse effects of noise can be explained by the masking of inner speech. Broadbent (1976, 1978) has countered Poulton's arguments but the debate continues.

7.5 Noise and health

Both short-term and long-term physiological changes may occur as a result of exposure to noise (Davies, 1968, 1976; Welch and Welch, 1970; Ahrlin and Ohrstrom, 1978) and such changes may have consequences for an individual's health and well-being. In this section we outline, extremely briefly, some of the effects on physical and mental health that exposure to loud noise at the workplace may produce. Studies of noise-induced sleep disturbance, of noise-induced stress in animals, and of reactions to aircraft and traffic noise are therefore excluded (recent views of these areas can be found in Shepherd, 1974; Griefahn and Jansen, 1977; Griefahn and Muzet, 1978; Moller, 1978; Tarnopolosky, 1978).

In the laboratory, measures of autonomic nervous system activity have been shown to adapt to loud intermittent noise (108 dB(A)), although it is less clear whether physiological adaptation to loud continuous noise (110 dB) occurs (Helper, 1957). It seems probable that long-term exposure to high noise levels produces adverse physiological effects, particularly on the cardiovascular system, and there are indications from some studies that long-term exposure to loud noise increases the risk of hypertension (Jansen, 1959, 1961; Parvizpoor, 1976; Jonsson and Hansson, 1977), although other studies have found no relation between noise exposure and cardiovascular disease (Drettner *et al.*, 1975).

The number of studies that have examined the effects of noise upon mental health is very small indeed, and the problems involved in establishing a clear link between exposure to noise at work and mental health symptoms are considerable. Jansen (1959) examined more than 1000 foundry workers in various parts of Germany. As well as the physiological observations mentioned above, Jansen also assessed mental health symptoms, by means of a clinical interview, and the factories from which the workers were drawn were classified in terms of the typical level of noise exposure. Jansen then compared workers exposed to lower noise levels and found that the former showed more emotional tension in the home and in the factory. Ingham (1970) made three points about this investigation. First, the individuals assessed for mental health symptoms were selected in a rather uncontrolled way; secondly, the methods of assessing symptoms were highly subjective; and thirdly, in any case only a few symptoms, from the large number assessed, distinguished between workers from high- and low-noise environments.

In a carefully controlled survey of personnel involved in aircraft launch operations aboard US Navy aircraft carriers, Davis (1958) found that there appeared to be a consistent tendency for men most exposed to aircraft noise to perform the least well on a variety of psychomotor tests. These included aspects of steadiness, reaction time and critical flicker frequency. However, overall there were no completely clear-cut differences between the most-exposed and the least-exposed groups. In addition, psychiatric examination by intensive tests and interview procedures revealed that the most exposed men appeared to experience greater feelings of anxiety. It is possible that the greater anxiety experienced by these men is attributable to the fact that their jobs were more difficult and more dangerous (Kryter, 1970). The presence of intense noise usually implies the presence of powerful machinery which requires highly skilled operation and where mistakes can have dangerous consequences. Thus, the expressed anxiety may relate more to these factors than to the high noise levels. In this connection it should be noted that the most exposed personnel did not rate the jet aircraft noise as any more disturbing than did personnel who were less exposed, but they did express more anxiety about the jobs they were doing.

Ingham (1970) carried out a preliminary survey of men exposed to different noise levels in a large steelworks. He examined new personnel coming into the works for symptoms of psychological disorder, using a battery of questionnaires validated in previous studies, and repeated these observations on the same personnel after varying periods of exposure to different noise levels. Although this work is based on only about 100 cases, Ingham did find evidence suggesting that on the Cornell Medical Index there was a small number of people who displayed a considerably larger number of symptoms after exposure to noise levels above 87 dB(A).

Although the conclusions of this last study are tentative, the methods used seem able to circumvent most of the main problems in research on noise and mental health. First, there is the problem of the duration of exposure to noise. Presumably in older, longer-exposed groups the psychological impairment, if any, would be greater. Secondly, there is the problem of selection out of and also into noisy jobs. Workers who find high noise levels disturbing and in whom symptoms of mental ill-health are perhaps provoked may well leave the job or the industry in question. Therefore, only if there is knowledge of the presence or absence of disability in this group can the effects of noise be fully assessed. It is also possible that noisy jobs attract a lower proportion of people who are unstable to begin with and so the effects of noise are masked. Thirdly, although noise may not produce any increase in symptoms, the psychological cost of maintaining apparent stability at work may be very high, and signs of instability may appear at home, or within the work situation in terms of reduced morale. Rodda (1967) suggested the interesting possibility that personnel in noisy industries, for example the motor-manufacturing and shipbuilding industries, may have lower morale, although he admitted that there could be many other reasons for this.

It thus appears that while there is insufficient evidence to assert that noise adversely affects mental health, some of the available evidence suggests that it may, but further research, along the lines indicated by Ingham's work, is necessary to establish more precise conclusions. In any event it seems probable that even if noise is not a direct contributor in itself to mental disturbance, it may, in some cases, act as a 'last straw' when combined with the other stresses of daily life.

References

Ahrlin, U. and Ohrstrom, E. (1978) Medical effects of environmental noise on humans. *Journal of Sound and Vibration*, **59**, 79–87.

Benignus, V.A., Otto, D.A. and Knelson, J.H. (1975) Effect of low-frequency random noises on performance of a numeric monitoring task. *Perceptual and Motor Skills*, **40**, 231–9.

Boggs, D.H. and Simon, J.R. (1968) Differential effect of noise on tasks of varying complexity. *Journal of Applied Psychology,* **52,** 148–53.
Broadbent, D.E. (1950) *The Twenty Dials Test Under Quiet Conditions.* Report, Applied Psychology Unit, No. 130/50. London: HMSO.
Broadbent, D.E. (1951) *The Twenty Dials and Twenty Lights Test Under Noise Conditions.* Report, Applied Psychology Unit No. 160/51. Cambridge : Applied Psychology Research Unit.
Broadbent, D.E. (1953) Noise, paced performance and vigilance tasks. *British Journal of Psychology,* **44,** 295–303.
Broadbent, D.E. (1954) Some effects of noise on visual performance. *Quarterly Journal of Experimental Psychology,* **6,** 1–5.
Broadbent, D.E. (1957*a*) Effects of noise on behavior. In *Handbook of Noise Control,* ed. C.M. Harris, chapt. 10. New York : McGraw-Hill.
Broadbent, D.E. (1957*b*) Effects of noises of high and low frequency on behaviour. *Ergonomics,* **1,** 21–9.
Broadbent, D.E. (1958*a*) *Perception and Communication.* London : Pergamon Press.
Broadbent, D.E. (1958*b*) Effect of noise on an 'intellectual' task. *Journal of the Acoustical Society of America,* **30,** 824–7.
Broadbent, D.E. (1963) Possibilities and difficulties in the concept of arousal. In *Vigilance: A Symposium,* ed. D.N. Buckner and J.J. McGrath, pp. 72–87. New York : McGraw-Hill.
Broadbent, D.E. (1970) Noise and work performance. In *Proceedings of the Symposium on the Psychological Effects of Noise,* ed. W. Taylor, chapt. 1.1. London: Research Panel of the Society of Occupational Medicine.
Broadbent, D.E. (1971) *Decision and Stress.* New York and London: Academic Press.
Broadbent, D.E. (1976) Noise and the details of experiments: a reply to Poulton. *Applied Ergonomics,* **7,** 231–5.
Broadbent, D.E. (1978) The current state of noise research: reply to Poulton. *Psychological Bulletin,* **85,** 1052–67.
Broadbent, D.E. (1979) Human Performance in noise. In *Handbook of Noise Control,* ed. C.M. Harris, chapt. 18. New York: McGraw-Hill.
Broadbent, D.E. and Gregory, M. (1963) Vigilance considered as a statistical decision. *British Journal of Psychology,* **54,** 309–23.
Broadbent, D.E. and Gregory, M. (1965) Effects of noise and signal rate upon vigilance analysed by means of decision theory. *Human Factors,* **7,** 155–62.
Broadbent, D.E. and Little, E.A.J. (1960) Effects of noise reduction in a work situation. *Occupational Psychology,* **34,** 133–40.
Bruce, R.D. (1976) The costs and benefits of implementing 90 and 85dB(A) noise exposure standards. In *Man and Noise,* ed. G. Rossi and M. Vigone, pp. 429–36. Turin: Edizioni Minerva Medica.
Bruel, P.V. (1976) Determination of noise levels. In *Man and Noise,* ed. G. Rossi and M. Vigone, pp. 189–220. Turin: Edizioni Minerva Medica.
Burns, W. (1973) *Noise and Man,* 2nd ed. London: John Murray.
Ceypek, T., Kuzniarz, J.J. and Lipowczan, A. (1973) Hearing loss due to impulse noise: a field study. In *Proceedings of the International Congress on Noise as a Public Health Problem.* Dubrovnik, Yugoslavia, ed. D. Ward, pp. 219–28. Washington, DC : US Environmental Protection Agency.
Cohen, A. (1973) Industrial noise and medical absence and accident record data on exposed workers. In *Proceedings of the International Congress on Noise as a Public Health Problem.* Dubrovnik, Yugoslavia, ed. D. Ward, pp. 441–54. Washington, DC : US Environmental Protection Agency.
Cohen, H.H., Conrad, D.W., O'Brien, J.F. and Pearson, R.G. (1973) Noise effects, arousal

and human information processing: task difficulty and performance. Unpublished report, Department of Psychology and Industrial Engineering, North Carolina State University, Raleigh.

Cohen, S. (1978) Environmental load and the allocation of attention. In *Advances in Environmental Psychology*, vol. 1, ed. A. Baum, J.E. Singer and S. Valins. Hillsdale, NJ : Erlbaum.

Cohen, S. (1980) After-effects of stress on human performance and social behaviour: a review of research and theory. *Psychological Bulletin*, **88**, 82–108.

Conrad, D.W. (1973) The effects of intermittent noise on human serial decoding performance and physiological response. *Ergonomics*, **16**, 739–47.

Corso, J.F. (1952) *The Effects of Noise on Human Behavior*. US Air Force WADC Technical Report 53–81. Ohio : Wright-Patterson Air Force Base.

Corso, J.F. (1976) Presbycusis as a complicating factor in evaluating noise-induced hearing loss. In *Effects of Noise on Hearing*, ed. D. Henderson, R.P. Hamernik, D.S. Dojanh and J.H. Mills, pp. 497–524. New York: Raven Press.

Corso, J.F. (1977) Auditory perception and communication. In *Handbook of the Psychology of Ageing*, ed. J.E. Birren and K.W Schaie, pp. 535–53. New York: Van Nostrand.

Daee, S. and Wilding, J.M. (1977) Effects of high intensity white noise on short-term memory for position in a list and sequence. *British Journal of Psychology*, **68**, 335–49.

Davies, A.D.M. and Davies, D.R. (1975) The effects of noise and time of day upon age differences in performance at two checking tasks. *Ergonomics*, **18**, 321–36.

Davies, D.R. (1968) Physiological and psychological effects of exposure to high intensity noise. *Applied Acoustics*, **1**, 215–33.

Davies, D.R. (1976) Noise and the autonomic nervous system. In *Man and Noise*, ed. G. Rossi and M. Vigone, pp. 157–63. Turin: Edizioni Minerva Medica.

Davies, D.R. and Gill, E.B. (1980) Noise, glucose and problem solving. Paper presented to a meeting of the Cognitive Section of the British Psychological Society, Bedford College, London.

Davies, D.R. and Hockey, G.R.J. (1966) The effects of noise and doubling the signal frequency on individual differences in visual vigilance performance. *British Journal of Psychology*, **57**, 381–9.

Davies, D.R. and Jones, D. M. (1975) The effects of noise and incentives upon attention in short-term memory. *British Journal of Psychology*, **66**, 61–8.

Davies, D.R., Lang, L. and Shackleton, V.J. (1973) The effects of music and task difficulty on performance at a visual vigilance task. *British Journal of Psychology*, **64**, 383–9.

Davis, H. (1958) *Project Anehin*. US Navy School of Aviation Medicine Project NM130100, Subtask 1, Report 7.

Drettner, B., Hedstrand, H., Klockhoff, G. and Svedberg, A. (1975) Cardiovascular risk factors and hearing loss. *Acta Otolaryngologica*, **79**, 366–71.

Easterbrook, J.A. (1959) The effect of emotion on cue utilization and the organization of behaviour. *Psychological Review*, **66**, 183–201.

Egan, J.P. and Hake, H.W. (1950) On the masking pattern of a simple auditory stimulus. *Journal of the Acoustical Society of America*, **22**, 622–30.

Eldredge, D.H. (1976) The problems of criteria for noise exposure. In *Effects of Noise on Hearing*, ed. D. Henderson, R.P. Hamernik, D.S. Dosanjh and J.H. Mills, 3–20. New York: Raven Press.

Elliot, L.L. (1971) Backward and forward masking. *Audiology*, **10**, 65–76.

Eschenbrenner, A.J. (1971) Effects of intermittent noise on the performance of a complex psychomotor task. *Human Factor*, **13**, 59–63.

Eysenck, M.W. (1977) *Human Memory*, Oxford: Pergamon Press.

Feddersen, W.E., Sandel, T.T., Teas, D.C. and Jeffress, L.A. (1957) Localization of high frequency tones. *Journal of The Acoustical Society of America*, **53**, 400–8.

Finkelman, J.M. and Glass, D.C. (1970) Reappraisal of the relationship between noise and human performance by means of a subsidiary task measure. *Journal of Applied Psychology*, **54**, 211–13.

Fleishman, E.A. (1975) Toward a taxonomy of human performance. *American Psychologist*, **30**, 1127–49.

Fletcher, J.L. and Loeb, M. (1967) Impulse duration and temporary threshold shift. *Journal of the Acoustical Society of America*, **44**, 1524–8.

Forster, P.M. and Grierson, A.T. (1978) Noise and attentional selectivity: a reproducible phenomenon? *British Journal of Psychology*, **69**, 489–98.

Fowler, C.J.H. and Wilding, J. (1979) Differential effects of noise and incentives on learning. *British Journal of Psychology*, **70**, 149–53.

Frankenhaeuser, M. and Lundberg, U. (1974) Immediate and delayed effects of noise on performance and arousal. *Biological Psychology*, **2**, 127–33.

Gebhardt, C.J. and Goldstein, D.P. (1972) Frequency discrimination and the MLD. *Journal of the Acoustical Society of America*, **51**, 1228–32.

Geldard, F.A. (1972) *The Human Senses*, (2nd ed). New York: Wiley.

Glass, D.C. and Singer, J.E. (1972) *Urban Stress*. New York and London: Academic Press.

Glorig, A. and Nixon, J. (1960) Distribution of hearing loss in various populations. *Annals of Otology, Rhinology and Laryngology*, **71**, 727–43.

Grether, W.F. (1970) *Noise and Human Performance*. Air Force Systems Command. Report AMRL-TR-70-29. Ohio: Wright-Patterson Air Force Base.

Grether, W.F. (1971) *Effects on Human Performance of Combined Environmental Stresses*. Aerospace Medical Research Laboratory Technical Report 70-68. Ohio: Wright-Patterson Air Force Base.

Grether, W.F., Harris, C.S., Mohr, G.C., Nixon, C.W., Ohlbaum, M., Sommer, H.C., Thaler, V.H. and Veghte, J.H. (1971) Effects of combined heat, noise and vibration stress on human performance and psychological functions. *Aerospace Medicine*, **42**, 1092–7.

Griefahn, B. and Jansen, G. (1977) Effects of noise on sleep: a literature review. In *Proceedings of the 9th International Congress on Acoustics*, Madrid.

Griefahn, B. and Muzet, A. (1978) Noise-induced sleep disturbances and their effects on health. *Journal of Sound and Vibration*, **59**, 99–106.

Grimaldi, J.V. (1958) Sensorimotor performance under varying noise conditions. *Ergonomics*, **2**, 34–43.

Hamilton, P. (1969) Selective attention in multisource monitoring tasks. *Journal of Experimental Psychology*, **82**, 34–7.

Harris, C.S. and Sommer, H.C. (1971) *Combined Effects of Noise and Vibration on Mental Performance*. Air Force Systems Command Report AMRL-TR-70-21. Ohio: Wright-Patterson Air Force Base.

Hartley, L.R. (1973) Effect of prior noise or prior performance on serial reaction. *Journal of Experimental Psychology*, **101**, 255–61.

Hartley, L.R. and Adams, R.G. (1974) Effect of noise on the Stroop test. *Journal of Experimental Psychology*, **102**, 62–6.

Hay, B. (1975) Occupational noise exposure: the laws in the EEC, Sweden, Norway, Australia, Canada and the USA. *Applied Acoustics*, **8**, 299–314.

Helper, M.M. (1957) *The Effects of Noise on Work Output and Physiological Activation*. US Army Medical Laboratory Report 270.

Hiroto, D.S. (1974) Locus of control and learned helplessness. *Journal of Experimental Psychology*, **102**, 187–93.

Hiroto, D.S. and Seligman, M.E.P. (1974) Generality of learned helplessness in man. *Journal of Personality and Social Psychology*, **31**, 311–27.

Hirsh, I.J. and Bilger, R.C. (1955) Auditory threshold recovery after exposure to pure tones. *Journal of the Acoustical Society of America*, **27**, 1186.

Hirsh, I.J., Bilger, R.C. and Detherage, B.H. (1956) The effect of auditory and visual background on apparent duration. *American Journal of Psychology*, **69**, 561–74.

Hockey, G.R.J. (1970*a*) Effect of loud noise on attentional selectivity. *Quarterly Journal of Experimental Psychology*, **22**, 28–36.

Hockey, G.R.J. (1970*b*) Signal probability and spatial location as possible bases of increased selectivity in noise. *Quarterly Journal of Experimental Psychology*, **22**, 37–42.

Hockey, G.R.J. (1970*c*) Changes in attention allocation in a multicomponent task under loss of sleep. *British Journal of Psychology*, **61**, 473–80.

Hockey, G.R.J. (1973) Changes in information selection patterns in multisource monitoring as a function of induced arousal shifts. *Journal of Experimental Psychology*, **101**, 35–42.

Hockey, G.R.J. and Hamilton, P. (1970) Arousal and information selection in short-term memory. *Nature*, **226**, 866–7.

Hood, J.D. (1950) Studies in auditory fatigue and adaptation. *Acta Otolaryngologica*, Supplement **92.**

Houston, B.K. (1969) Noise, task difficulty and Stroop color-word performance. *Journal of Experimental Psychology*, **82**, 403–4.

Houston, B.K. and Jones, T.M. (1967) Distraction and Stroop color-word performance. *Journal of Experimental Psychology*, **74**, 54–6.

Howell, K. and Martin, A.M. (1975) An investigation of the effects of hearing protectors on vocal communication in noise. *Journal of Sound and Vibration*, **41**, 181–96.

Ingham, J.G. (1970) Symptoms of psychological disorder in noisy environments. In *Proceedings of Symposium on Psychological Effects of Noise*, ed. W. Taylor. London: Research Panel of the Society of Occupational Medicine.

Jansen, G. (1959) Zur Ensebelung vegetativer Funktionsstornungen durch Larmeinwurkung. *Archiv für Gewerbepathologie und Gewerbehygiene*, **17**, 253–61.

Jansen, G. (1961) Adverse effects of noise on iron and steel workers. *Stahl and Eisen*, **81**, 217–20.

Jerison, H.J. (1959) Effects of noise on human performance. *Journal of Applied Psychology*, **43**, 96–101.

Jerison, H.J. and Arginteau, J. (1958) *Time Judgement, Acoustic Noise and Judgement Drift*. US Air Force, Wright Air Development Center Report WADC-TR-57-454. Ohio: Wright-Patterson Air Force Base.

Jerison, H.J. and Smith, A.K. (1955) *Effect of Acoustic Noise on Time Judgement*. US Air Force, Wright Air Development Center Report WADC-TR-55-358. Ohio: Wright-Patterson Air Force Base.

Jerison, H.J. & Wallis, R.A. (1957) *Experiments on Vigilance. II. Performance on a Simple Task in Noise and Quiet.* US Air Force, Wright Air Development Center Report WADC-TR-57-318. Ohio: Wright-Patterson Air Force Base.

Jerison, H.J. and Wing, S. (1957) *Effects of Noise and Fatigue on a Complex Vigilance Task.* US Air Force, Wright Air Development Center Report WADC-TR-57-14. Ohio: Wright-Patterson Air Force Base.

Jerison, H.J., Crannel, C.W. and Pownall, D. (1957) *Acoustic Noise and Repeated Time Judgements in a Visual Movement Projection Task.* US Air Force, Wright Air Development, Center. Report WADC-TR-57-54. Ohio: Wright-Patterson Air Force Base.

Jones, D.M. (1979) Stress and memory. In *Applications of Memory Research*, ed. M.M. Gruneberg and P.E. Morris, pp. 185–214. New York and London: Academic Press.

Jones, D.M. and Chapman, A.J. (1979) Stress after hours. In *Psychophysiological Response to Occupational Stress*, ed. C. Mackay, pp. 106–11. London: IPC.

Jones, D.M., Smith, A.P. and Broadbent, D.E. (1979) Effects of moderate intensity noise on the Bakan vigilance task. *Journal of Applied Psychology*, **64**, 627–34.

Jonsson, A. and Hansson, L. (1977) Prolonged exposure to a stressful stimulus (noise) as a cause of raised blood pressure. *Lancet*, **iii**, 86–7.

Karmy, S.J. and Coles, R.R.A. (1976) Hearing protection: factors affecting its use. In *Man and Noise,* ed. G. Rossi and M. Vigone, pp. 260–73. Turin: Edizioni Minerva Medica.

Kell, R.L., Pearson, J.C.G. and Taylor, W. (1970) Hearing thresholds of an island population in North Scotland. *International Audiology,* **9,** 334–9.

Kerr, W.A. (1950) Accident proneness of factory departments. *Journal of Applied Psychology,* **34,** 167–75.

Kirk, R.E. and Hecht, E. (1963) Maintenance of vigilance by programmed noise. *Perceptual and Motor Skills,* **16,** 553–60.

Kryter, K.D. (1970) *Effects of Noise on Man.* New York and London: Academic Press.

Kryter, K.D. (1973) A critique of some procedures for evaluating damage risk from exposure to noise. In *Proceedings of the International Congress on Noise as a Public Health Problem,* Dubrovnik, Yugoslavia ed. D. Ward. Washington, DC: US Environmental Protection Agency.

Lebo, C.P. and Reddell, R.C. (1972) The presbycusis component in occupational, noise-induced hearing loss. *Laryngoscope,* **82,** 1399–409.

Leonard, J.A. (1959) *Five-Choice Serial Reaction Apparatus.* MRC Report, Applied Psychology Unit, No. 326/59. Cambridge: Applied Psychology Research Unit.

Loeb, M. (1956) *The Effects of Intense Stimulation on the Perception of Time.* US Army Medical Research Laboratory, Fort Knox, Kentucky, Report 269.

Loeb, M. and Jones, P.D. (1978) Noise exposure, monitoring and tracking performance as a function of signal bias and task priority. *Ergonomics,* **21,** 265–72.

Loeb, M. and Richmond, G. (1956) *The Influence of Intense Noise on the Performance of a Precise Fatiguing Task.* US Army Medical Research Laboratory, Fort Knox, Kentucky, Report 268.

Loeb, M. and Smith, R.P. (1967) Relation of induced tinnitus to physical characteristics of the inducing stimuli. *Journal of the Acoustical Society of America,* **42,** 453–5.

Martin, A. (1976) The equal energy concept applied to impulse noise. In *Effects of Noise on Hearing,* ed. D. Henderson, R.P. Hamernik, D.S. Dosanjh and J.H. Mills, pp. 421–53. New York: Raven Press.

Miller, G.A., Heise, G.A. and Lichten, W. (1951) The intelligibility of speech as a function of the context of the test materials. *Journal of Experimental Psychology,* **41,** 329–35.

Miller, G.A. and Licklider, J.C.R. (1950) The intelligibility of interrupted speech. *Journal of the Acoustical Society of America,* **19,** 120–5.

Miller, J.D. (1974) Effects of noise on people. *Journal of the Acoustical Society of America,* **56,** 729–63.

Moller, A. (1978) Review of animal experiments. *Journal of Sound and Vibration,* **59,** 73–7.

Mosko, J.F. and Fletcher, J.L. (1971) Evaluation of the gundefender earplug: temporary threshold shift and speech intelligibility. *Journal of the Acoustical Society of America,* **49,** 1732–4.

Nixon, J.C. and Glorig, A. (1961) Noise induced permanent threshold shift at 2000 c.p.s. and 4000 c.p.s. *Journal of the Acoustical Society of America,* **33,** 904–8.

Nuckols, W.H. (1969) An investigation of the effect of music and noise on a prolonged intellectual task with environmental conditioning. Unpublished Master's Thesis, Department of Industrial Engineering, Texas A. and M. University.

Ollerhead, J.B. (1973) Noise: how can the nuisance be controlled? *Applied Ergonomics,* September, 130–8.

O'Malley, J.J. and Poplawski, A. (1971) Noise induced arousal and breadth of attention. *Perceptual and Motor Skills,* **33,** 887–90.

Parvizpoor, D. (1976) Noise exposure and prevalence of high blood pressure among weavers in Iran. *Journal of Occupational Medicine,* **18,** 739–1.

Passchier-Vermeer, W. (1974) Hearing loss due to continuous exposure to steady state broad-band noise. *Journal of the Acoustical Society of America,* **56,** 1585–93.

Patterson, R.D. (1969) Noise masking of a change in residue pitch. *Journal of the Acoustical Society of America,* **45,** 1520–4.

Pickett, J.M. (1956) Effects of vocal force on the intelligibility of speech sounds. *Journal of the Acoustical Society of America,* **28,** 902–5.

Plutchik, R. (1960) Effect of high intensity intermittent sound on compensatory tracking and mirror tracing. *Perceptual and Motor Skills,* **12,** 187–94.

Poulton, E.C. (1965) On increasing the sensitivity of measures of performance. *Ergonomics,* **8,** 69–76.

Poulton, E.C. (1966) Engineering psychology. *Annual Review of Psychology,* **17,** 177–200.

Poulton, E.C. (1970) *Environment and Human Efficiency.* Springfield, Ill. Charles C. Thomas.

Poulton, E.C. (1973) Unwanted range effects from using within-subject experimental designs. *Psychological Bulletin,* **80,** 113–21.

Poulton, E.C. (1976) Continuous noise interferes with work by masking auditory feedback and inner speech. *Applied Ergonomics,* **7,** 79–84.

Poulton, E.C. (1977*a*) Arousing stresses increase vigilance. In *Vigilance: Theory, Operational Performance and Physiological Correlates,* ed. R.R. Mackie, pp. 423–59. New York: Plenum Press.

Poulton, E.C. (1977*b*) Continuous intense noise masks auditory feedback and inner speech. *Psychological Bulletin,* **84,** 977–1001.

Poulton, E.C. (1978) A new look at the effects of noise: a rejoinder. *Psychological Bulletin,* **85,** 1068–79.

Poulton, E.C. and Edwards, R.S. (1974) Interactions and range effects in experiments on pairs of stresses: mild heat and low frequency noise. *Journal of Experimental Psychology,* **102,** 621–8.

Poulton, E.C. and Freeman, P.R. (1966) Unwanted asymmetrical transfer effects with balanced experimental designs. *Psychological Bulletin,* **66,** 1–8.

Rabbitt, P.M.A. (1966) Recognition memory for words correctly heard in noise. *Psychonomic Science,* **8,** 383–4.

Rabbitt, P.M.A. (1968) Channel capacity, intelligibility and immediate memory. *Quarterly Journal of Experimental Psychology,* **20,** 241–8.

Robinson, D.W. (1970) Relations between hearing loss and noise exposure, analysis of results of a retrospective study. In *Hearing and Noise in Industry,* ed. W. Burns and D.W. Robinson, pp. 100–51. London: HMSO.

Robinson, D.W. (1971) Estimating the risk of hearing loss due to exposure to continuous noise. In *Occupational Hearing Loss,* ed. D.W. Robinson, pp. 43–62. New York and London: Academic Press.

Robinson, D.W. (1976) Characteristics of occupational noise: induced hearing loss. In *Effects of Noise on Hearing,* ed. D. Henderson, R.P. Hamernik, D.S. Dosanjh and J.H. Mills, pp. 383–406. New York: Raven Press.

Robinson, D.W. and Dadson, R.S. (1956) A redetermination of the equal loudness relations for pure tones. *British Journal of Applied Physics,* **7,** 166–81.

Rodda, M. (1967) *Noise and Society.* Edinburgh: Oliver and Boyd.

Rosen, S., Plester, D., El Mofty, A. and Rosen, H.V. (1964) High frequency audiometry in presbycusis: a comparative study of the Mabaan tribe in the Sudan with urban populations. *Archives of Otolaryngology,* **79,** 18–32.

Samuel, W.M.S. (1964) Noise and the shifting of attention. *Quarterly Journal of Experimental Psychology,* **16,** 264–7.

Schouten, J.F. (1940) *Five Articles on the Perception of Sound.* Eindhoven: Institute for Perception. Cited by Green, D.M. (1976) In *An Introduction to Hearing.* New York: Wiley.

Schultz, T.J. (1972) *Community Noise Ratings.* Amsterdam: Applied Science Publishers.

Schwartz, S. (1974) Arousal and recall: effects of noise on two retrieval strategies. *Journal of Experimental Psychology,* **102,** 896–8.

Seligman, M.E.P. (1975) *Helplessness: On Depression, Development and Death.* San Francisco: Freeman.

Seligman, M.E.P. and Maier, S.F. (1976) Failure to escape traumatic shock. *Journal of Experimental Psychology,* **74,** 1–9.

Shepherd, M. (1974) Pollution and mental health, with particular reference to the problem of noise. *Psychiatria Clinica,* **7,** 226–36.

Simpson, G.C., Cox, T. and Rothschild, D.R. (1974) The effects of noise stress on blood glucose level and skilled performance. *Ergonomics,* **17,** 481–6.

Smith, A.P. (1980) The effects of noise on memory for order and location. Paper presented at a meeting of the Cognitive Section of the British Psychological Society, Bedford College, London.

Smith, A.P. and Broadbent, D.E. (1980) Effects of noise on performance on embedded figures tasks. *Journal of Applied Psychology,* **65,** 246–8.

Spoor, A. (1967) Presbycusis values in relation to noise induced hearing loss. *Laryngoscope,* **82,** 1399–409.

Stevens, S.S. (1956) Calculation of the loudness of complex noise. *Journal of the Acoustical Society of America,* **28,** 807–32.

Stevens, S.S. (1957) On the psychophysical law. *Psychological Review,* **64,** 153–81.

Stevens, S.S. (1972) Stability of human performance under intense noise. *Journal of Sound and Vibration,* **20,** 35–56.

Stevens, S.S. (1975) *Psychophysics.* New York: Wiley.

Tarnopolsky, A. (1978) Effects of aircraft noise on mental health. *Journal of Sound and Vibration,* **59,** 89–97.

Taylor, W., Pearson, J., Mair, A. and Burns, W. (1965) Study of noise and hearing in jute weaving. *Journal of the Acoustical Society of America,* **38,** 113–20.

Teas, D.C., Eldredge, D.H. and Davis, H. (1962) Cochlear responses to acoustic transients. *Journal of the Acoustical Society of America,* **34,** 1337–50.

Theologus, G.C., Wheaton, G.R. and Fleishman, E.A. (1974) Effects of intermittent moderate intensity noise stress on human performance. *Journal of Applied Psychology,* **59,** 539–47.

Thorton, J.W. and Jacobs, P.D. (1971) Learned helplessness in human subjects. *Journal of Experimental Psychology,* **87,** 367–72.

Tobias, J.V. and Irons, F.M. (1973) Perception of distorted speech. In *Proceedings of International Congress on Noise as a Public Health Problem, Dubrovnik, Yugoslavia,* ed. W.D. Ward, pp. 43–56. Washington, D.C: Environmental Protection Agency.

Van Dishoeck, H.A.E. (1948) The continuous threshold or detailed audiogram for recording simulation deafness. *Acta Otolaryngologica,* Supplement, **78,** 183–92.

von Békésy, G. (1960) *Experiments in Hearing,* ed. E.G. Wever. New York: McGraw-Hill. Hill.

Walker, T.G. (1970) Temporary threshold shift from impulse noise. *Annals of Occupational Hygiene,* **13,** 51–8.

Waltzman, S.B and Levitt, H. (1978) Speech interference level as a predictor of face to face communication in noise. *Journal of the Acoustical Society of America,* **63,** 581–90.

Ward, W.D. (1962) Damage risk criteria for line spectra. *Journal of the Acoustical Society of America,* **34,** 1610–19.

Ward, W.D. (1976) Transient changes in hearing. In *Man and Noise,* ed. G. Rossi and M. Vigone, pp. 111–20. Turin: Edizioni Minerva Medica.

Ward, W.D. (1977) Effects of noise on hearing. In *Handbook of Physiology: Reactions to*

Environmental Agents, ed. D.H.K. Lee, pp. 1–16. Bethesda, Maryland: American Physiological Society.

Warner, H.D. (1969) Effects of intermittent noise on human target detection. *Human Factors,* **11,** 245–9.

Warner, H.D. and Heimstra, N.W. (1971) Effects of intermittent noise on visual search tasks of varying complexity. *Perceptual and Motor Skills,* **32,** 219–26.

Warner, H.D. and Heimstra, N.W. (1972) Effects of noise intensity on visual target-detection performance. *Human Factors,* **14,** 181–5.

Warner, H.D. and Heimstra, N.W. (1973) Target-detection performance as a function of noise intensity and task difficulty. *Perceptual and Motor Skills,* **36,** 439–42.

Webster, J.C. (1969) SIL: past, present and future. *Journal of Sound and Vibration,* **3,** 22–6.

Webster, J.C. (1973) The effects of noise on hearing of speech. In *Proceedings of the International Congress on Noise as a Public Health Problem,* Dubrovnik, Yugoslavia, ed. W.D. Ward. Washington, DC: US Environmental Protection Agency.

Weinstein, A. and Mackenzie, R.S. (1966) Manual performance and arousal. *Perceptual and Motor Skills,* **22,** 498.

Welch, B. and Welch, A. (eds.) (1970) *Physiological Effects on Noise.* New York: Plenum Press.

Weston, H.C. and Adams, S. (1935) *The Performance of Weavers under Varying Conditions of Noise.* Industrial Health Research Board Report 70. London: HMSO.

Wilbanks, W.A., Webb, W.B. and Tolhurst, G.C. (1956) *A Study of Intellectual Activity in a Noise Environment.* US Naval School of Aviation Medicine, Research Project NN001 104 100, Report 1.

Wilkinson, R.T. (1963) Interaction of noise with sleep deprivation and knowledge of results. *Journal of Experimental Psychology,* **66,** 332–7.

Wilkinson, R.T. (1969) Some factors influencing the effect of environmental stressors upon performance. *Psychological Bulletin,* **72,** 260–72.

Wohlwill, J.F., Nasar, J.L., Dejoy, D.M. and Foruzani, H.H. (1976) Behavioral effects of a noisy environment: task involvement versus passive exposure. *Journal of Applied Psychology,* **61,** 67–74.

Woodhead, M.M. (1964) The effects of bursts of noise on an arithmetic task. *American Journal of Psychology,* **77,** 627–33.

Woodhead, M.M. (1966) An effect of noise on the distribution of attention. *Journal of Applied Psychology,* **50,** 296–9.

Zwicker, E. (1958) Über psychologische und methodische Grundlagen der Lautheit. *Akustishe Beihefte,* **1,** 237–58.

Zwislocki, J.J. (1965) Analysis of some auditory characteristics. In *Handbook of Mathematical Psychology,* ed. R.D. Luce, R.R. Bush and E. Galanter. New York: Wiley.

Zwislocki, J.J. (1971) Central masking and neural activity in the cochlear nucleus. *Audiology,* **10,** 48–59.

AUTHOR INDEX

Page numbers in roman type refer to the text and those in *italic type* refer to the lists of references.

SUBJECT INDEX

see also centre of gravity

hand-operated controls, location of, 139